T0142324

Lecture Notes on Data Engineering and Communications Technologies

Volume 56

The aim of the book series is to present cutting edge engineering approaches to data technologies and communications. It will publish latest advances on the engineering task of building and deploying distributed, scalable and reliable data infrastructures and communication systems.

The series will have a prominent applied focus on data technologies and communications with aim to promote the bridging from fundamental research on data science and networking to data engineering and communications that lead to industry products, business knowledge and standardisation.

Indexed by SCOPUS, INSPEC.
All books published in the series are submitted for consideration in Web of Science.

More information about this series at http://www.springer.com/series/15362

Nabendu Chaki · Jerzy Pejas ·
Nagaraju Devarakonda · Ram Mohan Rao Kovvur
Editors

Proceedings of International Conference on Computational Intelligence and Data Engineering

ICCIDE 2020

Editors
Nabendu Chaki
University of Calcutta
Kolkata, India

Jerzy Pejas
West Pomeranian University of Technology
Szczecin, Poland

Nagaraju Devarakonda
VIT-AP University
Amaravati, India

Ram Mohan Rao Kovvur
Vasavi College of Engineering
Hyderabad, India

ISSN 2367-4512 ISSN 2367-4520 (electronic)
Lecture Notes on Data Engineering and Communications Technologies
ISBN 978-981-15-8769-6 ISBN 978-981-15-8767-2 (eBook)
https://doi.org/10.1007/978-981-15-8767-2

This Springer imprint is published by the registered company Springer Nature Singapore Pte Ltd.
The registered company address is: 152 Beach Road, #21-01/04 Gateway East, Singapore 189721, Singapore

Preface

The AICTE Sponsored Third International Conference on Computational Intelligence & Data Engineering (ICCIDE 2020) took place on 8 and 9 August 2020 at Vasavi College of Engineering (VCE), Autonomous at Ibrahimbagh, Hyderabad, Telangana, India. ICCIDE is conceived as a forum for presenting and exchanging ideas, which aims at resulting in high-quality research work in cutting edge technologies and most happening areas of Computational Intelligence and Data Engineering.

The conference solicited latest research ideas on Computational Intelligence and Data Engineering, thus inviting researchers working in the domains of Machine Learning, Bayesian Networks, Computational Paradigms and Computational Complexity, Rough Sets, Semantic Web, Knowledge Representation, Fuzzy Systems, Soft Computing, Data Models, Ubiquitous Data Management, Mobile Databases, Data Provenance, Workflows, Cloud Computing, Bigdata Analytics, Scientific Data Management, and Security.

The sincere effort of the program committee members coupled with indexing initiatives from Springer have drawn a large number of high-quality submissions from scholars all over India and abroad. A thorough peer-reviewed process has been carried out by the PC members and by external reviewers. While reviewing the papers, the reviewers mainly looked at the novelty of the contributions, besides the technical content, the organization and the clarity of the presentation. The entire process of paper submission, review, and acceptance process was done electronically.

The Technical Program Committee eventually could identify 38 papers for publication out of 151 submissions. The resulting acceptance ratio is 25.16%, which is healthy and quite good in this third International conference.

The program also includes seven keynote addresses, by Prof. Rajkumar Buyya (The University of Melbourne, Australia), Dr. P. Sateesh Kumar (Department of Computer Science & Engineering, IIT Roorkee, India), Prof. Maode Ma (Nanyang Technological University, Singapore), Prof. Atul Negi (University of Hyderabad, India). Mr Aninda Bose (Senior Publishing Editor, Springer New Delhi, India), Dr. K. Raghavendra (Head High Performance Computing and Drones, Advanced Data

Processing Research Institute (ADRIN), ISRO), and Prof. Giuseppe Di Fatta (University of Reading, London). The Conference is structured with eight technical sessions which are chaired by distinguished faculty of outstanding institutions in the country.

We eventually extend our gratitude to all the members of the Program Committee and the external reviewers for their excellent and time-bound review work. We thank AICTE, who have sponsored this conference under the scheme of GOC. We are thankful to the entire management of Vasavi College of Engineering, Sri P. Balaji, CEO, and Prof. S.V. Ramana, Principal, for their patronage and continual support to make the event successful. We appreciate the initiative and support from Mr. Aninda Bose and his colleagues in Springer Nature for their strong support toward publishing this volume in the Lecture Notes on Data Engineering and Communications Technologies (LNDECT) series of Springer Nature. Finally, we thank all the authors without whom the conference would not have reached the expected standards.

Kolkata, India — Nabendu Chaki
Szczecin, Poland — Jerzy Pejas
Amaravati, India — Nagaraju Devarakonda
Hyderabad, India — Ram Mohan Rao Kovvur

Contents

About the Editors

Nabendu Chaki is a Professor at the Department Computer Science & Engineering, University of Calcutta, Kolkata, India. Dr. Chaki graduated in Physics from the legendary Presidency College in Kolkata and then in Computer Science & Engineering from the University of Calcutta. He completed his Ph.D. in 2000 at Jadavpur University, India. He shares 6 international patents, including 4 US patents with his students. Prof. Chaki has been active in developing international standards for software engineering and cloud computing as a part of the Global Directory (GD) for ISO-IEC. As well as editing more than 25 books, Nabendu has authored 6 textbooks and research books and more than 150 Scopus indexed research papers in journals and at international conferences. His areas of research include distributed systems, image processing and software engineering. Dr. Chaki has served as a member of the research faculty in the Ph.D. program in Software Engineering in the US Naval Postgraduate School, Monterey, CA. He is a visiting faculty member of many universities in India and abroad. In addition to being on the editorial board for several international journals, he has also served on the committees of over 50 international conferences. Prof. Chaki is the founder Chair of the ACM Professional Chapter in Kolkata.

Jerzy Pejas received the M.Sc. degree in Computer Science and Engineering from the Wroclaw University of Technology, and the Ph.D. degree in Control Systems from the Gdansk University of Technology. In 2013, he defended his habilitation thesis and obtained habilitation (D.Sc.) degree at West Pomeranian University of Technology in Szczecin. He is currently the Professor with the Faculty of Computer Science and Information Technology, West Pomeranian University of Technology, Szczecin, Poland. His main research interest topics include information security and computer networks, methods of secure electronic signature, as well as new trends in applied cryptography, in particular in pairing-based cryptography.

Nagaraju Devarakonda received his B.Tech. from Sri Venkateswara University, M.Tech. from Jawaharlal Nehru University (JNU), New Delhi, and Ph.D. from Jawaharlal Nehru Technological University, Hyderabad. He has published over 70 research papers in international conferences and journals. He is the Co-Editor of Proceedings of ICCIDE 2017 and 2018. The proceedings were published in Lecture

Notes on Data Engineering and Communication Technologies of SPRINGER. He published papers in ICCI*CC 2017 at OXFORD UNIVERSITY, UK, and ICCI*CC 2018 at UNIVERSITY OF CALIFORNIA, BERKELEY, USA. He is currently working as an Associate Professor in the School of Computer Science & Engineering at VIT-AP University and has 16 years of experience in teaching. His research areas are data mining, soft computing, machine learning and pattern recognition. He has supervised 25 M.Tech. students, guided 2 Ph.D.s and currently guiding 7 Ph.D.s.

Ram Mohan Rao Kovvur Obtained B.Tech. (Computer Technology) degree from Nagpur University in the year 1993, M.Tech. degree with the specialization "Computer Science and Engineering" from Osmania University in the year 1997 and received Ph.D. in Computer Science and Engineering from Jawaharlal Nehru Technology University (JNTU) in the year 2014 with research specialization as grid computing. He is having teaching experience of 25 years in various cadres such as Lecturer, Assistant Professor, Associate Professor and Professor at Vasavi College of Engineering (22 Years 6 Months) and Shadan Institute of Computer Studies (2 Years 6 Months). He was Functional Head (Systems Manager), Main Computer Center for a period of 15 years, i.e., from January 2001 to August, 2016. Presently, he is having responsibility as HoD & BoS Chairman of the Department of Information Technology, since August 2016. He had published and presented 15 number of technical papers in national and international journals and conferences. He obtained a grant of Rs 19.31 Lakhs from AICTE under MODROBS for Establishing Deep Learning Lab. As a part research work, he established a Grid Environment using Globus ToolKit (open source software toolkit used for building grid systems) in the Department of CSE, on 32 nodes, and also offered projects to UG students. Further, he also Established Cloud Lab, VCE using Aneka Platform (US Patented) of Manjrasoft Pvt. Ltd. at M.Tech. Lab, Computer Science Department on 12 nodes, and offered projects to UG/PG students. He had received award as one of the Best Nodal Coordinator from IIIT, Hyderabad (Virtual Labs), for promoting the use of Virtual Labs MHRD, Government of India Initiative, and received Appreciation from IIT Madras-QEEE (MHRD Initiative) as a Best QEEE Coordinator for three Times, i.e., in Phase VI, V and IV of Vasavi College of Engineering , in recognition of "Efforts towards making the Program Effective for both students and Institute" and also "Recognized as a Proactive Member in the overall QEEE Program" and also received long standing service award—Instructor Years of Service from Cisco Networking Academy (2016) in recognition of "15 Years of active Participation and Service" in the Cisco Networking Academy Program. He received citation from Vasavi College of Engineering (5th September 2003), in recognition of "Proactive Initiative" taken in upgrading the Computer centre and in the Introduction of Value Added Course in J2EE in Collaboration with Pramati Technologies.

Autonomous Obstacle Avoidance Robot Using Regression

Vakada Naveen, Chunduri Aasish, Manne Kavya, Meda Vidhyalakshmi, and Kl Sailaja

Abstract Obstacle avoidance is considered as one of the main features of autonomous intelligent systems. There are various methods for obstacle avoidance. In this paper, obstacle avoidance is achieved by the difference between left wheel velocity and right wheel velocity of differential drive robot. The magnitude of difference between the wheel velocities is used to steer the robot in the correct direction. Data is collected by driving the robot manually. Ultrasonic sensors are used for distance measurement and IR sensors are used to collect the data of wheel velocities. This data is used to build a linear machine learning model which uses sonar data as input features. The model is used to predict the wheel velocities of the differential drive robot. The model built is then programmed into Atmega328 microcontroller using Arduino IDE. This enables the mobile robot to steer itself to avoid the obstacles. Since all the components used for this robot are highly available and cost-effective, the robot is economically affordable.

Keywords Obstacle avoidance · Autonomous mobile robot · Arduino · Nodemcu · Raspberry pi · Machine learning · Regression · Stochastic gradient descent · Pseudoinverse

---

V. Naveen (✉) · C. Aasish · M. Kavya · M. Vidhyalakshmi · K. Sailaja
VR Siddhartha Engineering College, Kanuru 520007, India
e-mail: vakadanaveen@gmail.com

C. Aasish
e-mail: ch.aasish999@gmail.com

M. Kavya
e-mail: kavyamanne69@gmail.com

M. Vidhyalakshmi
e-mail: vidhyalakshmimeda@gmail.com

K. Sailaja
e-mail: sailaja0905@gmail.com

N. Chaki et al. (eds.), *Proceedings of International Conference on Computational Intelligence and Data Engineering*, Lecture Notes on Data Engineering and Communications Technologies 56, https://doi.org/10.1007/978-981-15-8767-2_1

1 Introduction

Robotics has transformed and revolutionized industries in various fields. One of the recent trends in robotics is autonomous robots. They have gained much popularity because manual intervention is not required to control the robot. These robots can take decisions by themselves. They use a variety of sensors to sense the changes in their environment. Machine learning and artificial intelligence allow these robots to take decisions by themselves and maintain high degree of autonomy.

Autonomous robots have been useful in developing self-driving cars, intelligent rovers, autonomous delivery systems, unmanned vehicles, surveillance robots etc. These wide arrays of applications led to the increase in research community of autonomous systems.

Obstacle avoidance is considered as one of the important features of autonomous mobile robots. Various algorithms have been successful in achieving obstacle avoidance for mobile robots autonomously. In the paper "The Obstacle Detection and Obstacle Avoidance Algorithm Based on 2-D Lidar" [1], Peng Yan et al. used a 2D LIDAR sensor with a measuring range of up to 80 m. They proposed a simple solution using visibility graph method. Using raw data from LIDAR sensor, the position and shape of the obstacle is found out. Then, optimum direction is selected by using a cost function.

Another method to avoid obstacles is by using image processing. Cheng-Pei et al. [2] proposed a solution for obstacle avoidance using single camera. They have used a camera and two laser projectors fixed on same base. They have used image-based distance measurement system (IBDMS) to find the location of obstacles from the image. It consists of simple image processing steps. Path planning is also done to achieve autonomous patrol.

One of the easiest ways to detect obstacles is by using ultrasonic sensor. It gives the range of obstacles in front of it. Jin, Yun et al. [3] developed an omnidirectional intelligent obstacle avoidance system. They used an ultrasonic sensor supported by PWM servo motor which helps it to rotate in any direction. The obstacle distances measured by the ultrasonic sensor are stored in an array for selecting an optimal path. The intelligent car will move in the direction obtained by minimizing the objective function [3].

Wu, Ter Feng et al. [4] also used ultrasonic sensors to implement a real-time object avoidance system for wheeled robots. They used six ultrasonic sensors to measure the distance between the robot and the obstacles. It uses a wall following method to achieve optimal path design. But the robot may or may not reach the target, which is a drawback of that system.

In the paper "Video surveillance robot control using smart phone and Raspberry pi," Bokade et al. [5] has created a robot which can be controlled by using the mobile phone. They have used raspberry pi to control the robot through wireless connection. The streaming speed of the video is 15 frames per second.

Singh et al. [6] proposed a wireless robot to live stream both video and audio. The robot can be used as a surveillance robot. A web application is developed to control the robot wirelessly. Arduino Uno R3 board is used to control the mobile robot.

Kadiam et al. [7] developed a robot that can be used for video surveillance using wifi and raspberry pi. They have used ARM 11 processor and a USB camera to capture the video. The USB camera is connected to Raspberry pi. The video is then used for extracting useful information using data mining techniques and pattern recognition. This process can be often termed as smart surveillance [8].

Lei Tai [9] et al. in his paper proposed a deep network solution for obstacle avoidance of the robot. They used convolutional neural networks for capturing images as input and for the decision-making process which is an output that gives commands to the robot in which direction it should move while avoiding the obstacles. The dataset that has been used for training the network has been collected by moving the robot manually by avoiding the obstacles. The robot is confined to avoid the obstacles in the indoor environment. The model shows high similarity between human decisions and robot decisions while avoiding the obstacles.

Shichao Yang [10] et al. predicted the trajectory of the robots to avoid the obstacles using deep networks. The detection of the obstacles and generating the commands to the robot is done from the monocular images. The dataset that is used for training the model is NYUv2 RGB-D. Convolutional neural networks are used to predict the depth and surface normal and also to predict the trajectory of the robot. The 3D cost functions are used which helps in choosing the best trajectory path.

Wilbert G [11] et al. proposed a system for obstacle avoidance for unmanned aerial vehicles. The input images are taken from the camera and compared to the images that are stored in the database. They include SURF which detects the obstacles and avoid them using the control law. The UAVs recovers the path after avoiding an obstacle. The proposed system is faster in detecting the obstacles and flexible.

Mihai Duguleana [12] et al. proposed a new methodology for solving the issue of autonomous development of robots that contain both static and dynamic obstructions. They proposed path planning with artificial intelligence approaches.They designed a trajectory planner algorithm so that robot can be maintained at any speed. They modeled the robot in VR and MATLAB and they also tested in both VR and in real environment.

Punarjay Chakravarty [13] et al. proposed CNN Architecture for predicting depth estimation from a single image. They proposed control algorithm for estimating depth for guiding a quadrotor away from obstacles. They collected data, that is, online images, trained the network, and controlled the drone. They calculated the performance of the depth network in navigating the drone in different environments.

Wilbert G. Aguilar [14] et al. proposed an algorithm for obstacle avoidance system for unmanned aerial vehicles using one-eyed camera. They also tested the execution system including the obstacle detection and obstacle avoidance. They proposed Speeded up Robust Features (SURF) algorithm that is matching across the image from database and real-time image at Unmanned Aerial Vehicles, this algorithm gives the high execution considering the accuracy.

2 Methodology

In this paper, ultrasonic sensors are used to collect data related to a range of the obstacles and IR sensors to measure the wheel velocities. Unlike other methods, we use the collected data to build a machine learning model, which takes ultrasonic sensor data as input feature and predicts the left wheel velocity and right wheel velocity. The entire process of building the autonomous obstacle avoidance robot is divided into three phases as shown in Fig. 1.

Data is required to build a machine learning model to achieve obstacle avoidance. We used two sensors for collecting the data. They are ultrasonic sensor and IR sensor. The two sensors are placed on the mobile robot. Raspberry pi is used to collect the data from these sensors.

The ultrasonic sensors are used to collect the range of obstacles in the front direction. IR sensors are used to collect the rpm of left wheel and right wheel. Rpm values can be converted to into speed in meters per second using formula 1.

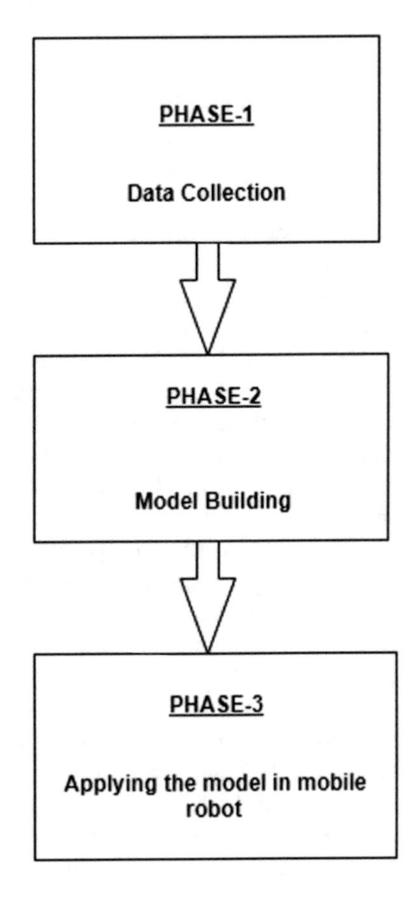

Fig. 1 Overview of building the autonomous object avoidance robot

$$speed = 3.14159 * \frac{D}{100} * \frac{RPM}{60} \tag{1}$$

The circuit diagram for the sensors and raspberry pi is shown in the Fig. 2. Note that the ultrasonic sensors is placed in the front side of the mobile robot and IR sensors are placed facing towards the wheels of the mobile robot as shown in the Fig. 3.

The mobile robot is controlled by NodeMCU [16]. The DC motors of the mobile robot are connected to NodeMCU as shown in Fig. 3. The circuit diagram for the DC motor connections with NodeMCU is shown in Fig. 4.

NodeMCU is connected to a mobile app which controls the mobile robot by changing the wheel velocity from the app. This helps in the manual training of the robot. The app which controls the mobile robot during training phase is should in

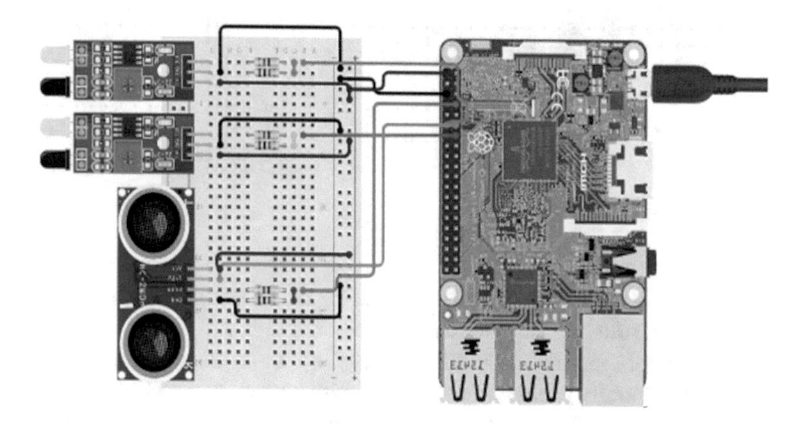

Fig. 2 Raspberry pi and sensors connection

Fig. 3 Position of sensors on the mobile robot

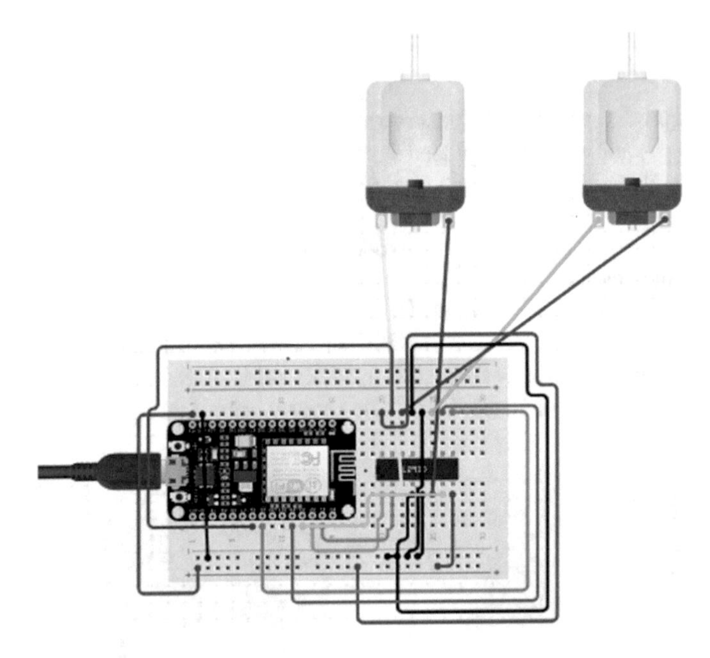

Fig. 4 DC motor and NodeMCU connections

Fig. 5. The android application can be built using MIT app inventor tool [17]. The connections of android app and Nodemcu are shown in Fig. 6.

Algorithm to train the mobile robot manually and the collect the sensor data:

1. Turn on the mobile robot by powering it with a battery or a power bank.

2. Connect to the mobile robot from the android application by using the IP address of NodeMCU.

3. Control the movement of the mobile robot manually by using the android application.

4. Avoid obstacles by controlling the mobile robot only in one direction, i.e., either left or right. Here, let us choose right direction only.

The data collected during the training phase is stored as a dataset in raspberry pi. The features stored in the dataset include obstacle distance in centimeters, left wheel speed and right wheel speed in terms of rpm. The sample dataset is shown in Table 1.

The flow chart for the data collection phase is shown in Fig. 7.

The dataset collected from the training phase is used to train the regression model. The input features for the training model is object range. We use this single feature to predict the left wheel velocity and right wheel velocity of the mobile robot. The flowchart for model building phase is shown in Fig. 8.

The machine learning model used for predictions is linear regression model. We use a simple linear model of degree 'd' to predict the left wheel velocity and right wheel velocity. The model used for the velocity prediction is shown below.

Model equations for wheel velocity:

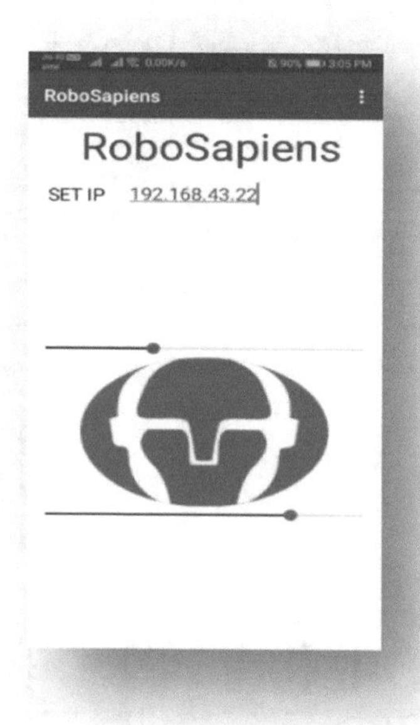

Fig. 5 Android app to control the mobile robot manually

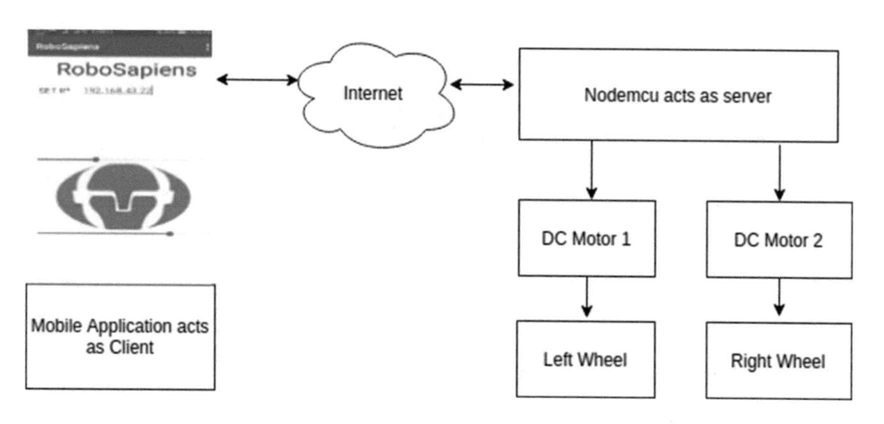

Fig. 6 Connections between android application and nodemcu

$$V[L] = W0 + W1 * x + W2 * x^2 + \ldots + Wd^d$$

$$V[L] = W'0 + W'1 * x + W'2 * x^2 + \ldots + W'd^d$$

V[l] = velocity of the left wheel
V[r] = velocity of the right wheel.

Table 1 Sample dataset for the machine learning model

Sonar Distance (cm)	Left wheel velocity(rpm)	Right wheel velocity(rpm)
30	1000	1000
25	950	950
23	900	855
21	850	800
15	800	500
10	800	200

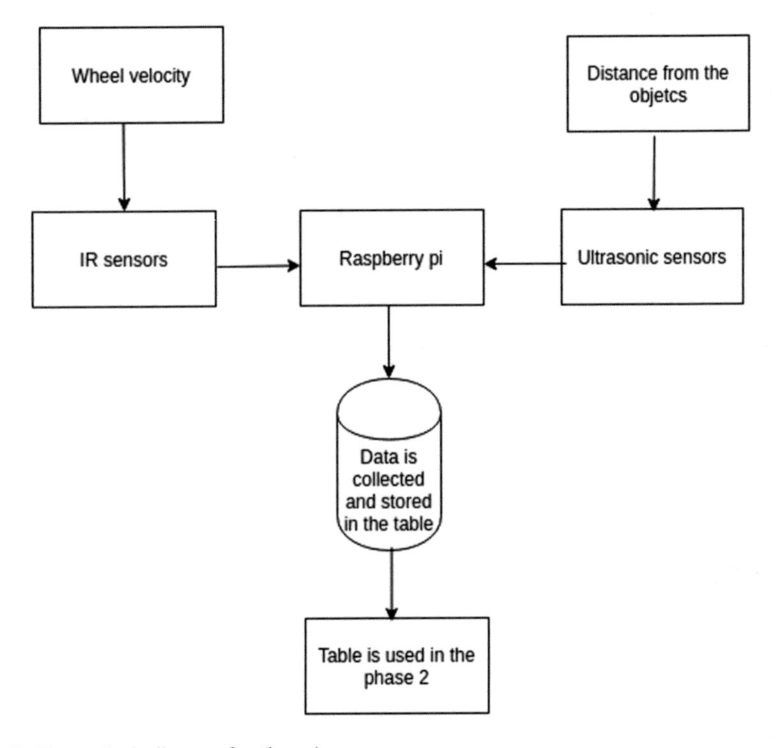

Fig. 7 The analysis diagram for phase 1

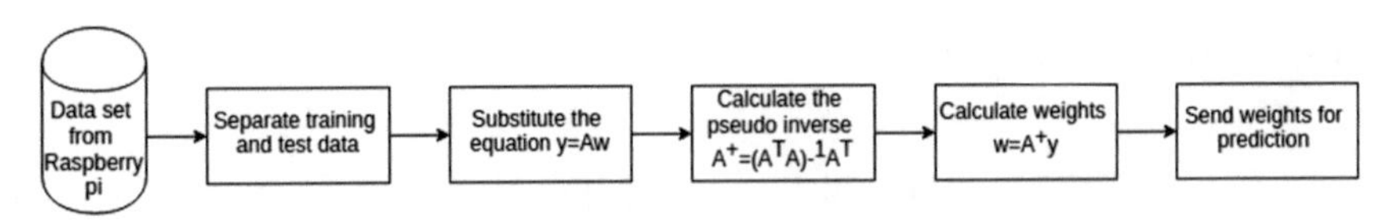

Fig. 8 Pseudo inverse algorithm to find weights of the model

where w0, w1,w2… wd, w'0, w'1, w'2… w'd are the parameters of the model which are found out using stochastic gradient descent algorithm. Pseudoinverse algorithm can also be used to find out the weights 'w.' The equation to find out w using stochastic gradient descent algorithms is shown below.

Updated rule for weights in stochastic gradient descent:

Error function $E = \frac{\sum E_i}{n}$

$E_i = (y - yd)^2$

Y = predicted velocity

Yd = desired velocity

w[i] = w[i] – η * dw[i]

where 'η' is a chosen parameter called learning rate which varies between 0 and 1 and dw[i] is the gradient of the error function corresponding to the weight w[i].

By iteratively updating the weights, the error function is minimized and the optimal weights are found out for the model.

The final model obtained by the optimal weights can be used in the mobile robot to achieve the obstacle avoidance algorithm.

The other way to find the weights of the model is using the pseudoinverse algorithm shown in Fig. 8.

3 Obstacle Avoidance Algorithm

There are three ultrasonic sensors placed on the mobile robot during testing phase on the front, left, and right sides.

The obstacle avoidance algorithm for the mobile robot is shown below:

1. Read the distance value from front side ultrasonic sensor of the mobile robot.
2. If the distance is greater than 25 cm, set the left wheel velocity and the right wheel velocity of the mobile robot as follows, otherwise go to step 3.

   V[l] = V[r] = 1000

   Here, 1000 is the pwm value sent to the DC motors.

3. If the distance is less than 25 cm, then read the distances values of ultrasonic sensors from left and right sides of the mobile robot.

   dl = left side distance of the obstacle

   dr = right side distance of the obstacle

(a) If dl < 10 cm and dr > 10 cm, then the robot should turn right side. We use the machine learning model to calculate the wheel velocities using right turn model.

(b) If dl > 10 cm and dr < 10 cm, then the robot should turn left side. We use the machine learning model to calculate the wheel velocities using the right turn model and interchange the left and right wheel velocities.

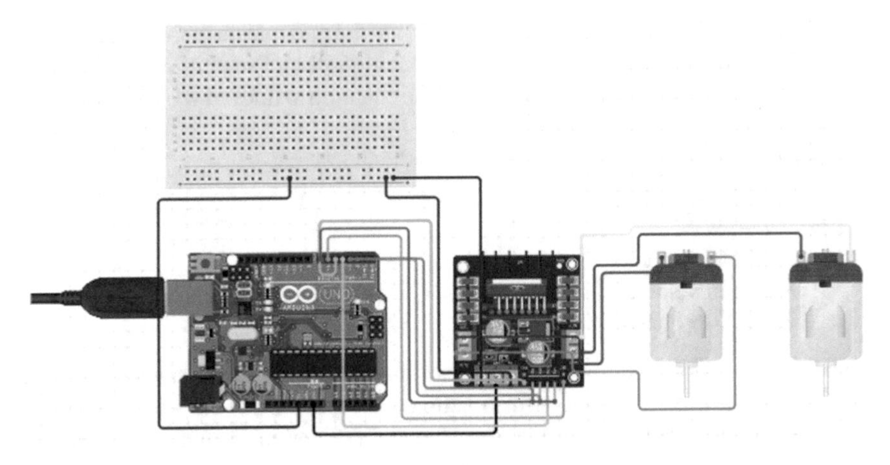

Fig. 9 Circuit diagram for implementing autonomous mobile robot

(c) If dl < 10 cm and dr < 10 cm, the robot turns backward. We set one of the wheel velocity to 0 and the other wheel velocity to 1000 for 2 s to turn the mobile robot in the backward direction.

The robot avoids the obstacle by changing its direction based on the magnitude of difference between the left wheel velocity and right wheel velocity. Thus, obstacle avoidance is achieved using the given object avoidance algorithm.

4 Implementing Obstacle Avoidance

The circuit for implementing the autonomous mobile robot is shown in Fig. 9.

The Arduino board [15] contains the code for implementing the linear regression model by substituting the distance value of the ultrasonic sensors. The arrangement of sensors over the mobile robot should as shown in Fig. 3.

5 Results and Analysis

The accuracy of the obstacle avoidance model depends on the accuracy of training data collected during the training phase. The model built using degree 3 linear equations performs better than the degree 2 model. The algorithm used is very simple to implement. The time complexity of the algorithm to predict the wheel velocities is O(1). We get the prediction instantaneously which makes it suitable for real-time obstacle avoidance.

The testing phase of the robot is shown in Figs. 10 and 11.

Fig. 10 Robot approaching the obstacle

Fig. 11 Robot avoiding the obstacle

6 Conclusion and Future Scope

In this paper, an autonomous obstacle avoidance robot is developed which is suitable for real-time applications. The robot developed serves as a prototype for implementation in real time applications. The robot can explore unknown environments and an IP cam can be placed on it for remote surveillance of the area. There are various applications for these types of robots in military vehicles, surveillance robots, unmanned vehicles, and self-driving cars.

An IP camera can be placed on the mobile robot which can be connected to a mobile application which enables live streaming of the surveillance area by connecting to the internet. The video data collected can be given as input to data mining algorithms and pattern recognition algorithms to extract useful information.

References

1. Peng, Yan, Dong Qu, Yuxuan Zhong, Shaorong Xie, Jun Luo, and Jason Gu. "The obstacle detection and obstacle avoidance algorithm based on 2-d lidar." In 2015 IEEE International Conference on Information and Automation, pp. 1648–1653. IEEE, 2015
2. Tsai, Cheng-Pei, Chin-Tun Chuang, Ming-Chih Lu, Wei-Yen Wang, Shun-Feng Su, and Shyang-Lih Chang. "Machine-vision based obstacle avoidance system for robot system." In 2013 International Conference on System Science and Engineering (ICSSE), pp. 273–277. IEEE, 2013
3. Jin, Yun, Shengquan Li, Juan Li, Hongbing Sun, and Yuanwang Wu. "Design of an Intelligent Active Obstacle Avoidance Car Based on Rotating Ultrasonic Sensors." In 2018 IEEE 8th Annual International Conference on CYBER Technology in Automation, Control, and Intelligent Systems (CYBER), pp. 753–757. IEEE, 2018
4. Wu, Ter Feng, Pu Sheng Tsai, Nien Tsu Hu, and Jen Yang Chen. "Use of ultrasonic sensors to enable wheeled mobile robots to avoid obstacles." In 2014 Tenth International Conference on Intelligent Information Hiding and Multimedia Signal Processing, pp. 958–961. IEEE, 2014
5. Bokade, Ashish U., and V. R. Ratnaparkhe. "Video surveillance robot control using smartphone and Raspberry pi." In 2016 International Conference on Communication and Signal Processing (ICCSP), pp. 2094–2097. IEEE, 2016
6. Singh, Diksha, Pooja Zaware, and Anil Nandgaonkar. "Wi-Fi surveillance bot with real time audio & video streaming through Android mobile." In 2017 2nd IEEE International Conference on Recent Trends in Electronics, Information & Communication Technology (RTEICT), pp. 746–750. IEEE, 2017
7. Kadiam, Vineela, and G. Pavani. "Smart Phone Controlled Two Axes Robot for Video Surveillance Using Wireless Internet & Raspberry Pi Processor." International Journal of Research in Advent Technology 2, no. 10 (2014): 97–100
8. Zhang, Lun, Stan Z. Li, Xiaotong Yuan, and Shiming Xiang. "Real-time object classification in video surveillance based on appearance learning." In 2007 IEEE Conference on Computer Vision and Pattern Recognition, pp. 1–8. IEEE, 2007
9. Tai L, Li S, Liu M (2016) A deep-network solution towards model-less obstacle avoidance. In: 2016 IEEE/RSJ International conference on intelligent robots and systems (IROS), pp 2759–2764. IEEE
10. Yang S, Konam S, Ma C, Rosenthal S, Veloso M, Scherer S (2017) Obstacle avoidance through deep networks based intermediate perception. arXiv preprint. arXiv:1704.08759
11. Aguilar WG, Casaliglla VP, Pólit JL (2017) Obstacle avoidance for low-cost UAVs. In: 2017 IEEE 11th international conference on semantic computing (ICSC), pp 503–508. IEEE
12. Duguleana M, Mogan G (2016) Neural networks based reinforcement learning for mobile robots obstacle avoidance. Expert Syst Appl 62:104–115
13. Chakravarty P, Kelchtermans K, Roussel T, Wellens S, Tuytelaars T, Van Eycken L (2017) CNN-based single image obstacle avoidance on a quadrotor. In: 2017 IEEE international conference on robotics and automation (ICRA), pp 6369–6374. IEEE

14. Aguilar WG, Casaliglla VP, Pólit JL, Abad V, Ruiz H (2017) Obstacle avoidance for flight safety on unmanned aerial vehicles. International work-conference on artificial neural networks, pp 575–584. Springer
15. https://www.arduino.cc/en/Main/Software
16. https://www.nodemcu.com/index_en.html
17. https://appinventor.mit.edu/explore/ai-with-mit-app-inventor

ABC-BSRF: Artificial Bee Colony and Borderline-SMOTE RF Algorithm for Intrusion Detection System on Data Imbalanced Problem

Pullagura Indira priyadarsini

Abstract In the current scenario, machine learning techniques are advantageous for making better decisions. They indeed are important and achieve better results in Intrusion Detection Systems (IDSs). Even though prominent classifiers may produce outstanding results regarding the majority classes, they are biased. Remarkably, the Imbalanced aspect of data leads to generate inaccurate results since the minority classes are not adjudicated properly and result in misclassification costs. In this paper, the classification problems due to imbalanced data are effectively addressed by the latest evolving techniques. To overcome this difficulty and improve the model performance, an acute algorithm, named Artificial Bee Colony Borderline SMOTE on Random Forests (ABC-BSRF) is proposed. It constitutes Artificial Bee Colony (ABC analysis) for Feature selection and Borderline SMOTE through random forests for oversampling. The results are compared with individual classifiers such as Support Vector Machines (SVMs), Decision Trees, and K-nearest neighbor (KNN). Observed results inferred from the experimentations done on KDD cup 99 dataset have proved that our proposed work can be excellently resolving the issue of imbalanced data. The ROC curve, F1 score, Precision, Recall, and AUC have shown noticeable results in contrast with other traditional methods.

Keywords Area under curve, artificial bee colony · Borderline SMOTE · Decision trees · K-nearest neighbor · Machine learning · Intrusion detection · Imbalanced data set · Oversampling algorithm · Random forest

1 Introduction

With the huge advancement of technology, enterprises are constantly stacking up massive information from numerous sources. Congregating, preprocessing, and then cataloguing these chock-full amounts of data are a toughened problem in recent times. In this respect, advanced machine learning techniques are robust to handle

---

P. I. priyadarsini (✉)
Vardhaman College of Engineering, Shamshabad, Hyderabad, Telangana, India
e-mail: indupullagura@gmail.com

N. Chaki et al. (eds.), *Proceedings of International Conference on Computational Intelligence and Data Engineering*, Lecture Notes on Data Engineering and Communications Technologies 56, https://doi.org/10.1007/978-981-15-8767-2_2

such overwhelming data. In the branch of data mining, Classifier learning with data sets that undergo imbalanced class distributions is an interesting point at issue [1]. Such a one arises if the number of examples that denote one class is much lesser than the ones of the other classes [2]. Its existence in several real-world applications such as text classification, fraud detection, and medical diagnosis has fetched the progress of attention from analysts. Considerably, a Classifier inclines to benefit the majority class. Whenever samples in the data set are uneven, there is a strong possibility that the classification function will end in false outcomes. Correspondingly in machine learning, the ensemble of one or more classifiers will augment the accuracy rather than neither of the individual classifiers and also solves the imbalanced class distribution hindrance. There are several novel ensemble learning algorithms that are instigated [3–6]. The performance of data mining algorithms deteriorates when the data set is imbalanced [7]. Thereupon, an apt model is a prerequisite to act on this difficulty.

Prevailing classification algorithms deteriorate if the data is skewed toward one class. One of the solutions to this dispute is sampling. It is a common action intended for exterminating the imbalance problem in various fields. Generally, sampling is of two types: oversampling and undersampling. There are several techniques to cope with unbalanced data. There is no technique which is good to work consistently better in all settings. At recent times, numerous oversampling techniques were proposed namely Synthetic Minority Oversampling Technique (SMOTE) [8], Borderline SMOTE [12], and random oversampling. Their predominance is that no data is lost. SMOTE is normally considered as a stereotype for oversampling algorithms [9–11]. SMOTE yields synthetic minority samples.

Furthermore, another issue which declines classification performance is data dimensionality. A data set with a plethora of features which are irrelevant or redundant leads to degradation of classification efficiency. Likewise, the Intrusion Detection System (IDS) data set is enormous and takes more time to get categorized. With this in mind, several feature selection methods are applied for preventing these irrelevant/redundant features without dropping IDS efficiency. Still noticing the features allied to the feature selection like simple to classify, shorten the processing time and improvise the classification proficiency and accuracy rate. Feature selection approach for IDS is determining [13, 14].

To overcome the abovementioned issues, Artificial Bee Colony Borderline SMOTE with Random Forests (ABC-BSRF) algorithm is proposed. It includes three principal phases: preprocessing, feature selection, and oversampling with classification for intrusion detection. The first phase is very essential for any data mining problem. The chosen data set KDD Cup 99 [15] is the typical data set for the Intrusion Detection System (IDS). The second phase is selecting features using Artificial Bee Colony optimization (ABC optimization). The third phase is applying oversampling technique Borderline SMOTE with random forests for classification of intrusion and normal data. Various experimentations, against three formal feature selection methods, are done to verify the observable results obtained. The Predictions are compared with individual classifiers Support Vector Machines (SVMs), K-nearest neighbor (KNN), and Decision Trees. This paper is well-arranged and supervened below. Section 2 will summarize the literature related to this area. The related algorithms are outlined in Sect. 3. And the propounded algorithm is given in Sect. 4. Experiments conducted and results inferred are depicted in Sect. 5. Conclusion and perspectives are drafted in Sect. 6.

2 Related Work

An Imbalanced problem in IDS is professed in lots of existing effective methods. In recent researches, Gaffer et al. [16] have given a novel fitness function for dealing with the problem of imbalanced data on KDD Cup 99 data set and have shown noticeable results. The works of Thomas et al. [17] have provided data-dependent decision fusion strategy and then different weights are remitted to the corresponding IDS. The fusion augments the weighted results to provide a single conclusion that is superior to existing IDSs and other existing ensemble techniques like OR, AND, ANN, and SVM. Mohammad et al. [18] have given a hybrid approach which is a combination of SMOTE and CANN (Cluster Center and Nearest Neighbor). It used Leave One Out method (LOO) for selecting important features in the NSL KDD data set.

Similarly, Ofek et al. [19] projected a fast clustering method built on the undersampling strategy for the binary-class imbalanced dilemma. An explicit classifier is accomplished for each cluster and outcomes are weighted. Verbeke et al. [20] demonstrated an oversampling method which copies the minority data to the training set. They indicated that simply oversampling the minority class with similar data (copied tuples) hasn't shown noticeable enhancement in the outcome of the classifier.

Some of the other oversampling techniques include Adaptive Synthetic Sampling (ADASYN), Mega-trend Diffusion Function (MTDF), and CUBE approach. ADASYN is also a powerful oversampling technique which progresses the learning in an effectual mode [21]. MTDF was first proposed by Li et al. [22] for estimating the domain range of a data set and obtaining artificial samples for training the Backpropagation Neural Network (BPNN). Its purpose is to balance the data set. MTDF oversampling is also combined with the Hybrid SVM data reduction technique [23]. CUBE is also an alternate oversampling procedure [24]. It is constructed as a geometric depiction of the sampling strategy. It is generalized for choosing proximately balanced samples with equal or unequal inclusion probabilities and any number of auxiliary variables. Japkowicz [25] has shown that both oversampling and undersampling are eminent in treating the problem of imbalanced data.

Based on the strategy that some specific features have more tendencies in improvising classification accuracy, feature selection is considered as a key step in data mining. In a research made by Wang et al. [26], a transformation of original features with logarithms of the marginal density ratios is done, obtaining novel, transformed features improving the execution of an SVM model. At recent times, Hajisalem et al. [27] illustrated a fusion of two methods such as artificial bee colony (ABC) and artificial fish swarm (AFS) along with the fuzzy C-means clustering (FCM) used for dividing the training data set and correlation-based feature selection (CFS) techniques for feature selection. Zhang et al. [28] have given a cost-based feature selection technique with multi-objective particle swarm optimization (PSO). They have compared multi-objective PSO with numerous multi-objective feature selection strategies tested on five benchmark data sets.

In a work done by Ren et al. [29], they have given a data augmentation method to construct IDS, named DO_IDS. In this, they used data sampling, iForest for sampling data, and fusion of Genetic Algorithm (GA) and Random Forest (RF) for optimizing sampling ratio. In the process of feature selection, a combination of GA and RF is done for selecting the optimum feature subset. Then DO_IDS is assessed by the intrusion detection data set UNSWNB15. Recently, Prudent Intrusion Detection System (PIDS) [30] was proposed using ensemble methodology which works with Ensemble Feature Selection (EFS) and Ensemble Classification (EC) algorithms and has shown that ensemble outperforms other classifiers.

3 Outline of ABC Analysis, SMOTE, and Random Forests

3.1 Feature Selection: ABC Analysis

The purpose of choosing this methodology is to find the finest features for IDS classification. For instance, in a colony, each type has its functions. It was first given by Karaboga [31], by simulating the foragers' behavior of obtaining food sources. According to the emphasis of the ABC analysis, a food source specifies a solution (for instance, food resource location) related to the problem, and the food source nectar amount shows the quality of the solution (for instance, fitness). The ABC procedure has four segments: initialization phase, employee bee phase, onlooker bee phase, and scout bee phase [32]. It is an iterative method. In the ABC analysis, a group of bee population is created where half of them are employee bees and the other half are onlooker bees. Its advantages are a few control parameters and fast convergence.

The common algorithmic organization of the ABC optimization method is given as follows:

Initialization Phase:

For each solution $x_i (i = 1, 2, S_N)$ where S_N is the number of solution cycle =1;

WHILE (cycle < = Maximum Cycle Number or a Maximum CPU time) DO

(1) Employed Bees Phase
(2) Onlooker Bees Phase
(3) Scout Bees Phase
(4) Memorize the best solution achieved so far
(5) cycle = cycle + 1; END WHILE

3.2 Oversampling: Borderline Synthetic Minority Oversampling Technique (B-SMOTE) Through Random Forests

In general, several models yielded a bias toward the classes with the maximum amount of samples in the case of imbalanced data. So, the model prediction cannot be accurate. Typical imbalanced data handling approaches include oversampling and undersampling. Yet, there are several shortcomings occurring in these two methods. The SMOTE (Synthetic Minority Oversampling Technique) approach is the latest one for oversampling and is given by Chawla et al. [19]. In essence, it intends to create the decision boundaries of the minority class more generally. It removes overfitting. The basis of this method is to produce new observations in the minority class by incorporating the existing ones. The process is as supervened as (1) For every observation x of the minority class, identify its K-nearest neighbor; (2) Choose randomly a few neighbors (the number depends on the rate of oversampling); and (3) Artificial observations are spread along the line joining the original observation x to its nearest neighbor.

There are some of the improved algorithms such as Borderline SMOTE [12], SMOTE–ENN [32], and so on, attaining enhanced proficiency than SMOTE. Batista et al. [33] announced a relative analysis of numerous sample schema (i.e., Edited Nearest Neighbor rule or ENN and SMOTE) to regulate the training set. They detached the redundant or irrelevant data from the training data, which enhanced the mean number of induced rules and also augmented the efficiency of SMOTE + ENN.

If minority class samples of data are copied, oversampling will be resulting in noise data that in turn increases the processing time, overfitting, and thus weakens the efficiency. Innumerable algorithms have been recommended for progressing SMOTE. Wang et al. [34] have stated that the SMOTE algorithm produces class overlapping or over-generation. Han [12] offered the Borderline SMOTE algorithm. It is the one which oversample examples which lie on the borderline of the minority classes. So for each minority sample, its K nearest neighbors of the same class are computed, then some samples are randomly chosen form them based on the oversampling rate. Then, new synthetic examples were made along the line between the minority example and their preferred nearest neighbors. Here in this approach, only borderline minority samples are oversampled. For achieving higher prediction, major classification algorithms endeavor the borderline of each class as precisely as possible in the training practice. The examples on the borderline and the ones nearby are more apt to be misclassified than the ones far from the borderline, and thus are more important for classification. Therefore, those examples far from the borderline may accord little to classification. Recently, two minority oversampling techniques namely borderline-SMOTE1 and borderline- SMOTE2 were proposed where the borderline samples of the minority class are oversampled.

These methods are disparate from the principal oversampling methods in which entire minority examples or an arbitrary subset of the minority class are oversampled

[35, 36]. Between borderline-SMOTE1 and borderline-SMOTE2, one of them can be operated.

Random Forest (RF) is an aggregation of supervised machine learning algorithms, which was first proposed by Breiman [37]. RF is capable of handling both binary and multi-class classification problems. Its classification efficiency is superior to other single classifier models. The foremost view of RF is to use random sampling with substitution to build several decision trees, and the outcome is attained by the method of voting.

The procedure for building RF is as follows:

(1) With random sampling, the substitution is done to extract samples from the data set and attain a training subset.
(2) For the training subset, features are randomly extracted from the feature set without substitution as the basis for splitting each node in the decision tree. From the root node, a complete decision tree is created from top to bottom.
(3) The decision trees are created by executing step (1) and step (2) reiterated "k" number of times.

Then finally, RF classifier is the one that is obtained by merging these decision trees. The outcome of classification is designated by these decision trees.

4 Proposed Methodology

In this methodology, the classification problems due to imbalanced data are effectively addressed by the latest evolving techniques. To overcome the difficulties of imbalanced data and improve the model performance, an acute algorithm named Artificial Bee Colony Borderline SMOTE on Random Forests (ABC-BSRF) is proposed. Globally, Ensemble modeling is an excellent approach theoretically and realistically producing better accuracy than any single classifier [38]. Since Ensemble learning/modeling have less overfitting, high variance, and improves prediction accuracy, the authors have used it in the proposed methodology. Ensemble learning in IDS is effective in enhancing the attack detection rate and lessening the FAR. Hence, the Ensemble methodology is promoted in this work done. The proposed work is given in three steps: preprocessing, feature selection, and oversampling with classification for intrusion detection. The first phase is indispensable for any data mining problem. Then, Feature selection is done using the Artificial Bee Colony algorithm. Later, Borderline SMOTE is applied with Random Forest for classification in Intrusion Detection System. Merging these three viewpoints is justified as they have rational pinpoints. Marking highlights of Artificial Bee Colony analysis are its faster convergence and usage of a few control parameters. It also outstrips the downsides of other evolutionary algorithms which eventually got stuck in their local minima resulting in irrelevant outcomes. Aiming at the ABC analysis can lead to the retrieval of the best features. The other one B-SMOTE centers on improvising the prediction competence of the minority class, while the last one Random Forests fixes on giving accuracy outputs and is a powerful aid for high-dimensional data. Its computational cost for training is a bit low. The whole process is done by using integrated ABC-BSRF algorithm. The mechanism of the proposed work is given below in Fig. 1.

Proposed ABC-BSRF Algorithm

In the ABC-BSRF algorithm, the crucial part is oversampling. Oversampling will gradually increase minority class performance. Random Forest is the ensemble technique which is robust to overfitting. In several applications, Random Forests were applied in unbalanced data sets. The proposed ABC-BSRF algorithm is given below in Fig. 2.

Figure 2 above describes the proposed ABC-BSRF algorithm. Steps 1 and 2 are employed for preprocessing the data. Normalization is customarily applied to the data set. And also duplicates are removed. In step 3 and step 4, Artificial Bee Colony is applied and their scores are obtained. On the scores, a threshold is kept. Accordingly, the best features are chosen. Ensemble classifier is created by steps 5 and 6. Firstly, the KDD data set is apportioned into a training data set and test data set. And then a model is built based on B-SMOTE and random forests. While Training data set is exploited for training the classifier, the test data set is exploited for testing the ensemble classifier. In step 7, testing has been done. In step 8, validation is done on the testing data. Step 9 is used for classifying attack and normal data. Finally, step 10 is used to quantify the proficiency of the classifier.

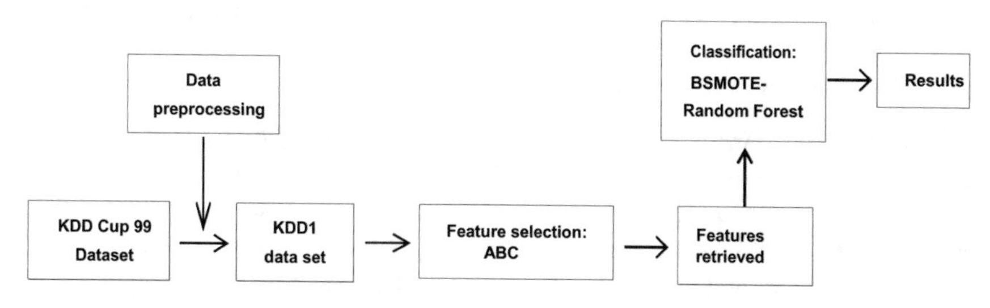

Fig. 1 Proposed flow of work

Algorithm: ABS-BSRF Algorithm
Input: KDD1 data set

Step 1: Convert non-numerical values to numerical
values Step 2: Apply normalization on the data set
Step 3: Perform Artificial Bee Colony for conducting Feature selection
Step 4: Select the best features
Step 5: Data set is scattered into training and testing
Step 6: Build ensemble classifier model: {Borderline SMOTE, Random forest}
Step 7: For each new sample F ∈ Test data
Step 8: Validate Ensemble classifier (test sample)
Step 9: Classify IDS attack data and normal data
Step 10: Evaluate the performance of ensemble classifier

Output: Results are evaluated using F-score, ROC curve, Precision and Recall

Fig. 2 Proposed ABC-BSRF algorithm

5 Experiential Work and Results Inferred

To perform the evaluation on Intrusion Detection System (IDS) using a proposed method, modeling is done on the KDD Cup 99 data set. This data set in our experiments is initiated from MIT's Lincoln Lab. It was extended for IDS evaluations by DARPA and is reflected as an emblematic for intrusion detection [15]. It is only 10 percent of the veritable data. It contains five million records. There are five classes and 41 features in it. From this KDD Cup 99 data set, a slice of it is retrieved for experimentation. It comprises 10230 instances and is named as "KDD" data set with a relative size of records as in the KDD Cup 99 data set with 1042 "Normal" instances and 9188 "attack" instances. It has discrete and continuous features. There are generally four types of attacks; they are DoS, Probe, U2R, and R2L. The five classes are mapped with 1, 2, 3, 4, and 5 for U2R, R2L, Probe, DoS, and Normal, respectively. Features are labeled as {A1, A2...., A41}.

The five classes in the KDD data set are defined as

(i) **DoS**: Denial of Service attack is the one where the attacker/intruder thwarts the authentic acquisition of the user over a machine.
(ii) **Probe**: A Probe attack is the one where the attacker searches for probable weakness on a computer system.
(iii) **Remote to local**: A Remote to local attack is the one where the attacker efforts to get illegitimate access to a computer system.
(iv) **User to root**: User to root attack is the one where an attacker tries to take advantage of Superuser privileges.
(v) **Normal**: It is normal data with good connections.

Now, Preprocessing is done to the KDD data set. Firstly, duplicates are removed. Then, Feature rescaling is accomplished for every feature individually. Necessarily, normalization has been applied to the data set. The KDD data set with its total instances of "attack" and "normal" is depicted in Fig. 3 below. It contains 1042 "normal" instances and 9188 "attack" instances. The KDD data set is separated as

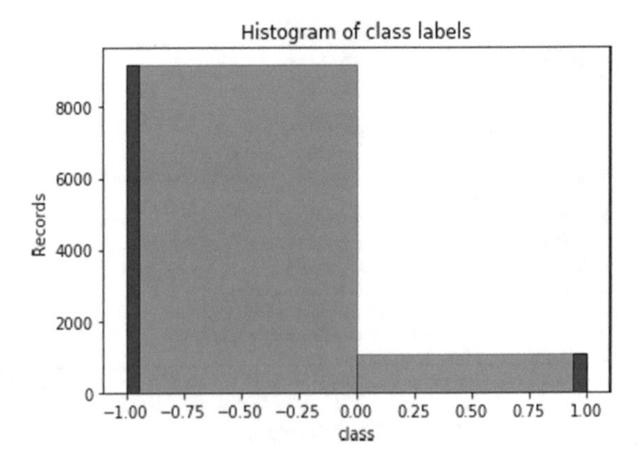

Fig. 3 Histogram of class distribution in the KDD data set

training and testing data. The training data is 20% and the testing data is 80% from the KDD data set.

The experiments on the proposed ABC-BSRF algorithm were conducted with R and Python interfaces in Anaconda 3.6 Environment [39]. The open-source form of Anaconda is an Excellency distribution of Python and embraces over 100 of the most common Python packages for data science. It is an open accessible tool with various modules such as Python and R language. The system with a processor of Intel i5, memory 8 GB, with system type 64-bit Operating System, and × 64 processor is exploited. It was usually employed in research areas like data science and applications of machine learning.

On the KDD data set, the ABC analysis is applied. The parameters for the ABC analysis are given as the total number of features N = 41, Max Limit = 20, Number of iterations = 100, and perturbation parameter MR is 0.1. As a result, based on the score obtained the features are selected. It is shown in Table 1 below. The ensuing features are 17.

Then KDD data set with 17 features is fed to the ensemble of Borderline SMOTE and Random Forests to construct a model. The feature selection of the ABC analysis is

Table 1 The result of ABC analysis on 41 features

S. no.	Feature #	Score obtained	S. no.	Feature #	Score obtained
1	A1	**0.04277964**	22	A22	−0.008572226
2	A2	**0.05071278**	23	A23	**0.005372601**
3	A3	−0.07364039	24	A24	**0.03184964**
4	A4	**0.1065986**	25	A25	−0.06492007
5	A5	**0.3156478**	26	A26	−1.063832
6	A6	**0.02174195**	27	A27	−0.007512986
7	A7	**0.1835877**	28	A28	−0.0004434346
8	A8	−0.4443892	29	A29	−0.6337133
9	A9	−0.616391	30	A30	−0.04900299
10	A10	−0.3542036	31	A31	−0.05859736
11	A11	−0.2304295	32	A32	**1.205842**
12	A12	**0.2442221**	33	A33	**0.7710653**
13	A13	−0.1079176	34	A34	−0.1081079
14	A14	**0.03973701**	35	A35	−0.1139069
15	A15	**0.6225633**	36	A36	**0.009961483**
16	A16	−.006361696	37	A37	**0.01144708**
17	A17	**5.243926**	38	A38	−0.0270871
18	A18	**0.006441499**	39	A39	−0.01110971
19	A19	−0.02183929	40	A40	−0.004748104
20	A20	−0.6958493	41	A41	−0.01068557
21	A21	−0.03246892			

Table 2 The number of features selected using various feature selection methods

S. no.	FS Method	Number of best features selected	Feature #
1	Chi-squared	15	A3, A4, A6, A7, A12, A13, A14, A24, A25, A33, A34, A35, A36, A37, A38
2	Information gain	13	A3, A4, A6, A7, A13, A24, A25, A33, A34, A35, A36, A37, A41
3	Gain ratio	12	A3, A4, A6, A7, A11, A13, A14, A23, A24, A25, A35, A41
4	ABC	17	A1, A2, A4, A5, A6, A7, A12, A14, A15, A17, A18, A23, A24, A32, A33, A36, A37

Table 3 The time taken to perform artificial bee colony analysis

	User	System	Elapsed
ABC analysis (ms)	0.61	0.39	1.02

compared with information gain, chi-square, and gain ratio. It is indicated in Table 2. The time taken to perform the ABC analysis is given in Table 3.

Then to assess the performance of the proposed methodology with BSRF (Borderline SMOTE with Random Forests), these two distinct data sets were used: Training and Testing.

(1) Training data set: The purpose of the training data set is to train the instances and construct a model.
(2) Testing data set: Testing data set is aimed to evaluate the performance of the ensemble classifier.

After the model building, the results are recorded. Several experiments are conducted on the KDD data set: (1) ABC-KNN, (2) ABC-SVM, (3) ABC-Decision Tree, and (4) ABC-BSRF. The error rate of three classifiers and BSRF concerning different feature selection methods like Information gain, chi-square, gain ratio, and ABC analysis is compared graphically. It is given in Fig. 4. It clearly illustrates the lower error rate for the proposed algorithm. The execution of the proposed ensemble is assessed using Precision, Recall, F1 Score, and Accuracy and Receiver Operating Characteristic (ROC) curve. AUC for ABC-BSRF algorithm is 1.0. The comparison of SVM, KNN, and Decision Tree classifiers with the proposed BSRF approach are given in Table 4. The Receiver Operating Characteristic (ROC) curve for the KNN, SVM, Decision Tree classifiers, and the proposed BSRF approach is depicted in Figs. 5, 6, 7, and 8, respectively, below. The Accuracy is 1.0 and Recall is 1.0 for the proposed algorithm. Figure 9 presents the accuracy and recall.

The assessment standards employed for making the examinations are illustrated as follows. (1) The F1 Score is specified as 2 * (Precision*Recall)/(Precision + Recall). (2) The accuracy of a classifier is professed as the fraction of the number of corrected predictions to the complete number of input samples. (3) The Area Under

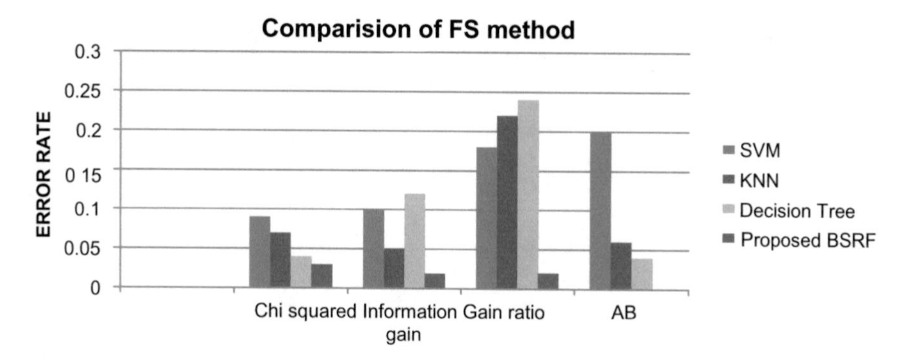

Fig. 4 The error rate of various classifiers with the chi-square, information gain, gain ration, and ABC analysis is given

Table 4 The results obtained for the proposed BSRF compared with other methods

S. no.	Classifier Used	Precision	Recall	F1 score	Accuracy	AUC
1	KNN	0.71	0.70	0.9	0.94	0.83
2	SVM	0.78	0.85	0.81	0.8	0.78
3	Decision TREE	0.67	0.83	0.9	0.96	0.90
4	Proposed BSRF	1.00	1.00	1.00	1.00	1.00

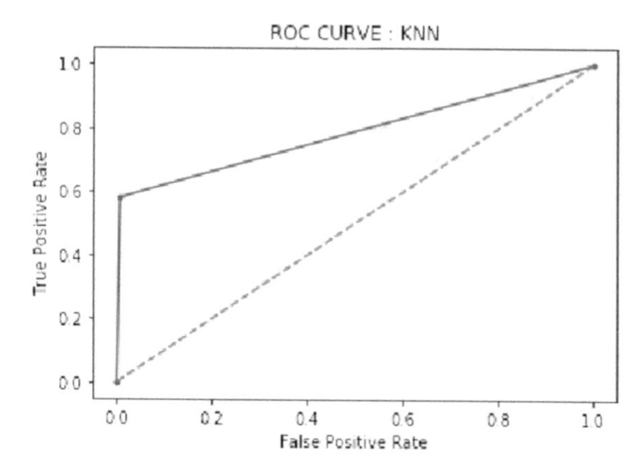

Fig. 5 The ROC curve for the KNN classifier with ABC analysis

the Curve is considered as AUC. It insists on the quality of the classification methodology. It is evaluated as the excellence of the model's estimates regardless of what classification threshold is taken. (4) Precision is expressed as the success likelihood

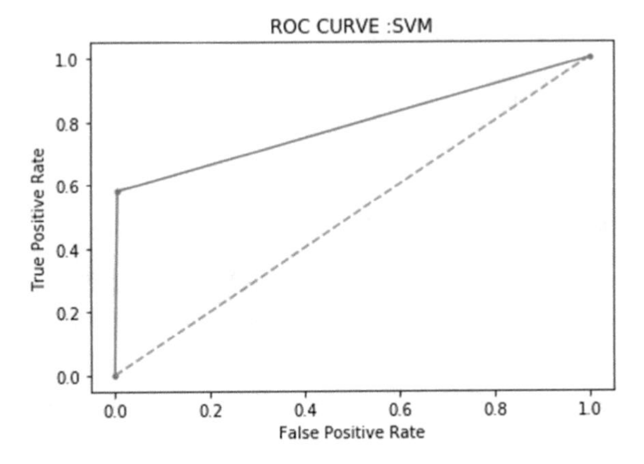

Fig. 6 The ROC curve for the SVM classifier with ABC analysis

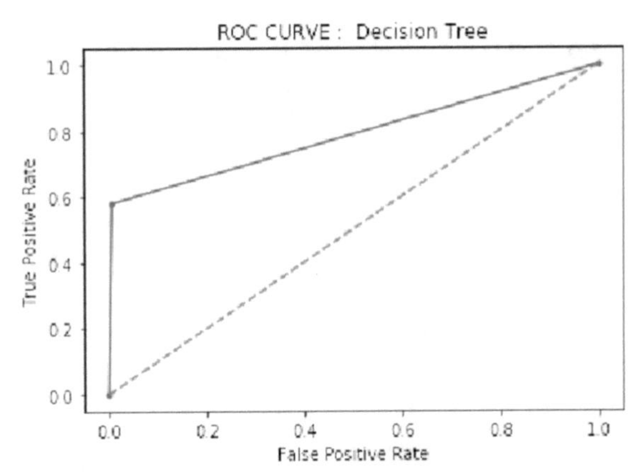

Fig. 7 The receiver operating characteristic curve obtained for decision tree classifier with ABC analysis

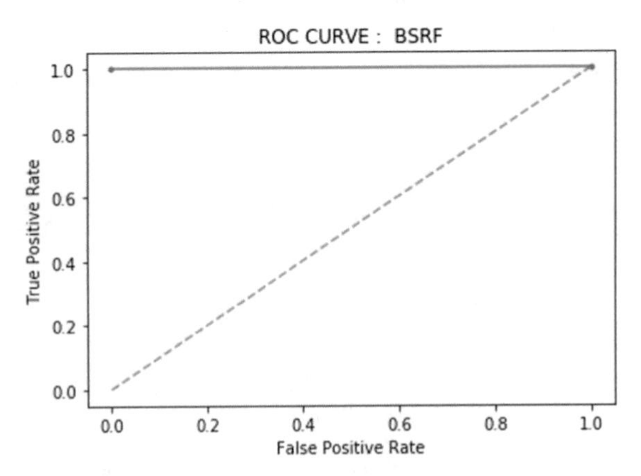

Fig. 8 The ROC curve for the proposed approach, the ABC-BSRF approach

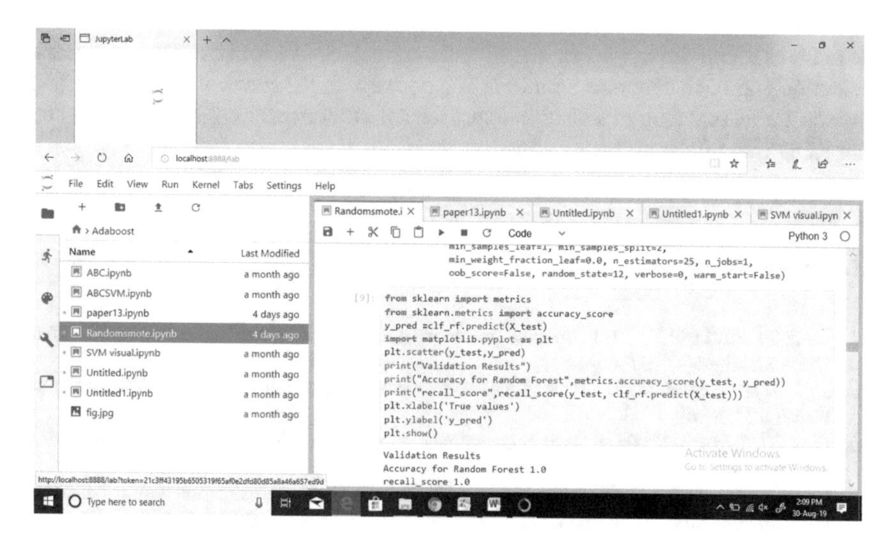

Fig. 9 The accuracy and recall of the proposed BSRF approach

of generating a correct positive-class classification. (5) A Recall is designated as the model's proficiency in finding out all the data points of interest in a data set.

6 Conclusions and Perspectives

Conventional imbalanced learning embraces oversampling and undersampling. Though, there are various delicacies occur in these two methods. There are some of the improved algorithms like Borderline SMOTE, SMOTE–ENN, and so on, attaining well than SMOTE. Undoubtedly, there are yet some problems when exploiting them in intrusion detection. In this paper, the classification problems due to an imbalanced data set are effectively addressed by cutting-edge techniques. To overcome this problem and improve the model performance, an innovative algorithm, namely ABC-BSRF based on Artificial Bee Colony (ABC algorithm) for feature selection and B-SMOTE on random forests is proposed. The results obtained through ABC-BSRF are compared against individual classifiers Support Vector Machines (SVMs), K-nearest neighbor (KNN), and Decision Trees. Feature selection approaches like Gain ratio, Information Gain, and Chi-Square are applied and validated for the above-said approach. Observed results on the KDD Cup 99 data set on the proposed algorithm have shown AUC to be 1.0 and F1 score to be 1.0. From the ROC curve, F1 score, Recall, and Precision, it is noticed that the proposed ABC-BSRF approach is an optimized one while comparing with other individual learning methods. On the other hand, Ensemble learning needs the minimum subset of features for attaining a better classification rate. Hence, identifying the apt feature selection approach becomes an interesting domain. In the forthcoming evaluations, the most effectual feature selection approach for other oversampling techniques for the KDD Cup 99 data set can be made.

References

1. Yang Q, Wu X (2006) 10 challenging problems in data mining research. Int J Inform Technol Decis Making 5(4):597–604
2. Jo T, Japkowicz N (2004) Class imbalances versus small disjuncts. ACM SIGKDD Explor Newsl 6(1):40–49
3. Oza NC, Tumer K (2008) Classifier ensembles: select real-world applications. Inform Fusion 9(1):4–20
4. Silva C, Lotric U, Ribeiro B, Dobnikar A (2010) Distributed text classification with an ensemble kernel-based learning approach. IEEE Trans Syst Man Cybern C 40(3):287–297
5. Yang Y, Chen K (2011) Time series clustering via RPCL network ensemble with different representations. IEEE Trans Syst Man Cybern C Appl Rev 41(2):190–199
6. Xu Y, Cao X, Qiao H (2011) An efficient tree classifier ensemble-based approach for pedestrian detection. IEEE Trans Syst Man Cybern B Cybern 41(1):107–117
7. Moayedikia A, Ong KL, Boo YL, Yeoh WG, Jensen R (2017) Feature selection for high dimensional imbalanced class data using harmony search. Eng Appl Artif Intell 57:38–49
8. Chawla NV, Bowyer KW, Hall LO, Kegelmeyer WP (2002) SMOTE: synthetic minority oversampling technique. J Artif Intell Res 16(1):321–357
9. Wang J, Xu M, Wang H, Zhang J (2006) Classification of imbalanced data by using the smote algorithm and locally linear embedding. In: Proceedings of the 8th international conference on signal processing, vol 3, pp 1–4
10. Ertekin CS (2013) Adaptive oversampling for imbalanced data classification. In: Proceedings of the 28th international sciences and systems, vol 264, pp 261–269
11. Chawla NV (2005) Data mining for imbalanced datasets: an overview. Data mining and knowledge discovery handbook, pp 853–867. Springer
12. Han H, Wang WY, Mao BH (2005) Borderline-SMOTE: a new over-sampling method in imbalanced data sets learning. In: Huang DS, Zhang XP, Huang GB (Eds.), ICIC 2005. LNCS, vol 3644, pp. 878–887. Springer, Heidelberg. https://doi.org/10.1007/1153805991
13. Sujitha B, Kavitha V (2015) Layered approach for intrusion detection using multi objective particle swarm optimization. Int J Appl Eng Res 10:31999–32014
14. Zorarpaco E, Ozel SA (2016) A hybrid approach of differential evolution and artificial bee colony for feature selection. Expert Syst Appl 62:91–103
15. Tavallaee M, Bagheri E, Lu W, Ghorbani AA (2009) A detailed analysis of the KDD CUP 99 data set. In: Proceedings of the 2009 IEEE symposium on computational intelligence in security and defense applications
16. Gaffer SM, Yahia ME, Ragab K (2013) Genetic fuzzy system for intrusion detection: analysis of improving of multiclass classification accuracy using KDD Cup-99 imbalance dataset. In: International conference on hybrid intelligent systems. IEEE press, pp 318–323
17. Thomas C (2012) Improving intrusion detection for imbalanced network traffic. Secur Commun Netw 6(3):309–324
18. Parsaei MR, Rostami SM, Javidan R (2016) A hybrid data mining approach for intrusion detection on imbalanced NSL-KDD dataset. (IJACSA) Int J Adv Comput Sci Appl 7(6):20–25
19. Ofek N, Rokach L, Stem R, Shabtai A (2017) Fast-CBUS: a fast clustering-based under sampling method for addressing the class imbalance problem. Neurocomputing 243(1):88–102

20. Verbeke W, Dejaeger K, Martens D, Hur J, Baesens B (2012) New insights into churn prediction in the telecommunication sector: a protdriven data mining approach. Eur J Oper Res 218(1):211–229
21. He H, Bai Y, Garcia EA, Li S (2008) ADASYN: adaptive synthetic sampling approach for imbalanced learning. In: Proceedings of the IEEE international joint conference on neural networks, IEEE world congress computational intelligence, pp 1322–1328
22. Li DC, Wu CS, Tsai TI, Lina YS (2007) Using mega-trend-diffusion and artificial samples in small data set learning for early flexible manufacturing system scheduling knowledge. Comput Oper Res 34(4):966–982
23. Teck CC, Xiang L, Junhong Z, Xiaoli L, Hong C, Woon D (2012) Hybrid rebalancing approach to handle imbalanced dataset for fault diagnosis in manufacturing systems. In: Proceedings of the 2012 7th IEEE conference on industrial electronics and applications (ICIEA), pp 1224–1229
24. Deville J-C, Tillé Y (2004) Efficient balanced sampling: the cube method. Biometrika 91(4):893–912
25. Japkowicz N (2000) The class imbalance problem: significance and strategies. In: Proceedings of the international conference on artificial intelligence, pp 1–7
26. Wang H, Gu J, Wang S (2017) An effective intrusion detection framework based on SVM with feature augmentation. Knowl-Based Syst 136:130–139
27. Hajisalem V, Babaie S (2018) A hybrid intrusion detection system based on ABC-AFS algorithm for misuse and anomaly detection. Comput Netw 136:37–50
28. Zhang Y, Gong DW, Cheng J (2017) Multi-objective particle swarm optimization approach for cost-based feature selection in classification. IEEE/ACM Trans Comput Biol Bioinf 14(1):64–75
29. Ren J, Guo J, Qian W, Yuan H, Hao X, Jingjing H (2019) Building an effective intrusion detection system by using hybrid data optimization based on machine learning algorithms. Secur Commun Netw 2019(7130868):11. https://doi.org/10.1155/2019/7130868
30. Priyadarsini PI, Nikhila K, Manvitha P (2018) Ensemble based framework for intrusion detection system. Int J Eng Technol 7(4)
31. Karaboga D (2005) An idea based on honey bee swarm for numerical optimization. Technical report-tr06, Erciyes university, engineering faculty, computer engineering department
32. Bansal JCH, Sharma H, Jadon SHS (2013) Artificial bee colony algorithm: a survey. Int J Adv Intell Paradigms 5:123–159
33. Batista GE, Prati RC, Monard MC (2004) A study of the behavior of several methods for balancing machine learning training data. ACM SIGKDD Explor Newslett 6(1):20–29
34. Putthiporn T, Chidchanok L (2013) Handling imbalanced data sets with synthetic boundary data generation using bootstrap re-sampling and adaboost techniques. Pattern Recogn Lett 34(3):1339–1347
35. Chawla NV, Japkowicz N, Kolcz A (2004) Special issue on learning from imbalanced data sets. SIGKDD Explor 6(1):1–6
36. Weiss G (2004) Mining with rarity: a unifying framework. SIGKDD Explor 6(1):7–19
37. Breiman L (2001) Random forest. Mach Learn 45:5–32
38. Ouyang MG, Wang WN, Zhang YT (2002) A fuzzy comprehensive evaluation based distributed intrusion detection. In: Proceedings 1st international conference on machine learning and cybernetics, pp 281–285. China, Beijing
39. Anaconda software. https://www.anaconda.com/download

A Novel Two-Layer Feature Selection for Emotion Recognition with Body Movements

M. M. Venkata Chalapathi

Abstract Programmed feeling acknowledgment from the investigation of body development can possibly change computer-generated reality, mechanical autonomy, conduct demonstrating, and biometric character acknowledgment spaces. A PC framework fit for perceiving human feeling from the body can likewise altogether change the manner in which we associate with the PCs. One of the critical difficulties is to recognize feeling explicit highlights from an immense number of descriptors of human body developments. Right now, we present a novel two-layer highlight choice structure for feeling order from an exhaustive rundown of body development highlights. We utilized the component choice structure to precisely perceive five essential feelings: satisfaction, pity, dread, outrage, and unbiased. In the main layer, one of a kind blend of Analysis of Variance (ANOVA) and Multivariate Analysis of Variance (MANOVA) was used to take out insignificant highlights. In the subsequent layer, a parallel chromosome-based hereditary calculation was proposed to choose a component subset from the significant rundown of highlights that expands the feeling acknowledgment rate. Score and rank-level combination was applied to additionally improve the exactness of the framework. The proposed framework was approved on restrictive and open datasets, containing 30 subjects. Diverse activity situations, for example, strolling and sitting activities, just as an activity autonomous case, were considered. In view of the exploratory outcomes, the proposed feeling acknowledgment framework accomplished a high feeling acknowledgment rate beating the entirety of the best in class strategies. The proposed framework accomplished acknowledgment exactness of 90.0% during strolling, 96.0% during sitting, and 86.66% in an activity free situation, exhibiting high precision and power of the created strategy.

Keywords Analysis of variance · Multivariate analysis of variance

M. M. Venkata Chalapathi (✉)
School of Engineering, Computer Science and Engineering Organization, Sri Satya Sai University of Technology and Medical Sciences Address Sehore, Bhopal, India
e-mail: mmv.chalapathi@gmail.com

N. Chaki et al. (eds.), *Proceedings of International Conference on Computational Intelligence and Data Engineering*, Lecture Notes on Data Engineering and Communications Technologies 56, https://doi.org/10.1007/978-981-15-8767-2_3

1 Introduction

Vocal inclination is a noteworthy factor in human correspondence. Right now, a machine to see sentiments from talk is an essential bit of the road manual to make correspondence among individuals and PCs progressively human-like. Notwithstanding, a great deal of investigate has been done to see sentiments subsequently from human discourse, low affirmation rate is up until now a significant issue. The general issue of customized talk feeling affirmation is that the acoustic indications of talk signals are affected by a combination of parts other than the inclination. The physiological complexities among all the speakers are the key trouble that prompts a low affirmation rate. Talks delivered by the condition of the vocal tract and its assortment with the time. Condition of vocal tract in which depends upon shape or size of vocal organs, unavoidably show solitary PCs.

In light of the above conversation, an expanding number of utilizations, that utilization body development data for emotionaltion acknowledgment, has risen. One of the ongoing works utilized a robot as a social arbiter to expand the nature of human–robot cooperation [7]. Feeling acknowledgment from body movement envelop countless applications including biometric security, medicinal services, gaming, and conduct modeling [5]. Instances of uses of feeling acknowledgment in biometric security area incorporate body development and outward appearance examination for video observation. Utilization of feeling acknowledgment in the clinical area incorporates identification of the mark conduct of patients having explicit mental conditions. In spite of a plentiful interest for an exact feeling acknowledgment from body development, this theme became slanting without a doubt, as of late.

Scientists have for the most part endeavored to perceive feelings from different modalities, for example, the face, head, and hand. Not many examinations have concentrated on entire body articulations for feeling investigation. Be that as it may, as expressed in, a PC model isn't just appropriate, however, may even surpass a human spectator capacity to perceive feeling, as it can recognize inconspicuous development changes not promptly obvious to the unaided eye. Additionally, body development data can be gotten noninvasively from a separation which might be gainful for some down to earth applications.

Past research concentrated uniquely on a set number of development highlights from countless calculable features. A fruitful endeavor to comprehend human feeling from entertainer's expressive body developments was vehicle ride out in. Creators presented a model dependent on Laban Movement Analysis (LMA) that coordinated Body, Effort, Shape, and Space highlights. Be that as it may, the rundown of highlights was exceptionally wide, unstructured, and a portion of the highlights were at no other time utilized for human feeling. Moreover, the pertinence factor of the highlights was rarely considered. The test is in this way to make a far-reaching rundown of movement includes that incorporate all nuanced development-related data applicable to the enthusiastic condition of an individual. At that point, the best blend of development highlights got utilizing a successful element choice calculation can be utilized to prepare AI calculations to perceive human feelings precisely. This paper explains all the previously mentioned challenges effectively.

Proposal of a one of a kind organizing of movement highlights into ten gatherings, each depicting an alternate part of human body development.

- Development of a two-layer include determination architecture that joins the intensity of a conventional channel-based methodology with a hereditary calculation.
- Identification of the most pertinent movement highlights for feeling acknowledgment from a far-reaching rundown of motion highlights. The significance factor was processed for a univariate situation where the highlights were considered freely, and a multivariate case, where highlights were considered as a feature of a gathering.
- Computation of highlight importance during two activity situations, which gives an extra understanding of the importance of highlights during feeling acknowledgment.
- Proposing a one of a kind blend of score and rank-level combination with two-layer highlight choice calculation to amplify the feeling acknowledgment precision.
- Introduction of a few new worldly highlights that exhibited upgrades over fleeting highlights, utilized pre-piously in the writing.

Primer work regarding this matter was done and distributed in.

2 Related Work

Feeling can be communicated through eye stare course, iris expansion, postural highlights, and development of the human body [5]. Pollick et al. [2] demonstrated that arm developments are altogether related to the enjoyableness measurement of the feeling model. Bianchi-Berthouze et al. presented a gradual learning model through gestural signs and a logical input framework to self-sort out postural highlights into discrete feeling classifications. In any case, those works were constrained to just pieces of the body. A few scientists endeavored to perceive feeling from move development.

Ca-murri et al. in extricated the amount of movement and constriction file from 2D video pictures delineating move developments of the subjects to perceive discrete feeling classifications. Recently, Durupinar et al. directed a perceptual report to set up a connection between the LMA (Laban Movement Analysis) highlights and the five-character attributes of a human. Senecal et al. broke down body movement articulation in theater execution dependent on LMA highlights. Specialists have likewise centered on perceiving feeling in arbitrary recording situations utilizing profound learning designs. In any case, those endeavors were restricted to explicit move developments.

Perhaps the greatest test of feeling acknowledgment is the high dimensionality portrayal of the movement highlights. Likewise, the writing gives next to no direction with respect to what kind of movement highlights are appropriate for feeling classification. A large portion of the current research has thought about a set number of highlights. Highlight importance was additionally not considered for feeling acknowledgment. Subsequently, the vast majority of the current research is one-sided towards a specific arrangement of movement highlights. For example, Glowinski et al. removed vitality, spatial degree, balance, and smoothness related highlights and afterward utilized Principal Component Analysis (PCA) to make a negligible portrayal of full of feeling motions. Saha et al. picked nine highlights identified with speed, acceleration, and rakish highlights to distinguish six feelings. This work effectively addresses the above lacks through the proposed far-reaching system for feeling acknowledgment, portrayed in subtleties in the following area.

3 Proposed Work

The main test is to make a total depiction of a human body development with feeling explicit recognizing data. This issue was overwhelmed by recognizing and figuring a thorough rundown of development highlights. These development highlights were then gathered into ten one of a kind categories so that every class spoke to an exceptional part of a body development (e.g., evenness, space, speed of movement, and so on.). The last rundown of highlights was processed dependent on the importance factor of these highlights utilizing a two-layer include choice calculation. The component choice system presented in segment III-A beats the trouble of identifying feeling explicit body development data.

A. OVERVIEW

The initial step of the proposed framework included the extraction of different geometric and kinematic highlights. A portion of these highlights was recently presented for 3D movement blend, grouping, and ordering. Scientists presently can't seem to build up an accord on the correct mix of different motion highlights. In this way, in the proposed feeling acknowledgment framework, an extensive rundown of movement highlights was separated expanding the accessible body development data. The movement highlights were figured either on a solitary casing or over an arrangement of edges traversed over a brief period. Accordingly, processed movement highlights portray different parts of human movement, for example, directions or geometric properties of the stances. These highlights were gathered into ten one of a kind restrictive gatherings which will be talked about in segment III-B.

Additionally, a fleeting profile was registered for every one of the highlights. The worldly profile comprises of 12, time arrangement capacities, as portrayed in segment III-C. A transient profile processed right now superior to a histogram with a fixed number of canisters. The quantity of receptacles of a histogram decides the degree of discretization of the determined features. A restriction of utilizing histogram is that the quantity of receptacles must be set observationally for the dataset. The estimations

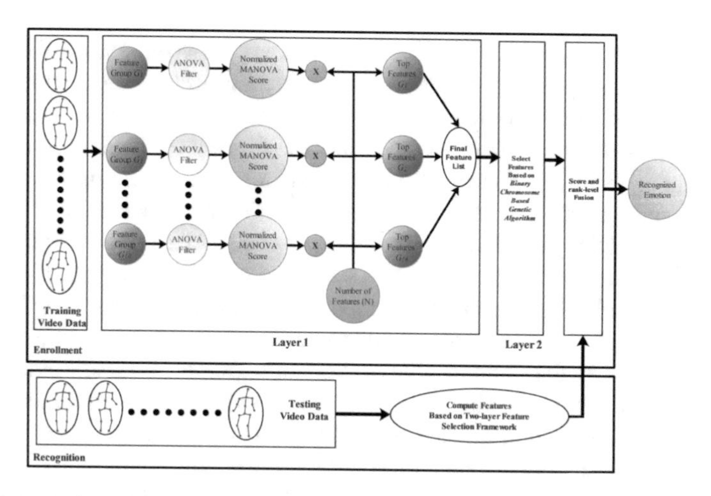

Fig. 1 An overview of the proposed framework for emotion recognition from body motion

of a histogram are additionally inadequate and the majority of the receptacles stay void after the histogram calculation.

The primary part of the proposed system includes a two-layer highlight choice procedure, as appeared in Fig. 1. In the principal layer, superfluous highlights are disposed of utilizing a mix of ANOVA and MANOVA. ANOVA is utilized to sort the highlights as indicated by their pertinence at perceiving feelings. ANOVA gives two measures: f-score and p-score to process the importance of a component. The f-score is a proportion of complete variety that exists among the number-crunching methods for the objective feelings. The p-score is a measure that decides the likelihood related to dismissing the f-score. The highlights that neglected to breeze through an essentialness assessment are disposed of right away. After the expulsion of these highlights, remaining highlights are considered as measurably pertinent for additional investigation.

MANOVA was utilized to figure bunch hugeness and to appropriate highlights among different component gatherings. The quantity of highlights considered from each gathering was inferred utilizing standardized MANOVA score processed for each gathering independently. The main layer may not be sufficient to accomplish ideal model execution dependent on the registered relevant highlights. The purpose behind this might be ascribed to the exhibition improvement of explicit component mix for certain master models. In this manner, a few top highlights from each element bunch were utilized as a contribution to the second layer of the system. The goal of the subsequent layer is to locate the best subset of highlights that amplifies the feeling acknowledgment pace of the master models.

As indicated by, the quantity of highlights can be chosen as a component of the example size, N, and the most extreme element size is N. In the proposed framework, the complete number of figured highlights was set depending on the example size of N. Since each gathering of highlights portrays an alternate part of human body development, the all-out number of highlights were dispersed among the element gatherings. Top ANOVA highlights were chosen from each element bunch dependent on the all-out number of highlights and the standardized MANOVA score registered for each gathering. MANOVA was utilized to measure bunch centrality, and the quantity of features registered from each gathering depended on the processed MANOVA score of the gathering. The gathering criticalness scores were standardized with the goal that each score ranges from 0 to 1 and their aggregate equivalents to 1. At that point, the processed MANOVA score was utilized to disperse the absolute number of highlights from each movement include gathering. Along these lines, just a portion of the top highlights from each movement includes bunch stayed for the ensuing advances. In the event that the quantity of highlights for a movement bunch surpassed the quantity of highlights that passed the ANOVA noteworthiness the highlights that created a p-score, which was higher than a predefined limit, was picked for the hereditary calculation. During the test, the p-score was picked as 0.005. This guarantees there exists a negligible possibility that the registered f-score was delivered from an alternate dissemination. Right now, the first layer utilized the importance of the highlights to get ready for the second layer of the proposed include choice structure. The subsequent layer utilizes the hereditary calculation that assesses the unmistakable capacity of the highlights to boost feeling acknowledgment exactness.

In the second layer of the two-layer structure, a twofold chromosome-based hereditary calculation was utilized to recognize the ideal element subset that amplifies the feeling recognition rate. The hereditary calculation utilized in the proposed system accomplished a level inside 80 ages. The transformation rate was set to 0.03, as portrayed in area.

B. MOTION FEATURE GROUPS

In view of a careful investigation of the current writing, all-inclusive rundown of 3D movement highlights was removed. These highlights were assembled into ten special classifications limiting the quantity of covering highlights depicting different body development types, however, much as could be expected.

- Group of Features 1 This gathering of highlights comprises of low-level element descriptors that measure the speed of the movement, for example, speed, increasing speed, and jolt. On the off chance that X characterizes a movement that is portrayed as n continuous postures, where X = x(t1), x(t2), x(t3), x(tn).

$$v^k(t_i) = \frac{X^k(t_{i+1}) - X^i(t_i)}{2\delta t}$$

$$\left\| v^k(t_i) \right\| = \sqrt{v_x^k(t_i)^2 + v_y^k(t_i)^2 + v_z^k(t_i)^2}$$

At that point, the speed is characterized in condition 1 and the magnitude of the speed is resolved utilizing the condition 2. In conditions 1 and 2, vk(ti) is the speed of the kth joint at time ti, vk(ti) is the x-segment of the speed of the kth joint at time ti, and δt alludes to a little portion of the time required for progressing between back to back casings. Normally, δt is set to an extremely little worth. During the test, the

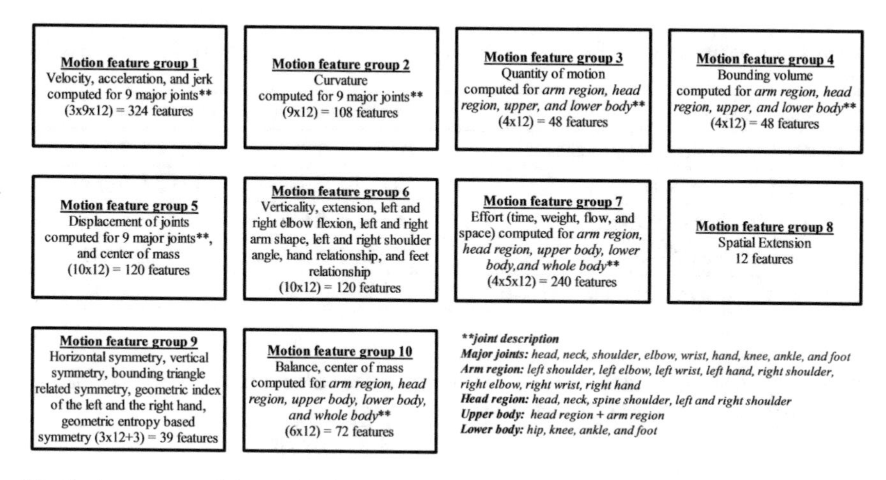

Fig. 2 A taxonomy of the motion features computed for emotion recognition from body motion features

worth was set to 1 s as Kinect v2 has an edge pace of 30 fps. test for that gathering, at that point, the entirety of the highlights that breezed through the assessment in that gathering was chosen (Fig. 2).

$$d_x = \left| \max_{k \varepsilon K} j_x - \min_{k \varepsilon K} j_x \right|$$

$$d_y = \left| \max_{k \varepsilon K} j_y - \min_{k \varepsilon K} j_y \right|$$

$$d_z = \left| \max_{k \varepsilon K} j_z - \min_{k \varepsilon K} j_z \right|$$

$$\text{Bounding Volume (BV)} = d_x * d_y * d_z$$

4 Experimental Results

The dataset gathered developments from 30 nonprofessional on-screen characters. As indicated by the specialists, the explanations behind picking nonprofessionals were to stay away from a precise exaggeration of developments and to build fluctuation of development articulation.

Every on-screen character performed developments communicating four unique feelings including impartial, joy, trouble, and outrage. Just verbal directions were given to the subjects. Feelings were seen during strolling scenarios. Two accounts were gathered from each subject for each enthusiastic walk. Hence, an aggregate of 240 examples of different enthusiastic strolling groupings was considered for the

Table 1 The table shows measured consistency scores for various filter-based feature selection algorithm

Filter-Based method	Walliing action	Sitting action
ReliefF (number of neighbors = 30)	0.403 ± 0.067	0.241 ± 0.072
Mutual Information (MI)	0.477 ± 0.051	0.311 ± 0.058
Analysis of Variance (ANOVA)	***0.262 ± 0.062***	***0.210 ± 0.0S5***
Chi-squared Score (CHI2)	0.476 ± 0.051	0 279 ± 0.061
Ensemble decision tree-based Method (EDT)	0.907 ± 0.059	0.772 ± 0.042

Table 2 The table shows measured monotonicity scores for various filter-based feature selection algorithm for expert models: LDA and SVM

	Walking action			Sitting action		
Filter-Based method	LDA SVM		Average correlation	LDA SVM		Average correlation
ReliefF (aumber of neighbors = 30)	−0.878	−0.931	−0.9045	−0.936	−0.965	−0.9505
Mutual Information (MI)	−0.873	−0.926	−0.8995	−0.951	−0.951	−0.951
Analysis of Variance (ANOVA)	***−0.908***	***−0.952***	***−0.93***	***−0.963***	***−0.969***	***−0.966***
Chi-squared Score (CHT2)	−0.841	−0.941	−0.891	−0.925	−0.916	−0.9205
Ensemble decision tree-based Method (EDT)	−0.834	−0.861	−0.8475	−0.917	−0.889	−0.903

test. Each subject wore a suit with retro-reflective markers. These markers had the option to reflect light spots to a two-dimensional space. By and large, 35 markers were set (Tables 1, 2).

5 Conclusion

Feeling acknowledgment utilizing body development is a rising territory of research. Critical advantages can be accomplished for biometric security, tolerant conduct observing, gaming, and mechanical technology with the production of a development-based feeling mindful PC framework. Body development data can give important signals identified with the passionate condition for each child. Regardless of demonstrating the incredible potential to be a basic indicator of saw feelings, body development data is one of the least investigated modalities for feeling acknowledgment. This paper has tended to the issue of production of a total framework that can precisely perceive five fundamental feelings: satisfaction, bitterness, outrage, dread, and unbiased dependent on body development highlights. The exploratory outcomes demonstrated that it is conceivable to assemble a PC framework fit for perceiving human feeling just dependent on body development data. Univariate and multivariate investigation of the movement highlights gave significant prompts in regards to apparent feeling. Experiment results gave basic data with respect to the apparent feeling in strolling and sitting activity situations. During strolling activity, the amount of development in the arm and the chest area locale were basic markers. Then again, body space usage, elbow point, and spatial expansion were fundamental signals to perceive feeling during the sitting activity. During activity free cases, movement highlights significant during all activity situations should be considered to amplify the feeling acknowledgment rate.

References

1. Hillman DR, Lack LC (2013) Public health implications of sleep loss: the community burden. Med J Aust 199(8):S7–S10
2. Lam JC, Sharma S, Lam B et al (2010) Obstructive sleep apnoea: definitions, epidemiology & natural history. Indian J Med Res 131(2):165
3. Obstructive sleep apnoea (osa; sleep apnea) information (2018) https://www.myvmc.com/diseases/obstructive-sleep-apnoea-osa-sleep-apnea/
4. Kee K, Naughton MT et al (2009) Sleep apnoea-a general practice approach. Aust Fam Physician 38(5):284
5. Lam B, Ip M, Tench E, Ryan C (2005) Craniofacial profile in asian and white subjects with obstructive sleep apnoea. Thorax 60(6):504–510
6. Barrera JE, Pau CY, Forest V-I, Holbrook AB, Popelka GR (2017) Anatomic measures of upper airway structures in obstructive sleep apnea. World J Otorhinolaryngol-Head Neck Surg 3(2):85–91
7. Schwab RJ, Pasirstein M, Pierson R, Mackley A, Hachadoorian R, Arens R, Maislin G, Pack AI (2003) Identification of upper airway anatomic risk factors for obstructive sleep apnea with volumetric magnetic resonance imaging. Am J Respir Crit Care Med 168(5):522–530

Nearest Neighbors via a Hybrid Approach in Large Datasets: A Speed up

Y. Narasimhulu, Raghunadh Pasunuri, and V China Venkaiah

Abstract A Spatial data structure such as kd-tree is a proven data structure in searching Nearest Neighbors of a query point. However, constructing a kd-tree for determining the nearest neighbors becomes a computationally difficult task as the size of the data increases both in dimensions and the number of data points. So, we need a method that overcomes this shortcoming. This paper proposes a hybrid algorithm to speed up the process of identifying k-nearest neighbors for a given query point q. The proposed algorithm uses lightweight coreset algorithm to sample K points. These points are then used as a seed to the K-Means clustering algorithm to cluster the data points. The algorithm finally determines the nearest neighbors of a query point by searching the clusters that are closest to the query point. While analyzing the performance of the proposed algorithm, the time consumed for constructing the coreset and K-Means algorithms is not taken in to account. This is because these algorithms are used only once. The proposed method is compared with two existing algorithms in the literature. We called these two methods as "general or normal method" and "without using coresets". The comparative results prove that the proposed algorithm reduces the time consumed to generate kd-tree and also K-Means clustering.

Keywords Nearest neighbors · kd-tree · Coresets · k-means · Clustering.

Y. Narasimhulu (✉) · R. Pasunuri · V. C. Venkaiah
SCIS, University of Hyderabad, Hyderabad, India
e-mail: narasimedu@gmail.com

R. Pasunuri
e-mail: raghupasunuri@gmail.com

V. C. Venkaiah
e-mail: venkaiah@hotmail.com

N. Chaki et al. (eds.), *Proceedings of International Conference on Computational Intelligence and Data Engineering*, Lecture Notes on Data Engineering and Communications Technologies 56, https://doi.org/10.1007/978-981-15-8767-2_4

1 Introduction

In Computational Geometry, objects considered are set of points in Euclidean space. Collection of points in a higher dimensional space is called multidimensional data, that represent locations and objects in space. Representing multidimensional data and accessing is an important issue in various fields that include computer graphics, computer vision, computational geometry, image processing, machine learning, pattern recognition, and more. Number of different representations and methods for accessing multidimensional data were proposed [1]. Some of these include Inverted Lists [2], Fixed Grid [3], Quad Tree [4], PR Quad-tree [5], EXCELL [6], Grid File [7].

Machine learning algorithms use multidimensional data to solve problems like classifying the data, predicting the values of dependent variables, inferring new knowledge, finding nearest neighbors in a range, and suggesting products to customers. These algorithms can be classified into 4 categories:

1. Supervised learning algorithms: These algorithms are given a training set of examples with the correct answers. These algorithms infer new knowledge from the data. This kind of learning is also called as learning from examples. Example algorithms are Find-S, List-then-eliminate, Candidate Elimination, Regression, and Classification.
2. Unsupervised learning algorithms: These algorithms are given a training set of examples with no responses, but instead the algorithm tries to identify commonalities between the inputs so that inputs that have something similar are categorized together. One example is clustering by K-Means algorithm.
3. Reinforced learning algorithms: They are told only when the answer is wrong but not how to correct it. The algorithm has to find out a way to get the answer right. These algorithms are always monitored and the answers are scored.
4. Evolutionary learning algorithms: They work on an idea of $fitness$, which corresponds to a score for how good the current solution is. Genetic algorithm is an example of evolutionary learning.

This paper concentrates on unsupervised learning which finds k-nearest neighbors [12], of a query point q. The proposed algorithm analyzes the common properties in the data, categorize the data, and finds the nearest neighbors. The next section presents the tools and their algorithms that are used in the work.

2 Preliminaries

2.1 Coresets

Machine learning algorithms accuracy increases as the input data size increases. Processing huge data by these algorithms brings a new kind of problem concerning the time complexity. Reducing the data size may cause the loss of valuable information.

Table 1 Algorithm for coreset construction

Algorithm 1: lightwieght-coreset-construction(X,K)
X : Unsupervised complete data set K : Number of points to be sampled. 1. μ = mean of X. 2. for $x \in X$ do $q(x) = \frac{1}{2}\frac{1}{\lvert x\rvert} + \frac{1}{2}\frac{d(x,\mu)^2}{\sum_{x' \in X} d(x',\mu)^2}$ 3. C = Sample K weighted points from X where each point x has weight $\frac{1}{K.q(x)}$ and is sampled with probability q(x) 4. Return set C with K points that were sampled.

One of the major challenges for the researchers is to bound the trade-off between reducing the data size and the loss of valuable information. Coresets are one such way of solving this trade-off problem.

A coreset is a reduced data set which can be used as a proxy for the full data set. Hence, they are known as summaries of the big data available [8]. Coresets can be computed in linear time and more intricate algorithms can be run on these sets to provide approximate results as a full data set. Models that are trained on these subsets are provably competitive in the results they produce with the models that are trained on full data. Roughly, Coreset is obtained by sampling the data while honoring the distribution.

Table 1 contains a procedure [10], to construct coresets. The algorithm calculates mean of the data and then uses it to compute the distribution $q(x)$ for each point and assigns it as a weight to each point. Finally, it samples K weighted points from X(complete data set) where each point $x \in X$ has weight $\frac{1}{K.q(x)}$ and is sampled with probability q(x). The function $d(x, \mu)$ is a distance function from x to mean μ. The time complexity of the algorithm is $O(nd)$, where n is the size of the data, d is the dimensions.

One of the advantages of using coresets is that the size of the coreset is independent of the size of the original data. An added advantage of the algorithm is that it can be implemented with ease. This paper uses lightweight coresets construction algorithm for K-Means clustering discussed in the Sect. 2.2.

2.2 K-Means

Unsupervised data does not contain target values, then the task of generalization becomes difficult and the algorithm has to completely rely on the data itself. The kind of algorithms that rely on data properties to learn are called unsupervised learning

Table 2 Algorithm for K-means

Algorithm 2: K-Means(X,K,C)

K : number of clusters to create.
X : Unsupervised complete data set.
C : Initial cluster centers.

1. Associate each point to a cluster center that is closest to it.
2. For each cluster, move the position of the center to the mean of the points in that cluster.
3. If the cluster centers did not change then goto step 4 else goto step 1.
4. Assign classification label to each point and return new cluster centers C

algorithms. In the proposed method, K-Means [8], is performed on unsupervised data to form clusters that have similar properties. One aspect to be specified while determining the similarity among the data is the distance measure.. If the Euclidean distance between the points $x \in X$ and $y \in X$ is minimum then they are considered to be similar. The data point x_i is assigned to a cluster K_j when the distance between the point x_i and cluster mean μ_j is minimum. The objective function to form the clusters is

$$min \sum_{j=1}^{K} \sum_{i=1}^{n} \sqrt{(x_i - \mu_j)^2}$$

K-Means algorithm specified in Table 2, assigns data points to the nearest cluster centers. Using the distance measure and mean, K-Means learns to find the cluster centers. The process of finding best cluster centers starts by selecting them randomly and fine-tuning until the cluster centers stop changing. The cluster centers stop changing when the error criterion is minimum, called converging time.

The algorithm's complexity is dependent on initial centroids that are considered. K-Means algorithm is relatively slow, because it has to calculate the Euclidean distance between each cluster center and each data point. When the centers change after an iteration, Euclidean distance has to be recomputed making the algorithm inefficient. The general K-Means algorithm is NP-Hard [15], which takes exponential time to converge. However, with a fixed "t" number of iterations, "c" centroids, "n" points, and "d" dimensions, K-Means takes $O(tcnd)$ time. In the Proposed Work to reduce the number of iterations for searching the best cluster centroids, we use the data points produced by the coreset construction algorithm as initial centroids.

In the proposed algorithm, coresets are not used as the summaries of the whole data, but they are used as the initial cluster centers. Coreset points as the initial centroids has taken less time for finding the final clusters in K-Means algorithm. The K-Means clustering time comparisons are presented in the results section. In order to reduce the time to search for the k-closest points to the given query, Quadtree, a spatial data structure, is used.

2.3 *Quadtree and kd-tree*

Quadtree [4], is a hierarchical spatial tree data structure. Quadtree represents two-dimensional data on the geometric space by recursively decomposing the space using separators parallel to the coordinate axis. The initial decomposed four regions correspond to four children of the root node, hence the term *quad*. Decomposition of the space into regions helps in solving problems efficiently such as, range query, spherical query, and nearest neighbors query. Range query finds all points that are present within a range. Spherical region query finds all the points that lie within a distance r from query q. Nearest neighbor query finds the nearest neighbors of a certain quantity k from the query q.

Because of the principle of equal subdivision, the height of the quadtree cannot be estimated as the data may fall more in any of the quadrants. Height of the tree can be in balance only when the data is distributed uniformly. Performance is mostly based on the height of the tree. If the tree is skewed, the performance degrades. Hence, the division point can be a median of all the data or it can be a midpoint of the data [9], if the data is known in advance.

A height balanced quadtree can be constructed in $O(dnlogn)$ runtime, where d is the number of dimensions and with $O(n)$ storage. Search in tree take $O(dh)$ run time, where h is the height of the tree. Insertion is restricted to $O(dh)$. It takes more time to readjust the tree after deleting the points from it.

The notion of quadtree can be extended to k, where k is the number of dimensions, and hence is called as a kd-tree [11]. In a two-dimensional case, where $k = 2$, each point has 2 values, x-coordinate and y-coordinate. In the construction of two-dimensional tree, the initial split is on x-coordinate, next on y-coordinate, then again on x-coordinate, and so on as shown in the Fig. 1. The root of the tree will split the space into two parts left and right subsets of roughly equal size on x-axis. The left and

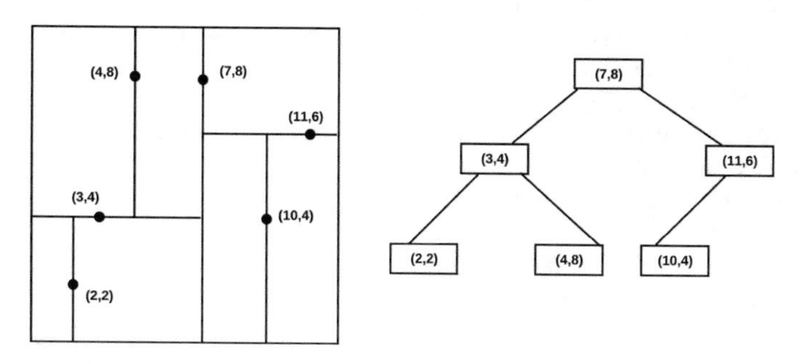

Fig. 1 kd-tree for 6 points in the cartesian plane

right subsets are further split according to y-axis and then again those are further split according to the axis until no further split is required. Depth of the tree is also based on which axis we divide on the first split. Another kd-tree based searching algorithm [14], which runs close to $O(logn)$ is proposed, which guarantees a theoretical proof of search accuracy as close as to Randomized Partitioning Tree (RPTree).

The next section presents the proposed algorithm that uses the above mentioned algorithms and data structures in retrieving the closest points to the query.

3 Proposed Work

In order to seek k-nearest neighbors, we may not require the entire data because we are not worried to return all points. Hence, we assume that $k < n$. Using this as the driving principle, the proposed algorithm presented in Table 3, considers only a subset of the data. This paper does not propose to classify or try to classify a query point to a particular class but returns its k nearest neighbors. The time complexity of the proposed algorithm is $O(nd) + O(tcnd) + O(dnlogn)$, which is asymptotically equal to $O(dnlogn)$, where d is the number of dimensions and n is the number of points.

Flow diagram of the entire work is presented in Fig. 2. The process starts by considering unsupervised data and constructing a coreset of size K. The value of K decides the number of clusters that are needed to form. It would be better if

Table 3 Algorithm for k-nearest neighbors

Algorithm 3: k-nearestneighbors(X,k,K,q)
k : Number of nearest points to be found. X : Unsupervised complete data set. K : Number of clusters to be created. q : Query point. 1. Let C be a set, of K coreset points. C = lightwieght-coreset-construction(X,K) 2. Using C as the initial centroids, K-Means constructs the new centriods that satisfy the criterian function and returns the new centriods. C = K-Means(X,K,C) 3. Identify the nearest cluster center from set C to the query point q. 4. Fetch the cluster data and store it in 'x'. 5. Construct kd-tree for data x and query the tree with q. $tree$ = kd-tree(x) $distance, knnIndices$ = tree.query(q) 6. return $distance, knnIndices$

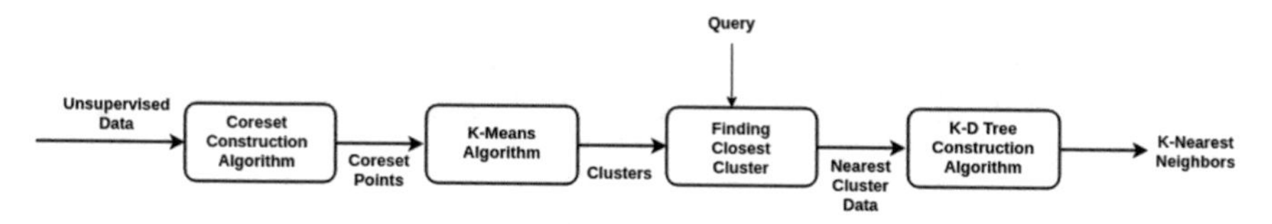

Fig. 2 Procedure for k-nearest neighbors

K is known in prior which helps in producing accurate results. When a wrong K is assumed the results could be wrong. So, the K value for the datasets that were considered in this paper is known in advance.

Using the K points as the initial centroids for K-Means, clusters are created. It is observed that the time taken for K-Means with some random points as initial centroids is much greater than the time taken for K-Means with coreset points as initial centroids. The results are shown in Sect. 4. Using the clusters generated by the K-Means, we find the closest cluster to the query point. The kd-tree is constructed using the closest cluster data. The kd-tree algorithm generates a tree and this tree is given a query point q and k value to produce the final k-nearest points to the query point. In the proposed method, data fed to kd-tree algorithm is less when compared with the normal method, hence reducing the time for construction.

In order to prove the results produced by the proposed method is better, a comparative study has been done by using two other approaches which we called them as "normal method" and "without using coresets". Another comparison is also done to prove that coreset points fed K-Means are better than straight forward K-Means. It is very clearly depicted that the proposed method outperforms the standard method.

Next section provides the complete results and a comparative study of the following three methods.

1. `Normal method`: Uses kd-tree on full data and query it.
2. `Without using coresets`: Uses K-means, kd-tree on clustered data and query the kd-tree
3. `Proposed method`: Uses K-Means using coreset data and kd-tree on cluster data.

Normal method does not require K-Means, hence K-Means comparison for the normal method is not presented.

4 Results

Algorithms k-nearestneighbors(X,k,K,q), K-Means(X,K,C), kd-tree [11], were implemeted in python. Details of the datasets [13], that are used for experimentation are given in the Table 4.

Table 4 Datasets and their properties

Name of the Dataset	No. of dimensions	No. of instances	No. of classes
Breast cancer data	30	569	2
Digits data	64	1797	10
CovType data	54	581012	7
Smartphone data	562	10299	6
Kddcup data	36	494020	23
Miniboone data	50	130062	7

Table 5 K-Means construction time for breast cancer data

Name of the Dataset	Name of the method	K-means construction time
Breast cancer data	Without using Coresets	0.021596901
	Proposed Work	0.00489233

Table 6 kd-tree construction time for breast cancer data

Name of the Dataset	Name of the method	KD-tree construction time
Breast cancer data	Normal method	0.000900756
	Without using Coresets	0.000555496
	Proposed Work	0.000510688

Table 7 Datasize variation for breast cancer data

Name of the Dataset	Initial data size	Data size at search time
Breast cancer data	569	438

Comparisons on kd-tree construction time, K-Means time, input data size initially, and at kd-tree construction point for all datasets that are present in the Table 4, are displayed in the following tables.

Table 6 presents variations in the times of constructing kd-tree for the three methods. Table 5 presents the K-Means time for random points as initial centroids and coreset points as initial centroids. Table 7 displays the initial data size considered and final data size drawn from Breast Cancer Data. Figure 3 is the graphical representation for the Tables 5, 6 and 7 provided above. It is clearly observed from the above figure that the proposed algorithm performed better than the other methods (Tables 8, 9, 10 and 11).

As in the case of Breast Cancer Data, the Tables 9, 12, 15, 18, and 21 present the comparison of kd-tree construction time on Digits data, Smartphone data, Kddcup data, and Miniboone data, respectively (Tables 12, 13, 14, 15 and 16).

Table 8 K-means construction time for digits data

Name of the Dataset	Name of the method	K-means construction time
Digits data	Without using Coresets	0.177046333
	Proposed Work	0.018532348

Table 9 kd-tree construction time for digits data

Name of the Dataset	Name of the method	kd-tree construction time
Digits data	Normal method	0.002147111
	Without using Coresets	0.000293409
	Proposed Work	0.000220433

Table 10 Datasize variation for digits data

Name of the Dataset	Initial data size	Data size at search time
Digits data	1797	252

Table 11 K-means construction time for Cov type data

Name of the Dataset	Name of the method	K-means construction time
CovType Data	Without Using Coresets	58.320359221
	Proposed Work	3.571427582

The Tables 8, 11, 14, 17, and 20 present the comparison of K-Means construction time on Digits data, Smartphone data, Kddcup data, and Miniboone data, respectively (Table 17, 18, 19, and 20).

The Tables 10, 13, 16, 19, and 22 present the comparison of initial data size and final data size for tree construction on Digits data, Smartphone data, Kddcup data, and Miniboone data, respectively.

Table 12 kd-tree construction time for Cov type data

Name of the Dataset	Name of the method	kd-tree construction time
Cov type data	Normal method	2.81545425
	Without using Coresets	0.557287733
	Proposed Work	0.528879423

Table 13 Datasize variation for Cov type data

Name of the Dataset	Initial data size	Data size at search time
Cov type data	581012	159981

Table 14 K-means construction time for smartphone data

Name of the Dataset	Name of the method	K-means construction time
Smartphone data	Without using Coresets	5.235056832
	Proposed Work	0.746759341

Table 15 kd-tree construction time for smartphone data

Name of the Dataset	Name of the method	kd-tree construction time
Smartphone data	Normal method	0.402041996
	Without using Coresets	0.01072642
	Proposed Work	0.0068364

Table 16 Datasize variation for smartphone data

Name of the Dataset	Initial data size	Data size at search time
Smartphone data	10299	686

Table 17 K-means construction time for Kddcup data

Name of the Dataset	Name of the method	K-means construction time
Kddcup data	Without using Coresets	51.760765204
	Proposed Work	8.746508095

The Figs. 4, 5, 6, 7, and 8 presents the pictorical representation of the comparisons done for Digits data, Smartphone data, Kddcup data, and Miniboone data, respectively (Tables 21 and 22).

Table 18 kd-tree Construction time for Kddcup data

Name of the Dataset	Name of the method	kd-tree construction time
Kddcup data	Normal method	1115.123546068
	Without using Coresets	1071.968181827
	Proposed Work	198.936803207

Table 19 Datasize variation for Kddcup data

Name of the Dataset	Initial data size	Data size at search time
Kddcup data	494020	190107

Table 20 K-means construction time for miniboone data

Name of the Dataset	Name of the method	K-means construction time
Miniboone data	Without using Coresets	3.630400168
	Proposed Work	1.522944811

Table 21 kd-tree construction time for miniboone data

Name of the Dataset	Name of the method	kd-tree construction time
Miniboone data	Normal method	0.765435536
	Without using Coresets	0.412943829
	Proposed Work	0.332265506

Table 22 Datasize variation for miniboone data

Name of the Dataset	Initial data size	Data size at search time
Miniboone data	130062	85649

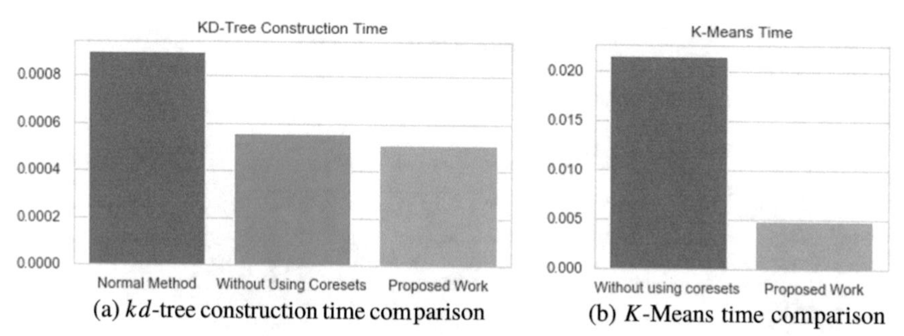

(a) kd-tree construction time comparison (b) K-Means time comparison

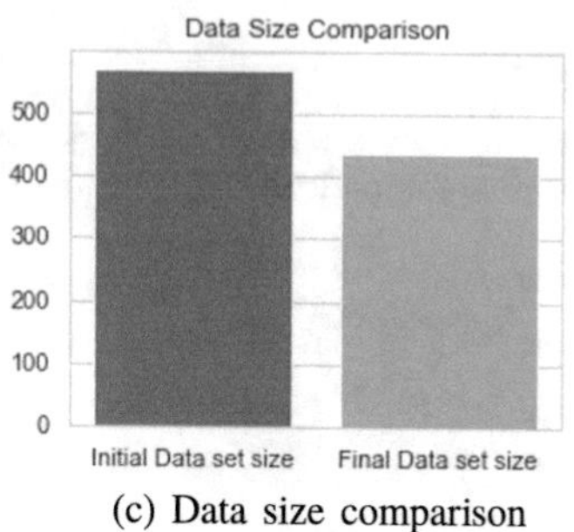

(c) Data size comparison

Fig. 3 Comparisons on breast cancer data

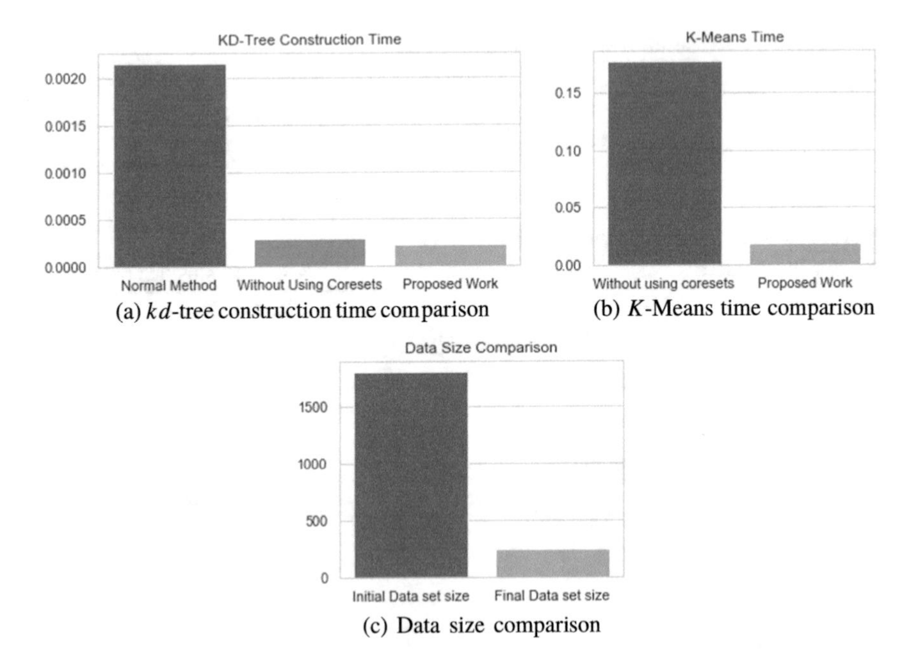

(a) kd-tree construction time comparison (b) K-Means time comparison

(c) Data size comparison

Fig. 4 Comparisons on digits data

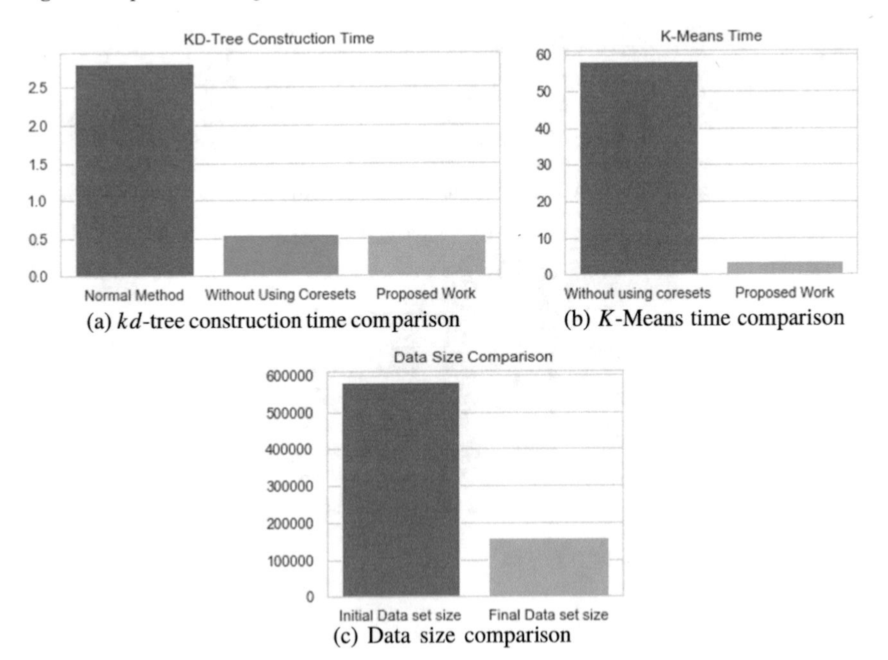

(a) kd-tree construction time comparison (b) K-Means time comparison

(c) Data size comparison

Fig. 5 Comparisons on Cov type data

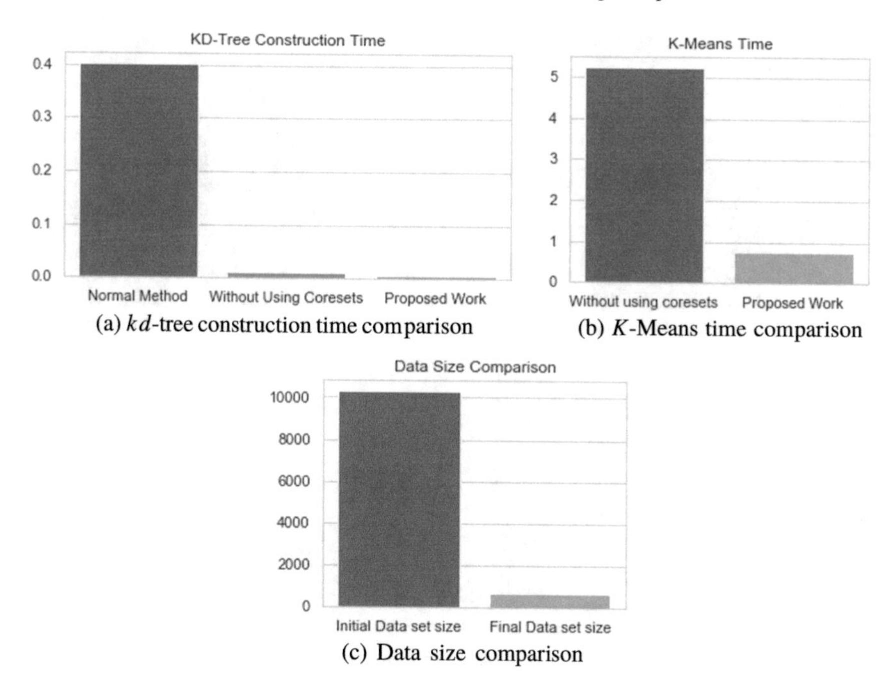

(a) kd-tree construction time comparison

(b) K-Means time comparison

(c) Data size comparison

Fig. 6 Comparisons on smartphone data

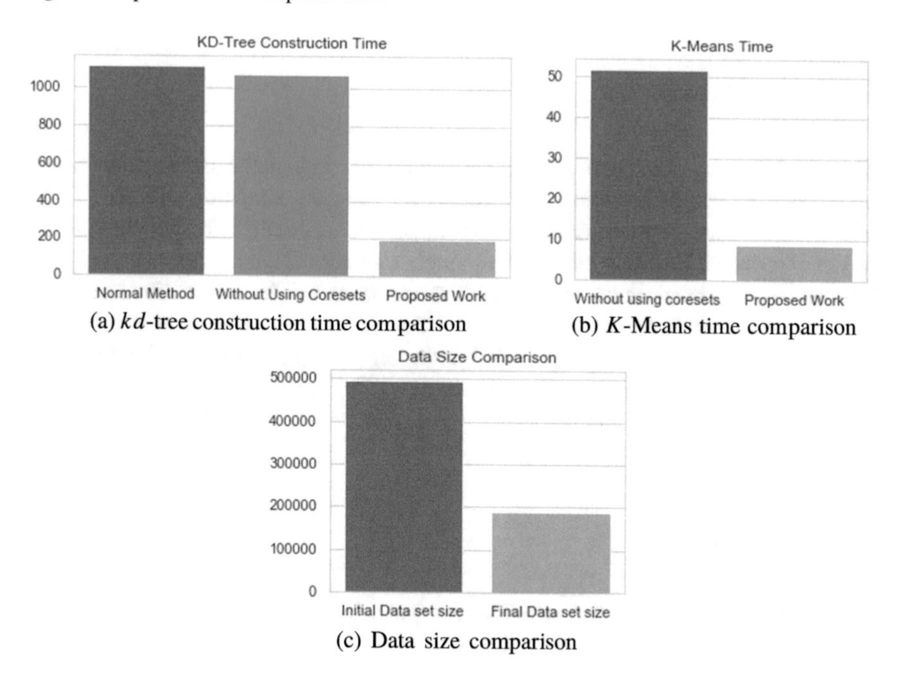

(a) kd-tree construction time comparison

(b) K-Means time comparison

(c) Data size comparison

Fig. 7 Comparisons on Kddcup data

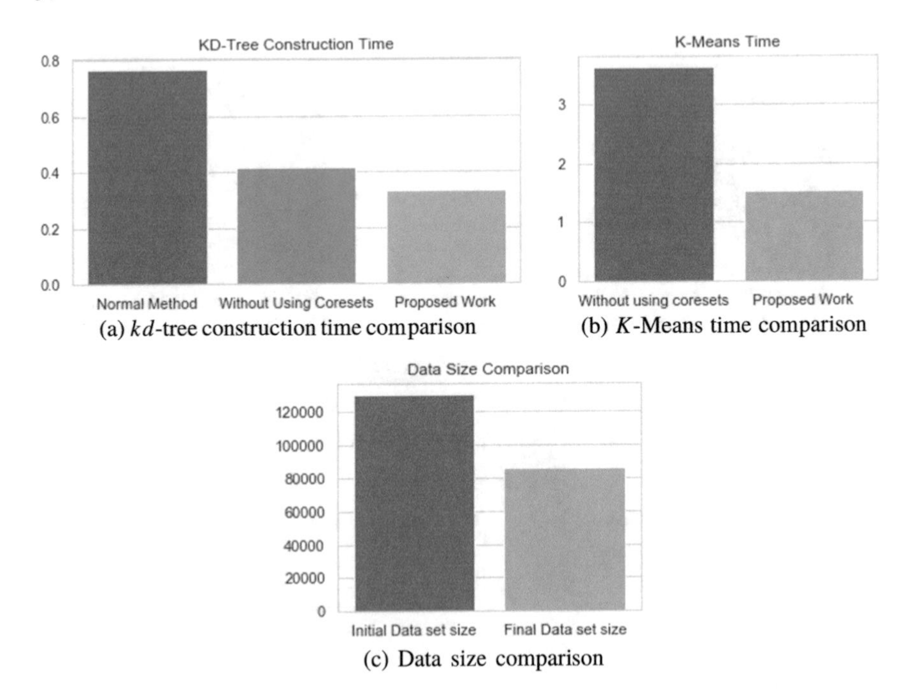

(a) kd-tree construction time comparison

(b) K-Means time comparison

(c) Data size comparison

Fig. 8 Comparisons on Kddcup data

5 Future Work

The methods considered here for comparison have produced the same nearest points on all datasets and these were proved to be correct when compared with the ground truth results. Size of the data has been reduced to a good factor but dimensions were not reduced. Building kd-tree for higher dimensional data is more complex and time-consuming because kd-tree suffers from the curse of dimensionality. If data can be reduced in length and dimensions, then results can be achieved in much lesser time. Algorithm 3 can also be extended to classify the query point q. The proposed algorithm can be applied to the supervised data. The issue of cluster size being less than k is not addressed in the paper.

6 Conclusion

The kd-tree is a standard data structure for searching nearest neighbors, yet proves to take more time when the data size increases. This paper proposes a k-nearest neighbors algorithm, which reduces the kd-tree time by performing two constant operations, coreset construction, and K-Means. Comparative results proved that the proposed method has performed better in terms of time to construct kd-tree and also to retrieve the nearest neighbors. It also showed that the size of the data for the kd-tree is reduced.

References

1. Samet H (1989) The design and analysis of spatial data structures. Addison-Wesley, Reading, MA
2. Knuth DE (1973) The art of computer programming. Sorting and Searching, vol 3. Addison-Wesley, Reading, MA
3. Bentley JL, Friedman JH (1979) Data structures for range searching. ACM Comput Surv 11(4):397–409
4. Finkel RA, Bentley JL (1974) Quadtrees: a data structure for retrieval on composite keys. Acta Informatica 4(1):1–9
5. Orenstein JA (1982) Multidimensional tries used for associative searching. Inform Process Lett 14(4):150–157
6. Tamminen M (1981) The EXCELL method for efficient geometric access to data. Series Acta Polytech Scand, Math Comput Sci 34
7. Nievergelt J, Hinterberger H, Sevcik KC (1984) The grid file: an adaptable, symmetric multikey file structure. ACM Trans Database Syst 9(1):38–71
8. Bachem O, Lucic M, Krause A (2018) Scalable k-means clustering via lightweight coresets. In: Proceedings of the 24th ACM SIGKDD international conference on knowledge discovery and data mining (KDD '18), pp 1119–1127. ACM, New York, NY, USA. https://doi.org/10.1145/3219819.3219973
9. Maneewongvatana S, David M (1999) Mount: its okay to be skinny, if your friends are fat. In: 4th annual CGC workshop on comptutational geometry
10. Ashok, Kumari OG, Pasunuri R, Vadlamudi CV, Subba Rao YV, Rukma Rekha N (2019) CKD tree: an improved KD-tree construction algorithm using coresets for k-NN based classification. Communicated
11. Spatial KD-tree. https://docs.scipy.org/doc/scipy-0.14.0/reference/generated/scipy.spatial.KDTree.html
12. Bentley JL (1979) Decomposable searching problems. Inform Process Lett 8:244–251
13. Datasets. https://archive.ics.uci.edu/ml/datasets.php, https://archive.ics.uci.edu/ml/datasets.php
14. Ram P, Sinha K (2019) Revisiting kd-tree for nearest neighbor search, KDD '19, Anchorage. AK, USA
15. Mahajana Meena, Nimbhorkara Prajakta, Varadarajanb Kasturi (2012) The planar k-means problem is NP-hard. Theor Comput Sci 442(13):13–21

Impact of Hyperparameters on Model Development in Deep Learning

Humera Shaziya and Raniah Zaheer

Abstract Deep learning has revolutionized the field of computer vision. To develop a deep learning model, one has to decide on the optimal values of various hyperparameters such as learning rate. These are also called as model parameters which are not learned by the model rather initialized by the user. Hyperparameters control other parameters of the model such as weights and biases. Parameter values are learned effectively by tuning the hyperparameters. Hence, hyperparameters determine the values of the parameters of the model. Manual Tuning is a tedious and time-consuming process. Automating the selection of values for hyperparameters results in the development of effective models. It has to be investigated to figure out which combinations yield the optimum results. This work has considered scikit-optimize library functions to study the impact of hyperparameters on the accuracy of MNIST dataset classification problem. The results obtained for different combination of learning rate, number of dense layers, number of nodes per dense layer, and activation function showed that a minimum of 8.68% and a maximum of 98.96% for gp_minimize function, 8.68% and 98.74% for forest_minimize function and gbrt_minimize generates 9.24% and 98.94% for lowest and highest accuracy, respectively.

Keywords Convolutional Neural Networks (CNN) · Deep learning · gp_minimize · forest_minimize · gbrt_minimize · Hyperparameters · Skopt

H. Shaziya (✉)
Department of Informatics, Nizam College, Osmania University, Hyderabad, India
e-mail: humera.shaziya@gmail.com

R. Zaheer
Department of CS, Najran University, Najran, Saudi Arabia

N. Chaki et al. (eds.), *Proceedings of International Conference on Computational Intelligence and Data Engineering*, Lecture Notes on Data Engineering and Communications Technologies 56, https://doi.org/10.1007/978-981-15-8767-2_5

1 Introduction

Hyperparameters determine the values for various attributes of network structure and training process. Few examples of hyperparameters are learning rate, number of layers, number of hidden layers, number of filters, dropout rate, number of iterations, batch size, activation function, optimizer, and regularization. Tuning of these parameters requires an additional step in the deep learning model development. Hyperparameter tuning can be done manually or automatically. Manual is not an effective way of choosing values as it requires more time. Automatic methods are of three kinds, they are grid search, random search and bayesian optimization. Grid search considers every combination of the specified parameters which often leads to an enormous number of configurations. Sometimes it's not feasible to look for every combination as it affects the performance drastically. Further to overcome this problem, random search was introduced. As the name suggests, hyperparameters are chosen randomly and experimented. Finally, bayesian method uses an objective function and keeps track of the past results to determine the probability of the next best combination of hyperparameters. Additionally, it is preferred to follow a systematic method to check on the combinations of hyperparameters. Scikit-optimize or skopt library is a popular one to automate the task of choosing hyperparameters. Skopt has significant functions to automatically select hyperparameters such as gp_minimize, forest_minimize and gbrt_minimize. In the present work, these three functions have been experimented on MNIST data set and the results are analyzed.

The paper is organized as follows: Sect. 2 will review the related work. The skopt hyperparameter library used in the current work has been discussed in Sect. 3. Environment and data of experiments carried out in the present study is outlined in Sect. 4. Results of the investigations are described in the Sect. 5, and the outcomes are explored and elaborated. Finally, the paper concludes in Sect. 6, with the discussion of key points of the present study.

2 Literature Review

Domhan et al. [6] state that automatic hyperparameter is far better than manual ones, and proposed a probabilistic model to determine the combination of values that are not leading to poor result and when that happens the learning process stops and the learning process is resumed with different set of parameters. The process of finding optimal hyperparameter speeds up because of this early stopping. It considers a number of steps of stochastic gradient descent as the function to minimize. The learning curve is observed using a probabilistic model to determine the performance of the model. Experiments were conducted on MNIST, CIFAR10, and CIFAR100 datasets. The implemented architectures are small and large convolutional neural networks and the result was 19.22% error rate. It is proposed by Dougal Maclauriny and others in [11], that reversing the values of stochastic gradient descent with momentum helps in finding the optimal hyperparameters. Bergstra et al. [4] has proposed two new methods for determining hyperparameters automatically and compared those methods with the random search technique of finding optimal hyperparameters without any intervention of human. The proposed methods are based on greedy sequential methods. Deep Belief Networks (DBN) are employed to test the developed greedy sequential methods and they have outperformed the random search. Ilievski et al. [8] proposed a method based on radial basis function named HORD that is deterministic and outperforms gaussian process. Fabolas, a bayesian optimization scheme was developed in [9], by Klein and others. Training time and loss is modeled as a function of data set size. It was tested on support vector machine and found to be 10–100 times faster than bayesian methods. Hyperparameter optimizations are performed using genetic algorithms by Young [14]. It is based on the concept of evolutionary computing wherein the best hyperparameter configuration is chosen by going through the natural like process of evolution. Loshchilov developed CMA-ES, the covariance matrix adaptation evolution strategy technique [10], to evaluate the hyperparameter selection in parallel. The method devised has been compared with bayesian strategy to find the optimal hyperparameter configurations on 30 GPUs. Hyperopt, a python based library for configuring hyperparameters automatically is reported in [3], by Bergstra and others. The library is comprehensively described in the article and design method is also presented. Snoek [12] states that the kernel type of gaussian process and proper hyperparameter treatment will assist in finding a better optimizer. In the study, authors developed an algorithm with bayesian optimization to be executed on parallel implementations. Zela [15] demonstrates that both the neural architecture and hyperparameter configurations can be done together in the same network. It is indicated in the work that separating architecture and configurations is suboptimal, whereas doing it combined will give good results. Feurer and others [7] address the question of AutoML, i.e., automated machine learning. Several methods including Bayesian and multi-fidelity optimizations were examined. Akiba et al. [2] developed a hyperparameter optimization framework, which follows a design-by-run principle. Tsai et al. [13] investigated the problem of predicting the number of passengers in a bus and simulated annealing technique is used to determine the hyperparameter of the model.

3 Skopt: Library for Hyperparameter Optimization

Scikit-optimize or Skopt is a library for implementing various optimizations [1]. It provides number of built-in functions to achieve optimizations.

3.1 gp_minimize

The function gp_minimize is based on the process of gaussian using Bayesian optimization. Optimizing hyperparameters using cross validation method for different folds is not an optimal process. It would take forever to determine optimal hyperparameters. The other way is to use the gaussian process to approximate the values. It is assumed that multivariate gaussian is being followed by the function, and thus the next values will be generated by the gaussian process kernel using the function values covariance. The acquisition function is then employed to assess the next value of the hyperparameter, which is relatively faster than cross validation approach.

3.2 forest_minimize

Decision trees are employed in forest_minimize to determine the values for the hyperparameters. It is a tree-based regression model, that is used to assess the values at regular intervals to determine the optimal hyerparameters. It works by looking into the function values one after the other. The point where it gets the minimum configuration is chosen.

3.3 gbrt_minimize

gbrt_minimize is a function of skopt that finds the optimal hyperparameters based on gradient boosted trees. This method essentially improves the performance by evaluating the function sequentially to determine the right parameters with the minimum number of evaluations. The function accepts the parameters that need to be learned to optimize the objective function.

4 Environment

The model has been developed on NVIDIA GPU based windows 10 machine. Python 3.6 has been used through anaconda. Anaconda is an open source software that provides an environment for the development of python programs. Tensorflow and Keras have been utilized for deep learning APIs. CUDA and cuDNN softwares of NVIDIA facilitate to harness the power of graphic processing unit (GPU).

4.1 Dataset

The experiments have been performed on MNIST dataset [5]. It is a defacto dataset for investigating machine and deep learning methods. This dataset is a collection of images of handwritten digits 0 through 9. Total images in MNIST are 70 k. Training set comprises of 60 k images and testing set has 10 k images. There are 10 classes (0–9) and each image belongs to any one of the classes. Images are of gray scale with 28×28 pixels dimensions.

5 Results and Discussions

The three functions gp_minimize, forest_minimize and gbrt_minimize of skopt library have been experimented on MNIST dataset to determine which function produces the optimal hyperparameters. The chosen hyperparameters to train are learning rate, activation function, number of dense layers, and number of dense nodes. The ranges for each hyperparameter is specified. The low and high of learning rate is 1e–6 to 1e–2, the number of layers is between 1 and 5, number of nodes has 5 and 512 as low and high values and the activation functions specified are relu and sigmoid. The initial values are 1e–5, 1, 16 and relu for learning rate, number of layers, number of nodes in each layer, and activation function, respectively. After initializing hyperparameters with the default values, a model is developed using which training and testing is done to know about the accuracy of different combination of hyperparameter values. A convolutional neural network is created with 4 layers. The first one being the input layer followed by two 2D convolutional layers and the last one is a fully connected or dense layer. There are five kernels and 16 filters employed in the first convolutional layer and the second one has five kernels and 36 filters. The dense layer uses softmax activation function and adam is the optimizer specified. Accuracy is the performance metric.

The acronyms used in the figures are LR: Learning Rate, DL: Number of Dense Layers, DN: Number of nodes for each dense layer, AT: Activation Function and ACC: Accuracy.

Figure 1 shows accuracy obtained on training gb_minimize function on MNSIT dataset using convolutional neural network. It is observed that the lowest accuracy is 8.68% for the learning rate, number of layers, number of nodes per layer, and activation function combination of 2.3e–06, 4, 69, sigmoid, respectively. 3.9e–04, 5, 512, and relu combinations of hyperparameters gives the highest accuracy of 98.96%.

Figure 2 presents accuracy obtained on training forest_minimize function on MNSIT dataset using convolutional neural network. It is showed that the lowest accuracy is 8.68% for the learning rate, number of layers, number of nodes per layer and activation function combination of 9.0e–03, 1, 305, sigmoid, respectively. 4.4e–03, 1, 319 ,and sigmoid combinations of hyperparameters gives the highest accuracy of 98.74%.

Figure 3 indicates accuracy obtained on training gbrt_minimize function on MNSIT dataset using convolutional neural network. It is listed that the lowest accuracy is 9.24% for the learning rate, number of layers, number of nodes per layer, and activation function combination of 2.8e–06, 5, 18, sigmoid, respectively. 6.4e–04, 1, 501, and relu combinations of hyperparameters gives the highest accuracy of 98.94%.

Figure 4 gives different values selected in gp_minimize for various epochs. It can be seen in the histogram shown in Fig. 4a, 10^{-3}–10^{-4} range of learning rate have highest selection count upto 12 times. As shown in Fig. 6b, relu activation function has been choosen for almost 30 times, whereas sigmoid for only 10 times. Maximum

LR	DL	DN	AT	Acc(%)	LR	DL	DN	AT	Acc(%)
1.0e-05	1	16	Relu	65.08	3.9e-04	2	512	Sigmoid	95.08
2.5e-05	1	325	Sigmoid	11.26	4.0e-04	1	512	Relu	98.80
9.3e-06	2	293	Sigmoid	11.26	**3.9e-04**	**5**	**512**	**Relu**	**98.96**
9.6e-03	1	172	Sigmoid	9.24	3.9e-04	4	512	Relu	98.74
5.3e-04	1	332	Sigmoid	94.32	3.9e-04	5	512	Relu	98.66
7.5e-03	1	398	Sigmoid	11.26	2.2e-05	1	512	Relu	94.10
2.3e-06	**4**	**69**	**Sigmoid**	**8.68**	3.3e-05	5	5	Relu	27.56
1.2e-05	3	459	Relu	98.12	5.5e-04	5	5	Sigmoid	11.26
1.8e-05	3	312	Relu	93.30	3.4e-04	5	217	Relu	98.60
6.1e-04	2	147	Relu	98.74	6.4e-06	5	381	Relu	89.24
1.7e-04	5	435	Relu	98.22	5.9e-04	5	5	Relu	79.80
7.0e-03	5	509	Relu	11.26	6.3e-04	1	262	Relu	98.82
1.4e-06	1	11	Relu	24.18	1.0e-06	2	512	Relu	66.16
3.9e-05	5	512	Relu	97.28	4.3e-04	1	421	Sigmoid	94.66
2.3e-04	1	5	Relu	86.54	5.3e-04	1	425	Sigmoid	95.34
4.9e-04	1	512	Relu	98.82	1.6e-04	1	512	Relu	97.98
5.5e-04	5	512	Relu	98.72	4.7e-04	1	181	Relu	98.42
5.1e-04	5	452	Relu	98.64	1.1e-03	5	213	Relu	98.88
5.1e-04	1	512	Relu	98.90	1.7e-04	1	250	Relu	97.88
2.3e-04	1	512	Sigmoid	91.48	7.9e-04	5	201	Relu	98.76

Fig. 1 Shows the accuracy obtained for gp_minimize

LR	DL	DN	AT	Acc(%)	LR	DL	DN	AT	Acc(%)
1.0e-05	1	16	Relu	65.62	4.2e-03	1	475	Sigmoid	98.52
2.9e-03	3	367	Sigmoid	97.64	4.4e-03	1	324	Sigmoid	97.98
2.3e-06	3	164	Sigmoid	9.86	**4.4e-03**	**1**	**319**	**Sigmoid**	**98.74**
3.6e-04	4	91	Sigmoid	91.04	5.0e-03	1	326	Sigmoid	98.04
2.1e-05	4	194	Sigmoid	11.26	7.8e-03	1	327	Sigmoid	96.96
3.3e-04	5	38	Relu	97.22	6.4e-03	1	331	Sigmoid	9.86
3.0e-06	2	11	Sigmoid	10.02	4.2e-03	1	314	Sigmoid	97.92
3.4e-04	2	412	Sigmoid	94.68	4.9e-03	1	336	Sigmoid	97.92
1.6e-06	1	296	Sigmoid	11.26	4.8e-03	1	306	Sigmoid	97.44
1.1e-04	2	52	Relu	96.20	5.3e-03	1	324	Sigmoid	98.42
1.2e-05	2	31	Relu	80.40	4.7e-03	1	284	Sigmoid	98.16
2.8e-04	2	310	Sigmoid	92.94	5.5e-03	1	359	Sigmoid	97.64
2.3e-04	3	300	Sigmoid	92.24	5.9e-03	1	325	Sigmoid	95.86
1.8e-04	4	454	Sigmoid	88.94	5.8e-03	1	336	Relu	98.58
1.4e-04	3	74	Sigmoid	23.44	5.7e-03	1	305	Sigmoid	97.94
3.9e-03	1	248	Sigmoid	97.90	6.2e-03	1	321	Sigmoid	97.94
9.2e-03	1	204	Sigmoid	9.86	6.1e-03	1	310	Sigmoid	97.88
5.8e-03	1	328	Sigmoid	98.72	6.4e-03	1	315	Sigmoid	98.12
5.3e-03	1	325	Sigmoid	96.80	6.4e-03	1	289	Sigmoid	97.32
9.0e-03	**1**	**305**	**Sigmoid**	**8.68**	6.4e-03	1	298	Sigmoid	94.48

Fig. 2 Shows the accuracy obtained for forest_minimize

LR	DL	DN	AT	Acc(%)	LR	DL	DN	AT	Acc(%)
1.0e-05	1	16	Relu	71.10	1.2e-03	5	479	Relu	98.38
2.1e-06	1	410	Relu	76.12	6.5e-05	4	16	Relu	87.88
6.5e-06	5	480	Relu	89.10	1.2e-03	1	45	Relu	98.92
2.9e-03	5	102	Sigmoid	11.26	9.8e-04	1	510	Sigmoid	97.18
3.0e-05	3	508	Relu	96.08	3.4e-03	1	33	Sigmoid	98.18
2.7e-06	2	143	Sigmoid	11.00	3.4e-03	1	507	Sigmoid	97.62
9.2e-03	3	339	Relu	97.24	6.2e-03	1	512	Sigmoid	98.68
2.0e-03	1	143	Relu	98.68	3.1e-03	2	468	Relu	98.82
2.8e-06	**5**	**18**	**Sigmoid**	**9.24**	7.7e-03	5	13	Sigmoid	11.26
2.4e-04	2	213	Relu	98.46	6.2e-03	1	509	Sigmoid	96.90
4.6e-06	2	256	Sigmoid	11.26	4.7e-03	1	10	Relu	98.28
1.2e-03	1	235	Relu	98.92	2.9e-03	1	512	Sigmoid	97.62
6.4e-04	**1**	**501**	**Relu**	**98.94**	2.2e-03	1	357	Sigmoid	98.08
3.9e-04	1	506	Relu	98.74	3.3e-03	5	271	Sigmoid	11.26
5.3e-04	1	196	Relu	98.66	3.6e-06	3	415	Relu	81.02
7.8e-04	2	500	Relu	98.76	1.3e-03	5	509	Relu	98.88
1.6e-03	1	108	Relu	98.70	6.3e-06	4	506	Relu	89.04
4.2e-05	1	63	Relu	92.88	1.4e-06	4	510	Relu	63.42
2.4e-05	5	493	Sigmoid	9.86	5.4e-06	4	477	Relu	88.20
1.3e-03	5	456	Relu	98.16	1.0e-03	5	510	Sigmoid	95.96

Fig. 3 Shows the accuracy obtained for gbrt_minimize

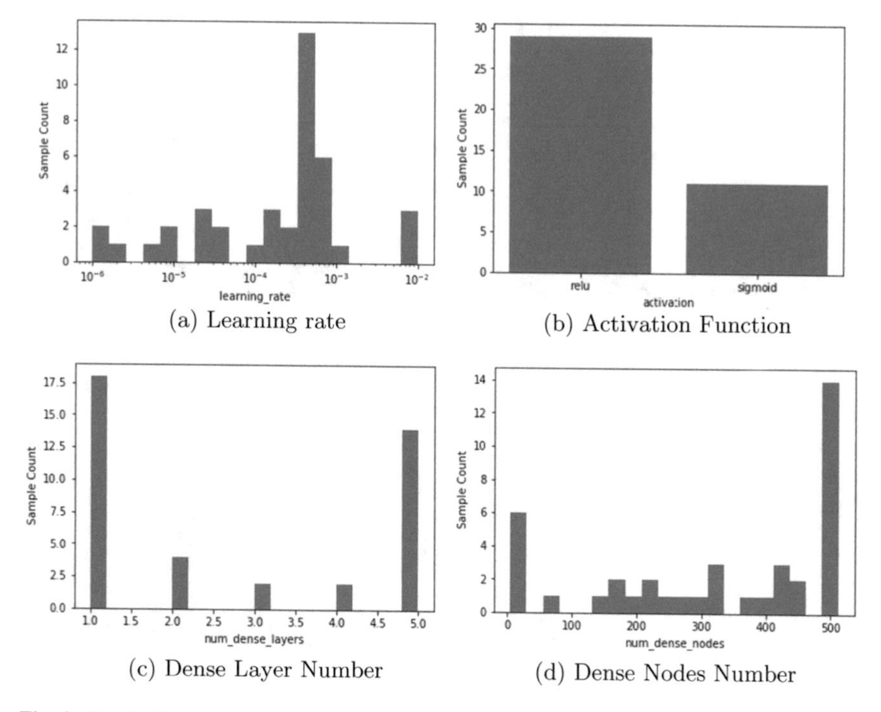

(a) Learning rate

(b) Activation Function

(c) Dense Layer Number

(d) Dense Nodes Number

Fig. 4 Count of hyperparameter selected in gp_minimize function

time 1 and 5 dense layers has been selected as given in Fig. 6c, and count of dense nodes is 500 most of the time presented in Fig. 6d.

Figure 5 presents different values chosen for different epochs while execution. Figure 5a represents the choice of learning rate, as can be seen, 10^{-2} was the rate selected for almost 16 times which is the highest count for learning rate. Preference for sigmoid activation function has been evident from the histogram in Fig. 5b, for almost 35 times and relu was selected only form 5 times. In Fig. 5c, it is clear that one dense layer was chosen for more than 25 times which is greater than any other value. Number of dense nodes range is 300–400 as shown in Fig. 5d.

Figure 6 illustrates various chosen values of gbrt_minimize in terms of different combinations for number of epochs. It can be noted from the histogram shown in Fig. 6a, that 10^{-3} and 10^{-2} were chosen for maximum times. Relu activation function has been preferred by gbrt_minimize shown in Fig. 6b, for more times than sigmoid. Relu was the choice for 25 times, whereas sigmoid selected for only 15 times. As for dense layers in Fig. 6c, the pattern is the same as other two functions and one dense is chosen for 18 times. Figure 6c histogram shows that 14 times 500 value was initialized to dense nodes.

Figure 7 gives the comparison of the low and high accuracy values for the three functions.

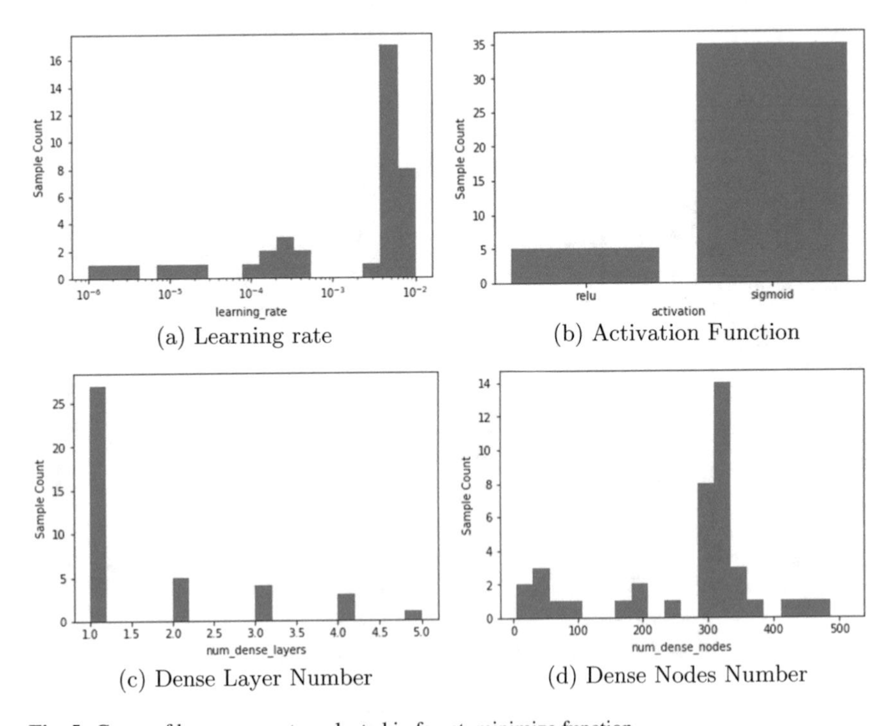

(a) Learning rate (b) Activation Function

(c) Dense Layer Number (d) Dense Nodes Number

Fig. 5 Count of hyperparameter selected in forest_minimize function

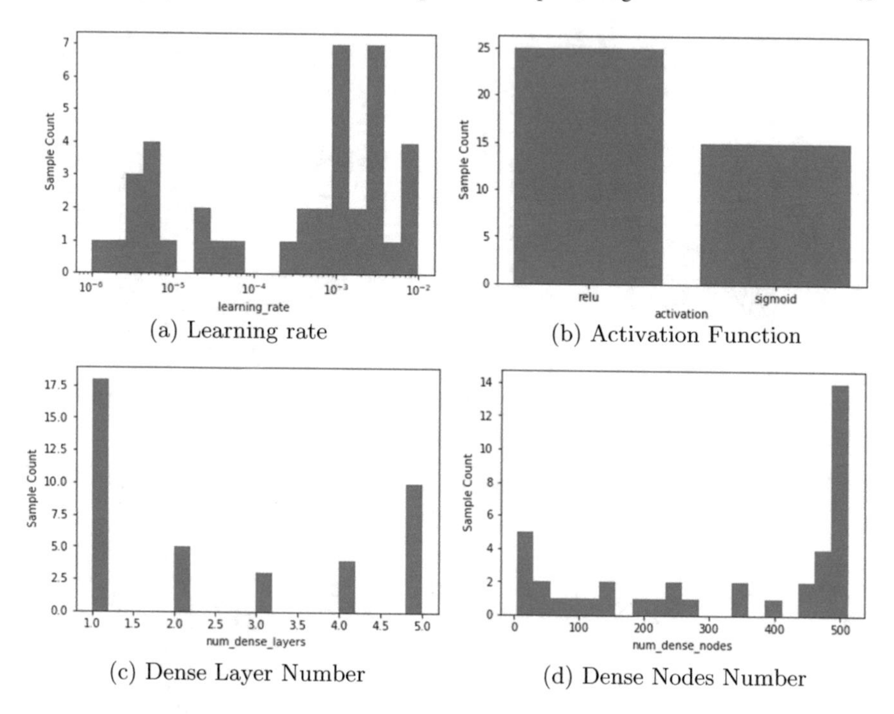

Fig. 6 Count of hyperparameter selected in gbrt_minimize function

Type	L/H	LR	DL	DN	AT	Acc(%)
gp_minimize	L	2.3e-06	4	69	Sigmoid	8.68
	H	3.9e-04	5	512	Relu	**98.96**
forest_minimize	L	9.0e-03	1	305	Sigmoid	8.68
	H	4.4e-03	1	319	Sigmoid	98.74
gbrt_minimize	L	2.8e-06	5	18	Sigmoid	9.24
	H	6.4e-04	1	501	Relu	98.94

Fig. 7 Analysis of gp_minimize, forest_minimize and gbrt_minimize

Figure 8 shows the best three and least two accuracy values for the three functions. The analysis of results of the top three and last two configurations shows that relu activation function in combination with other parameters produces the best results. It is observed that there is no specific pattern about the number of dense layers in obtaining low or high outcomes. However, it is noted that the accuracy increased drastically when the number of nodes per each dense layer is more. When there are

	Parameter	gp_minimize	forest_minimize	gbrt_minimize
Top 3	LR	3.9e-04	4.4e-03	6.4e-04
	DL	5	1	1
	DN	512	319	501
	AT	Relu	Sigmoid	Relu
	ACC	**98.96**	98.74	98.94
	LR	5.1e-04	5.8e-03	1.2e-03
	DL	1	1	1
	DN	512	328	235
	AT	Relu	Sigmoid	Relu
	ACC	98.90	98.72	**98.92**
	LR	1.1e-03	5.8e-03	1.2e-03
	DL	5	1	1
	DN	213	336	45
	AT	Relu	Relu	Relu
	ACC	98.88	98.58	**98.92**
Bottom 2	LR	9.6e-03	9.2e-03	2.4e-05
	DL	1	1	5
	DN	172	204	493
	AT	Sigmoid	Sigmoid	Sigmoid
	ACC	**9.24**	9.86	9.86
	LR	2.3e-06	9.0e-03	2.8e-06
	DL	4	1	5
	DN	69	305	18
	AT	Sigmoid	Sigmoid	Sigmoid
	ACC	**8.68**	**8.68**	9.24

Fig. 8 Showing first three and last two accuracies

few nodes, the accuracy seems to decrease and when the nodes increase, accuracy also elevates. It can be seen that Bayesian optimization is better than tree-based and gradient-based, in fact, all the three are so close in terms of high and low accuracy.

6 Conclusion

Hyperparameter tuning is an essential task in machine learning and deep learning models. Essentially, hyperparameters determine the accuracy of the model. An optimal configuration gives the best results whereas suboptimal choices may result in worst accuracy. Generally, by using trial and error basis hyperparameters are chosen. However, it is time-consuming to test for various combinations without any scientific reason. Hence, automatic schemes were proposed to facilitate selection of hyperparameters by the system itself. Bayesian optimizations are mostly employed to automate the process of hyperparameter tuning. In the present study, three functions based on gaussian process, decision trees, and gradient boosted methods have been considered to evaluate on MNIST dataset. The current work has developed a two layered convolutional neural network to experiment with the gp_minimize, forest_minimize and gbrt_minimize functions of skopt library. It is indicated in the results that the gp_minimize function generates 98.96% accuracy. It is specified that gp_minimize is based on bayesian optimizations which has outperformed the other methods of decision tree and gradient boosted schemes.

References

1. skopt api documentation. https://scikit-optimize.github.io/. Accessed on 11 February 2019
2. Akiba T, Sano S, Yanase T, Ohta T, Koyama M (2019) Optuna: a next-generation hyperparameter optimization framework. In: Proceedings of the 25th ACM SIGKDD international conference on knowledge discovery & data mining, pp 2623–2631
3. Bergstra J, Komer B, Eliasmith C, Yamins D, Cox DD (2015) Hyperopt: a python library for model selection and hyperparameter optimization. Comput Sci & Discov 8(1):014008
4. Bergstra JS, Bardenet R, Bengio Y, Kégl B (2011) Algorithms for hyper-parameter optimization. In: Advances in neural information processing systems, pp 2546–2554
5. Deng L (2012) The mnist database of handwritten digit images for machine learning research [best of the web]. IEEE Sig Process Mag 29(6):141–142
6. Domhan T, Springenberg JT, Hutter F (2015) Speeding up automatic hyperparameter optimization of deep neural networks by extrapolation of learning curves. In: Twenty-fourth international joint conference on artificial intelligence
7. Feurer M, Hutter F (2019) Hyperparameter optimization. In: automated machine learning. Springer, pp 3–33
8. Ilievski I., Akhtar T, Feng J, Shoemaker CA (2017) Efficient hyperparameter optimization for deep learning algorithms using deterministic rbf surrogates. In: Thirty-first AAAI conference on artificial intelligence
9. Klein A, Falkner S, Bartels S, Hennig P, Hutter F (2016) Fast Bayesian optimization of machine learning hyperparameters on large datasets. arXiv preprint arXiv:1605.07079
10. Loshchilov I, Hutter F (2016) CMA-ES for hyperparameter optimization of deep neural networks. arXiv preprint arXiv:1604.07269
11. Maclaurin D, Duvenaud D, Adams R (2015) Gradient-based hyperparameter optimization through reversible learning. In: International conference on machine learning, pp 2113–2122
12. Snoek J, Larochelle H, Adams RP (2012) Practical Bayesian optimization of machine learning algorithms. In: Advances in neural information processing systems, pp 2951–2959
13. Tsai CW, Hsia CH, Yang SJ, Liu SJ, Fang ZY (2020) Optimizing hyperparameters of deep learning in predicting bus passengers based on simulated annealing. Appl Soft Comput 106068
14. Young SR, Rose DC, Karnowski TP, Lim SH, Patton RM (2015) Optimizing deep learning hyper-parameters through an evolutionary algorithm. In: Proceedings of the workshop on machine learning in high-performance computing environments. ACM, p 4
15. Zela A, Klein A, Falkner S, Hutter F (2018) Towards automated deep learning: Efficient joint neural architecture and hyperparameter search. arXiv preprint arXiv:1807.06906

Strategies to Effectively Integrate Visualization with Active Learning in Computer Science Class

Humera Shaziya and Raniah Zaheer

Abstract Educational technology has disrupted the teaching learning process. Additionally students involvement during the class work accelerates the learning rate. Hence, integrating visualizations with active learnings (VAL) can transform the education system. The three strategies discussed in this study are Predict, Calculate, and Explain. Each strategy consists of three phases. The first phase is partial explanation of the topic by instructor followed by students activity. Second phase requires students to predict the outcome, calculate the result or provide conceptual explanation for predict, calculate, and explain strategies, respectively. Third phase deals with discussions among the instructor and students over the topic. It has been observed through this study that active learning is far beneficial than traditional method of teaching and when it is integrated with appropriate visualization tool further develop the comprehension of the subject. Experimental outcome of this investigation proved that indeed VAL achieves the goal of engaging students with the use of technology. The results with VAL is 85% and without VAL is 68%. Thus, VAL strategies are more efficient than conventional classroom approach in a computer science class.

Keywords Active learning · Animation · Calculate · Explain · Predict · Simulation · VAL · Video · Visualization

1 Introduction

Innovative learning techniques are the demand of the present education system. Traditional method of teaching follows an instructor-centric approach wherein an instructor presents the content material to the students with little or no involvement of students during the lecture. However, there is a need to engage students during

---

H. Shaziya (✉)
Department of Informatics, Nizam College, Osmania University, Hyderabad, India
e-mail: humera.shaziya@gmail.com

R. Zaheer
Department of CS, Najran University, Najran, Saudi Arabia

N. Chaki et al. (eds.), *Proceedings of International Conference on Computational Intelligence and Data Engineering*, Lecture Notes on Data Engineering and Communications Technologies 56, https://doi.org/10.1007/978-981-15-8767-2_6

the lecture to ensure better understanding through participation. Active learning is a technique that addresses the problem of one-way communication where every student is required to participate in the learning activity. With the advent of the most advanced technology in the education field, the teaching learning process can be made more effective. The visualization tools such as videos, animation, and simulation can be used in a classroom. Choosing a suitable visualization tool for the explanation of a topic further improves the clarity of the topic. Integrating the visualization tools with active learning methods enhance the teaching activity. The learning outcomes may be measured effectively by considering the performance of the students during the lecture session. In this paper, effective integration strategies are discussed to incorporate in any classroom setting. The use of visualization and the engagement of students lead to better understanding of the subject.

The subjects like algorithms, programming languages, operating systems, computer networks, database management systems, can be explained with illustrations and demonstrations by using the appropriate visualization tool. Visualization tool when used properly can enhance the quality of teaching and provides the students more knowledge easily. There is a need to figure out whether to use visualization, and based on the suitability of the visualization the most appropriate technique can be chosen. Traditional method of teaching wherein the teacher is to speak and students are to listen does not ensure whether students really comprehend the topic. This method also does not provide any mechanism to determine the achievement of the learning objectives, except assessments that are conducted in the mid and end of the semester. Active learning is about engaging students in various activities in a class to make sure they learn by doing themselves. Active learning requires an instructor to create carefully designed activities. The research question addressed in this work is, whether the integration of visualization with active learning prove to be efficient for computer science courses at university level. The Contributions of this work are: to integrate visualizations with active learning, to implemented the VAL strategies in computer science class, to assess the impact of VAL by conducting an experiment and reporting the performance of students in computer science class.

The paper is organized as follows: Sect. 2 will report the related work. Section 3 introduces the idea of active learning and its methods. Visualization need and its appropriate choice is outlined in Sect. 4. Section 5 presents the integration of visualization and active learning. Section 6 explains the investigations performed with two groups of students and results are elaborately analyzed. The paper concludes in Sect. 7, giving insights about the current work and experiments.

2 Literature Review

According to Banerjee et al. [7], students predicting the outcome with visualization shows significant improvement of learning over students who simply view the visualization. They have also showed through their study that students who had prior training of active learning benefit a lot with the use of visualization. The research questions addressed in the paper are whether there is a better behavioral engagement and cognitive achievements when visualization is combined with the active learning and what are the student perception about the effectiveness of learning with visualization. Two experiments were conducted to study the impact of using visualization with active learning. Experiment one investigated the student group not trained in active learning and experiment two was carried out with student group trained in active learning. The results from experiment one showed that there are statistically better results of visualization with active learning in contrast to passive learning. However, there were null results with respect to cognitive achievements. Experiment two results were positive in terms of behavioral engagement, effective learning, and cognitive achievements. In [8] Banerjee et al., conducted an investigation to study the effects of using visualization to explain the concept of pointers to first year graduate students and found that there were no significant improvements in the post-test results and concluded that there are several other factors that have impact on results such as topic complexity, learner characteristics, and challenge level of assessment questions. Rodger developed visual and interactive tools for teaching and learning computer science concepts. Two such tools are JAWAA [5], used to create animations and JFLAP [20], for investigating about the automata and grammar. Hundhausen et al.[15] has done an extensive study of evaluating the effectiveness of algorithm visualization (AV) and their significant findings state that the effectiveness of AV depends greatly on how it is used by students rather than what AV shows them. Banerjee et al. have come up with a mapping strategy between instructional objectives and instructional strategies with visualization. They have also specified an implementation plan for the chosen strategy. Banerjee et al. [24] discusses about visualization and how it helps students to develop a mental model. Suresh et al. [22] developed a system to simulate data structures which facilitate quick comprehension of topics. Pegu [19] Information and Communication Technology in Higher Education in India: Challenges and Opportunities, discusses the use of information technology in the class that has changed the way teachers and students interact and communicate with each other. The paper also talks about different new and innovative ways to disseminate information to the learners. Finally, the paper explores the transformative potentials of ICT in higher education in India. Stoltzfus and Libarkin [21] examined the scale-up classroom setting and a regular classroom environment and conducted experiments to conclude that active learning is all about engaging students in activities and not indulging in expensive technological infrastructure. Grissom et al. [12] performed a survey to determine the rate of adoption of student-centric learning methods. It has reported that computer science faculty use active learning moderately. Bass [9] proved a hypothesis, active learning is better than lecture learning in social studies class, to be true through tests conducted on two groups. Active learning group performance escalated relative to lecture learning group. Abrahams and Singh [4] presented e-commerce class model that facilitated students to create an e-commerce website and training material. This specific environment enables students to learn the concepts by doing and implementing. García-Holgado et al. [11] investigated a comprehensive survey about the software engineering introductory course taught using active learning techniques and described the results that suggest there is an improvement in the students comprehension of abstract topics through learning by doing projects. Houseknecht et al. [14] incorporated active learning methodologies during workshops of organic chemistry, measured and compared the results from preceding workshop, and found that it was quite encouraging to consider the use of active learning even for workshops and seminars. Deslauriers et al. [10] discusses cognitive element involved in active learning. It differentiates between feeling of learning (FOL) with test of learning (TOL) by performing experiments to give students concrete results, and thus clarify their perception of learning through tests.

3 Active Learning

Active learning approach to teaching engages students during the class work. In [6], ArmbrusterIt introduced active learning in the class of basic biology. Students were given questions or problems and after solving them, they are made to discuss their answer with their peers to bring the active participation of every student in the class activities. Thus, it is shown in the paper that it is advantageous to make students participate in the learning process. Some of the strategies of active learning are (i) think pair share (ii) one minute paper (iii) collaborative learning (iv) learning by teaching (v) infographics.

3.1 Think Pair Share

In this method, a question or problem or topic is given to a student to think about it. One to two minutes will be provided to write down the solution. Then the student is asked to discuss her answers with her peer adjacent to her. Both the students share their responses and discuss about the topic. Kothiyal [16] have done experiments of think pair share in CS1 class for a period of ten weeks by doing total 13 activities over that period. It was found that 83% of the students were engaged in the activities.

3.2 One Minute Paper

A topic is provided to the students and they are asked to write their responses in a minute [13]. This method can be used at the end of the class to determine whether students have understood the concept. The teacher can give the right answer and the students can compare their responses with the correct answers. Murphy [18] have used this technique in a data structure lab to obtain the students responses. Traditional way of taking responses was pencil and paper, however, the authors have created electronic method and thus students provided longer responses compared to traditional method which has facilitated instructor to gain insights about the students comprehension.

3.3 Collaborative Learning

Groups are created from the class and projects or tasks can be given to the group. Students split the task into subtasks and each one is performed by a specific student. Later everyone's contribution is collected and combined to form the solution to the project. In [23], Tsai et al. have taught Web 2.0 course of computer science using wiki pages. This class was conducted for UG students and those who have participated in the wiki project to learn about the subject showed excitement and enthusiasm. The paper also outlines a methodology to incorporate wiki projects in the classroom.

3.4 Learning by Teaching

New topics which are not discussed in the class are given to students. They come prepared to the class and explain the topic to the entire class. So the students learn about the concept by teaching it to the class. Abrahams et al. [3] implemented learning by teaching in an e-commerce class wherein students have prepared teaching material and also taught the course concepts to the class.

3.5 Infographics

Infographics is about representing the concepts or ideas or topics in graphical forms. Instructor can explain the concept and ask the students to create graphical representations of the topic. This method facilitate development of students critical thinking by involving and participating in the process of comprehending a topic. This simplifies the learning process as the complex topic can be showed in a simple graphical form. Krauss [17] discussed about infographics and its use in the classroom environment.

4 Visualization

4.1 Visualization and Its Need

Visualization has been used since ages in the teaching of different subjects. Traditional ways to represent the topic visually is to use various kinds of diagrams, plans, graphs, flow charts, and so on. Visual cue of a jargon or important term helps students to remember and recall it later.

4.2 How to Choose Appropriate Visualization for Your Topic

To determine the appropriate tool to be used with the topic, one has to first figure out whether the topic is suitable to be explained with the help of visualization. Not all topics in a subject require the use of a tool. Figure 1a shows the subject structure is divided into three different types of topics: procedures, concepts, and facts. Typically, a procedure consists of number of steps or cycles that can be best explained by means of a visual tool. It turns out that the concepts also need some kind of visual to describe clearly or in a lucid manner. It is more intuitive to interpret facts and so visual tool might not be that effective in its explanation. There are several purposes of using visuals. They can be mainly categorized into 3 classes: instructor centric, topic centric, and student centric. An instructor motives to use a visualization to complete the syllabus faster or to avoid writing on the board are not to be considered for choosing visualization. Visualization use for such topics is recommended and the approach is known as topic centric use of visualization. It turns out that a student-centric approach is the most appropriate one for the usage of visualization as it aims at making students comprehend the topic in a relatively easier manner. Figure 1b

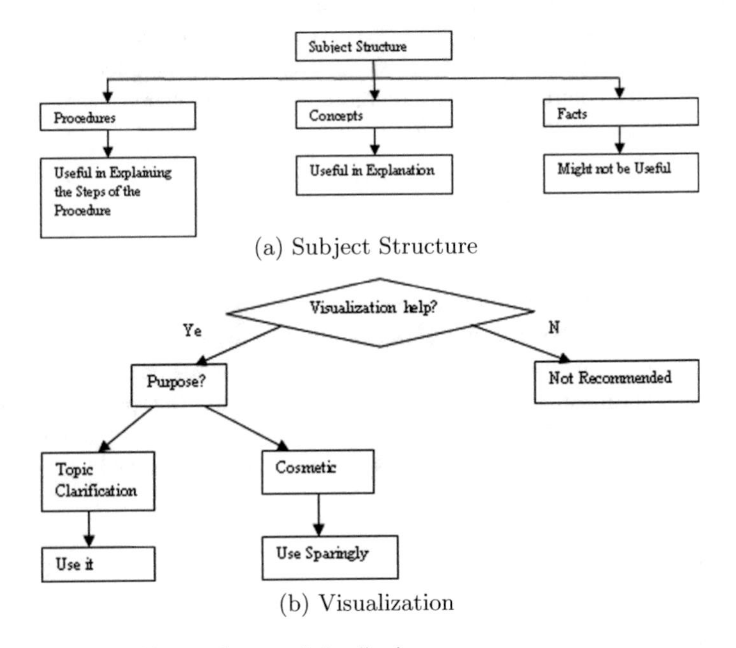

Fig. 1 Structure of subjects and usage of visualization

shows the process of making a decision about the purpose of using a visualization. Sometimes a tool may not bring about any difference in level of understanding of a topic, in such situation it is not recommended to use it. However, there are several topics that can be benefited by the use of it. Finally, use of a tool just to make a class or topic fancy does not satisfy the intended purpose of using it, so in that case it has to be used occasionally.

5 Strategies for Integration of Visualization with Active Learning

There are three strategies to integrate visualization with active learning: Predict, Calculate, and Explain. These active learning methods are combined with visualization and are executed in three phases: Observe, Predict or Calculate or Explain, and Correct. Essentially, the idea is to engage students with some sort of activity with the help of visualization. Typically, the observe phase to let students view the visualization in order to gain an understanding of concept. The second phase is concerned with students activity in which they come up with their results. Finally, in the correct phase, the instructor discusses the actual result and students results. In the current study, these strategies are used to teach computer science (CS) subjects in the classroom setting. It turns out that the use of visualization in a passive mode of teaching does not provide higher level of perception of concepts in contrast to the use of visualization in an active learning mode where there is a higher students engagement in the learning process.

5.1 Predict

This strategy focuses on prediction of outcome of an experiment. A video or animation or simulation of a concept is showed to students up to a particular point and is paused. The students are then asked to make a prediction of what happens next. The three phases corresponding to predict strategy are observe, predict, and correct. As mentioned earlier, choice of appropriate visualization for a topic is an important aspect of improving the understanding level of the topic. First, there is a need to figure out whether the topic is suitable to be explained with some kind of visualization. To demonstrate the use of predict strategy in CS, the topic chosen is CPU scheduling from operating system subject. There are various algorithms for scheduling jobs to processor. The setup for the demonstration includes a simulator wherein couple of input variables have to be chosen such as number of jobs, algorithm, speed, and so on as shown in Fig. 2a. Instructor choose the appropriate input values and run the simulation up to a particular point and then it is paused. Instructor asks the students to make a prediction of next job to schedule to cpu. After students write down their

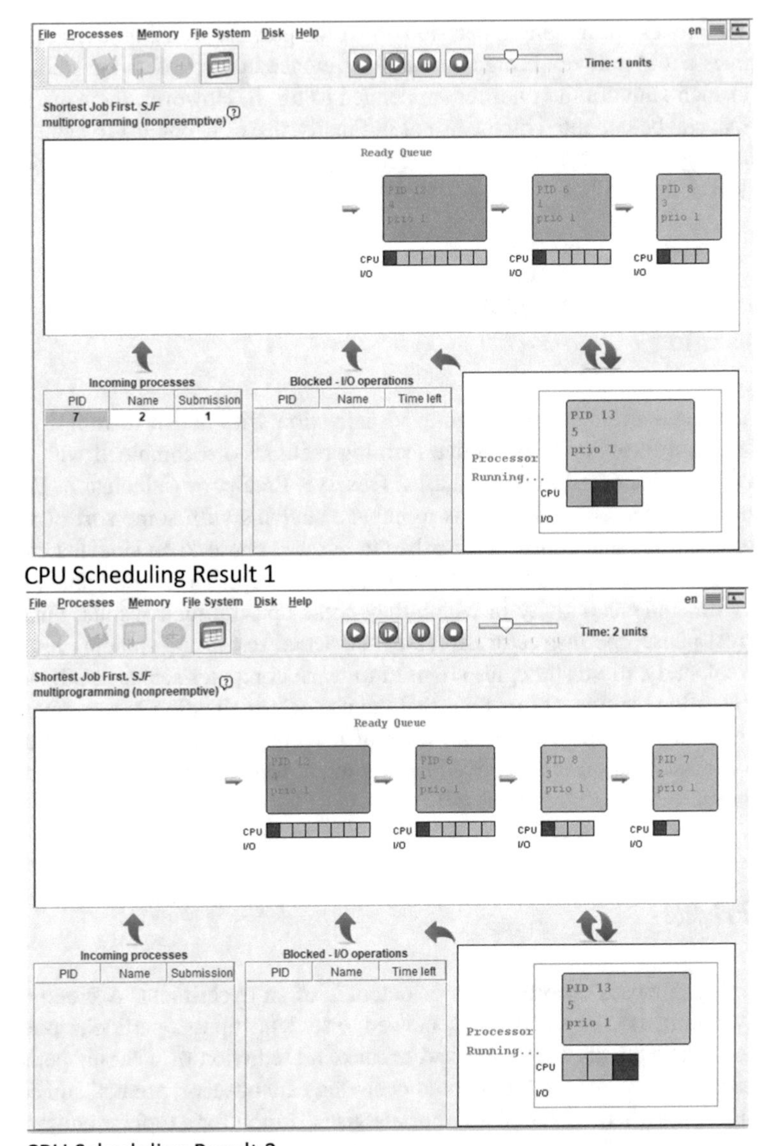

Fig. 2 Shows implementation of predict

predictions, instructor resumes the simulation and a comparison of student result with the actual result given in Fig. 2b, is done and the discussion follows.

5.2 *Calculate*

Calculate strategy is applied to topics that involve computations or mathematical calculations. This strategy consists of three phases: observe, calculate, and correct. In the observe phase, a video or simulation performing calculation of first step of a topic is played, students observe how it is being done. The video or simulation is paused just before the execution of second step and students are asked to perform calculations and write the results down. The third phase is about discussing and correcting the outcome of the calculation phase. During the correct phase, video or simulation is played again to get to know the correct output of the calculation, followed by the discussion on how to arrive to the actual output. Students have to state the logic for arriving at a particular result. For instance, Fig. 3a, shows pictorial implementation of searching of a key in a hash. This simulator can be found at [2]. A hash is created with few elements in it. A key value is given as input to be searched in the hash. The simulation in Fig. 3b, demonstrates step by step the search procedure in a hash. The result is displayed after the simulation of the process.

5.3 *Explain*

Explain strategy is used to illustrate a concept with the help of visualization. This can be done in the first class of a new topic or a concept. The phases in this strategy are observe, explain, and discuss. In the observe phase, a video or simulation of the entire concept is showed without giving any explanation of it. The students are then asked in the explain phase to come up with the explanation of that particular concept. This can be combined with think pair share strategy of active learning. Students are grouped into pairs, these pairs watch to the video or simulation in the observe phase and discuss to arrive to the explanation of the concept and share their interpretation with the class. Different groups present their understanding about the concept and later the teacher discusses scientifically correct explanation and argues the pros and cons of various explanations. Figure 4a simulates the concept of paging from operating system (OS) course. Simulation shows the behavior of OS when the pages are brought in and out of the memory and it helps learners to form a cognitive image of how an invisible concept works. For instance, when a page is brought in the memory and swapped out of the memory, changes are made to the page table that is shown in Fig. 4b.

6 Results and Discussions

To showcase the effectiveness of active learning in a classroom environment, B+ tree topic from database management system (DBMS) has been chosen. A simulator for

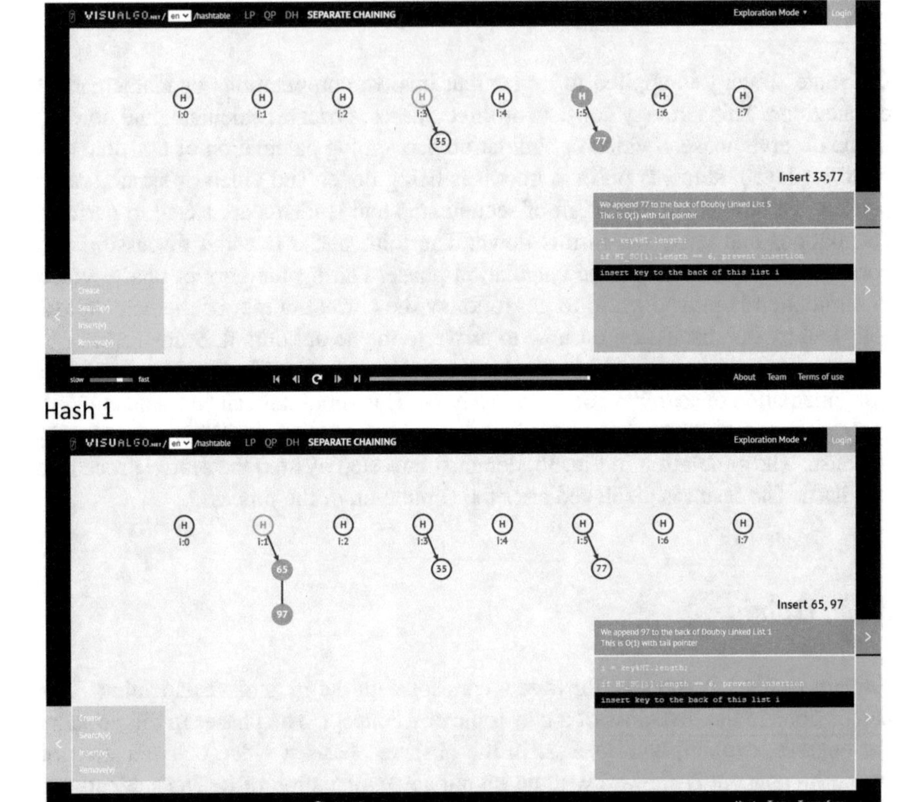

Fig. 3 Presents how to leverage calculate approach

B+ trees [1], is leveraged to perform the study in the class of PG Sem III. Students of this class had basic knowledge of data structures and relational database concepts. Total number of students attended the experiment is 20. The students were split in two groups, first group was taught the concept of B+ trees using conventional teaching style, whereas the other group has been made to learn the same concept using active learning integrated with visualization. The strategies explained in the active learning section and visualization section have been incorporated to disseminate the B+ topic of DBMS. Think pair share strategy of active learning and predict method is employed to engage students in the teaching learning process. The activity begins when the instructor explains the concepts and rules of B+ trees. After detailing the rules for constructing an example tree using the visualization technique, students were asked to involve in the activity. The activity starts by giving a series of numbers for the development of a B+ tree. The numbers given are 4, 2, 8, 3, 9, 11, 18, 5, 12, 6. Each student first creates a tree using pen and paper, then discuss it with his/her

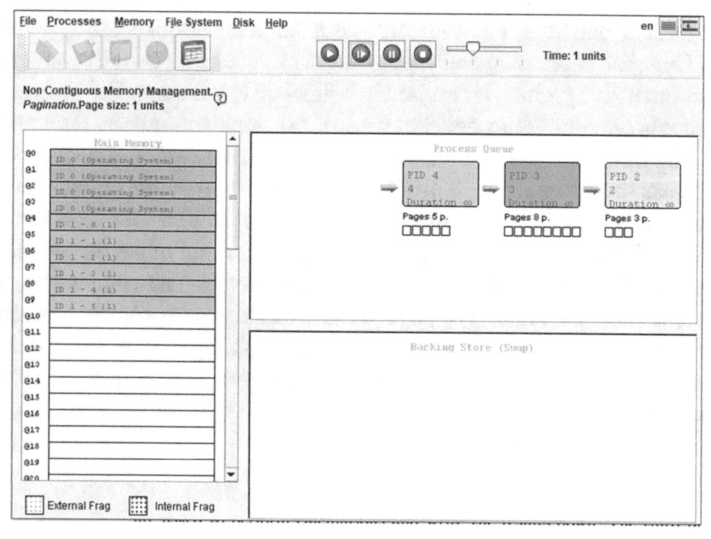

Paging result 1

Paging result 2

Fig. 4 Demonstrates explain method

adjacent student, and lastly create the tree with the same numbers using visualization method. Thus, active learning and visualization has been successfully integrated into the classroom activity which is resulted in better participation and involvement. This gives students opportunity to develop the critical thinking abilities. The number of students who have correctly created the tree using active learning and visualization is 85%, whereas only 68% students from other group could do the construction of the B+ tree correctly. Another experiment was carried out from operating system course for the determination of number of page faults. The concept of paging and page fault was explained with conventional means of chalk board. The students are given a problem to find out the number of page faults. Second group of students were taught by means of visualization and animation of page fault topic, from the test results of these students it turns out the performance was relatively better than the first group. Thus, the results proves that integrating visualization with active learning improves the performance of the students.

7 Conclusion

Information and communication technology in higher education has brought many changes in the manner teaching is conducted and thus improves the level of understanding of difficult concepts. Certain topics can be explained well with the help of visualization. Active learning engages students in the teaching learning process. In this paper, various strategies to integrate active learning with visualization have been discussed. The strategies have been applied concretely to computer science concepts. The active learning techniques such as predict, calculate, and explain are discussed in detail with the example concepts from the area of computer science. The visualization methods such as video, animation, and simulation are integrated effectively with the active learning strategies to enhance teaching learning process. It is shown that the effective utilization of visualization comes from the nature of topic of discussion. There are some topics that are explained well and cognitive achievement will be more if they are explained using the appropriate visualization. Passive learning does not yield better acquisition of knowledge compared to active learning that enhances comprehension of subject material. It is discussed in this paper about integrating active learning with visualization to produce better results in terms of knowledge acquisition and performance in exams. The students who could finish the activity without active learning and visualization are 68% whereas 85% students could do the assigned task using active learning and visualization.

References

1. B+ tree visualization. https://www.cs.usfca.edu/~galles/visualization/BPlusTree.html. Accessed on 11 March 2019

2. Visualgo: visualising data structures and algorithms through animation. https://visualgo.net/en. Accessed on 11 March 2019
3. Abrahams AS, Singh T (2010) An active, reflective learning cycle for e-commerce classes: learning about e-commerce by doing and teaching. J Inform Syst Edu 21(4):383
4. Abrahams AS, Singh T (2019) An active, reflective learning cycle for e-commerce classes: Learning about e-commerce by doing and teaching. J Inform Syst Edu 21(4):6
5. Akingbade A, Finley T, Jackson D, Patel P, Rodger SH (2003) Jawaa: easy web-based animation from cs 0 to advanced cs courses. In: ACM SIGCSE Bulletin, vol 35. ACM, pp 162–166 (2003)
6. Armbruster P, Patel M, Johnson E, Weiss M (2009) Active learning and student-centered pedagogy improve student attitudes and performance in introductory biology. CBELife Sci Educ 8(3):203–213
7. Banerjee G, Murthy S, Iyer S (2015) Effect of active learning using program visualization in technology-constrained college classrooms. Res Pract Technol Enhanced Learn 10(1):15
8. Banerjee G, Patwardhan M, Mavinkurve M (2013) Teaching with visualizations in classroom setting: mapping instructional strategies to instructional objectives. In: 2013 IEEE fifth international conference on technology for education (t4e 2013). IEEE, pp 176–183 (2013)
9. Bass B (2018) Action research study of classical teaching methods vs. active learning methods in the middle school social studies classroom. In: Culminating experience action research projects, vol 18, part 2. Spring, p 26 (2018)
10. Deslauriers L, McCarty LS, Miller K, Callaghan K, Kestin G (2019) Measuring actual learning versus feeling of learning in response to being actively engaged in the classroom. Proc Natl Acad Sci 116(39):19251–19257
11. García-Holgado A, García-Peñalvo FJ, Rodríguez-Conde MJ (2018) Pilot experience applying an active learning methodology in a software engineering classroom. In: 2018 IEEE global engineering education conference (EDUCON). IEEE, pp 940–947 (2018)
12. Grissom S, Mccauley R, Murphy L (2017) How student centered is the computer science classroom? A survey of college faculty. ACM Trans Comput Educ (TOCE) 18(1):1–27
13. Harwood WS (1996) The one-minute paper. J Chem Educ 73(3):229
14. Houseknecht JB, Bachinski GJ, Miller MH, White SA, Andrews DM (2020) Effectiveness of the active learning in organic chemistry faculty development workshops. Chem Educ Res Pract 21(1):387–398
15. Hundhausen CD, Douglas SA, Stasko JT (2002). J Vis Lang Comput 13(3):259–290
16. Kothiyal A, Majumdar R, Murthy S, Iyer S (2013) Effect of think-pair-share in a large cs1 class: 83% sustained engagement. In: Proceedings of the ninth annual international ACM conference on International computing education research. ACM, pp 137–144
17. Krauss J (2012) Infographics: more than words can say. Learn Lead Technol 39(5):10–14
18. Murphy L, Wolff D (2005) Take a minute to complete the loop: using electronic classroom assessment techniques in computer science labs. J Comput Sci Colleges 21(1):150–159
19. Pegu UK (2014) Information and communication technology in higher education in India: challenges and opportunities. Int J Inform Comput Technol 4(5):513–518
20. Rodger SH, Finley TW (2006) JFLAP: an interactive formal languages and automata package. Jones & Bartlett Learning
21. Stoltzfus JR, Libarkin J (2016) Does the room matter? active learning in traditional and enhanced lecture spaces. CBELife Sci Educ 15(4):ar68
22. Suresh N, Kottangodan SS, Niyas P (2014) Simulation of recursion and data structures
23. Tsai WT, Li W, Elston J, Chen Y (2010) Collaborative learning using wiki web sites for computer science undergraduate education: a case study. IEEE Trans Educ 54(1):114–124
24. Wong L et al (2013) Program visualization: Effect of viewing vs. responding on student learning

Supervised Learning-Based Classifiers in Healthcare Decision-Making

Barasha Mali, Chandrasekhar Yadav, and Santosh Kumar

Abstract Supervised classifiers are machine learning classifiers that can predict the present categorical class label if similar information from the past is given. The simple and easy nature of such classifier makes it popular in different applications. Healthcare is one such application area which has recently adopted this computerized decision-making approach to assist the experts and speed up the process. Also advances in healthcare electronics is generating a lot of data and making it easily available. The reference for designing a reliable decision-making system can be obtained from these easily available datasets. This paper briefs the recently adopted machine learning techniques in healthcare with special focus on the supervised classifier algorithms, its application to healthcare decision-making and their evaluation. These classifiers are applied to a Parkinson's disease benchmark dataset for validation.

Keywords Supervised classifier · Healthcare · Machine learning · Decision-making · Ensemble · Support vector machines

B. Mali (✉)
National Institute of Technology, Silchar, India
e-mail: barashamali@gmail.com

Sant Longowal Institute of Engineering and Technology, Longowal, India

C. Yadav · S. Kumar
Standardization Testing and Quality Certification, Ministry of Electronics and Information Technology, Hyderabad, India
e-mail: chandrtech15@gmail.com

S. Kumar
e-mail: santoshkumar7461@gmail.com

N. Chaki et al. (eds.), *Proceedings of International Conference on Computational Intelligence and Data Engineering*, Lecture Notes on Data Engineering and Communications Technologies 56, https://doi.org/10.1007/978-981-15-8767-2_7

1 Introduction

Supervised learning-based decision-making has proved its worth in different fields including healthcare and is gaining popularity. Recently [10], have reviewed machine learning applications in neuroscience and suggested it as the handy toolbox for the future neuroscientists. Supervised machine learning helps to accurately predict the performance of different methods used by neuroscientists to determine the correlated variables or to get some standard and simple models related to neuroscience. Esteva et al. [9] has cited that the improvement in computational and electronic technology has made a large amount of data available leading to the increasing popularity of deep machine learning in the healthcare sector, especially in the fields of medical imaging, robotic-assisted surgery and genomics.

Such computational techniques that make human-like decision are used to design smart healthcare environment in [15], to monitor and analyze the health of the user. Table 1 will help us visualize that various effective applications in healthcare have

Table 1 Recent application of computation in healthcare

Author and year	Method	Application
Carnevale et al. 2020 [6]	Supervised learning	Identifying critical patients based on the posts in a network
Glaser et al. 2019 [10]	Supervised Machine learning	Predictive variable identification and benchmarking brain models in neurobiology
Subasi et al. 2019 [15]	Machine learning	Designing smart healthcare for recognizing human activities
Esteva et al. 2019 [9]	Deep learning	Impact of computational techniques in medical imaging, robotic- assisted surgery and genomics
Bansal et al. 2019 [5]	Machine learning	Medical department identification of visiting patients
Jiang et al. 2017 [12]	Artificial intelligence	Reviews that cancer, cardiology and neurology as the major AI application areas for early detection, diagnosis and treatment
Miotto et al. 2017 [13]	Deep learning	Reviews the deep learning applications in medical applications and suggests it for improving the health sector
Dua et al. 2014 [8]	Machine learning	Challenges and solutions related to healthcare informatics are reviewed and applied
Dua et al. 2014 [7]	Supervised learning	Surveys the supervised machine learning models for fraud detection in healthcare
Soleimanian et al. 2012 [14]	Decision Tree	Determining the appropriate medical operation method for pregnant women
Zhang et al. 2015 [18]	Semi-supervised kernel learning	Classification and prediction of pulmonary embolism and breast cancer

come up utilizing supervised learning, machine learning and deep learning-based decision making classifiers.

The authors surveyed the recent healthcare applications using supervised learning and related methods. Decision making in healthcare is mostly benefitted in this sector. Therefore, to see how effective the supervised learning methods are, the different classifiers were implemented on a real dataset of Parkinson's disease. The method of data collection and dataset description is explained in Sect. 5. The supervised classifiers were compared using the different evaluation parameters.

Section 1 gives an introduction about supervised learning and related fields in recent healthcare applications, Sect. 2, briefs the classifier algorithms in general, Sect. 3, discusses the parameters used for evaluating the performance of the classifiers, Sect. 4, applies the supervised classifier algorithms to a healthcare dataset as a case study and the paper is concluded in Sect. 5, giving an idea about the future of supervised classifiers in healthcare.

2 Classification Algorithms

Classification is the method of forming a subset of data based on some similarities among them. It finds application in various areas to distinguish between different categories of available information. For supervised learning classifiers some subset of the data must be known [16]. Depending on the application and data availability, different classifiers may perform better than the other.

It basically consists of two phases: (i) training phase where the subset of data with known classification is used to construct a model for the next phase and (ii) testing phase where the subset of data with unknown classification is assigned classes based on the training phase.

Some commonly used classifier algorithms are as follows:

1. Decision trees: Decision trees are used to classify different classes based on split criteria such as gini-index or entropy starting from the beginning or root note to the decision or leaf node [17]. The indices give a measure of skew of different classes which is inversely proportional to the value of the indices. If $p_1, p_2, \ldots p_k$ are the fraction of classes in node N belonging to k different classes, gini-index and entropy is given by Eqs. 1 and 2. Gini-index ideally should lie between 0 and $1 - \frac{1}{k}$ and entropy between 0 and $\log(k)$.
   Gini-index,
   $$G(N) = 1 - \sum_{i=1}^{k} p_i^2 \tag{1}$$
   Entropy,
   $$E(N) = -\sum_{i=1}^{k} p_i.log(p_i) \tag{2}$$

Decision trees are subclassified as: deep, medium and shallow trees based on the depth of the trees.

2. Nearest neighbour: Nearest neighbour classifier is a simple instance-based classifier. The top k values of the training data are considered as the given test instance and majority class data found among the neighbouring k values is the predicted class. Improved performance of the classifier can be obtained by using a nearest neighbour index like centroid, mahalanobis distance, fisher's discriminant, etc. This can be easily converted to the multiclass nearest neighbour classifier [3]. Based on the index the nearest neighbour classifiers are classified as fine, medium, coarse, cubic, cosine, weighted, etc.
3. Support vector machines: Support vector machine or SVM is basically a binary linear classifier based on a single separating hyperplane that separates two different classes optimally [3]. It is modified into multiclass support vector machine to classify dataset with more than two classes. The generalized equation of SVM classification is as in Eq. 3.

$$Y(x) = W^T \phi(x) + b \tag{3}$$

where $Y(x)$ is the output of the classifier, b is the bias and W^T is the weight and is obtained from the training dataset. Data x is mapped to space S using $\phi(x)$. Minimizing $||W||$ and satisfying condition $y_i Y(x_i) \geq 1$ a suitable hyperplane is obtained for separating the different classes [4, 11].

A linear hyperplane is used for linear SVM. Complex datasets are not always linear and kernel functions are used to transform a low dimensional space to a high dimensional space. Therefore, based on the kernel function SVM classifiers are further classified as fine gaussian, medium gaussian, coarse gaussian, cubic, quadratic, etc.
4. Ensemble: Ensemble methods combine the information obtained from more than one classifier, and hence better performance can be expected from such classifiers. Based on the selection of base classifiers and specific combination strategies, ensemble methods are classified as boosted trees, bagged trees, subspace kNN, subspace discriminant, etc.

Classification algorithms are based on the type and amount of data available. Also, each model has its advantages and limitations.

3 Methodology

Recent applications of supervised learning method in making decisions in various healthcare applications were surveyed. The classification algorithms were analyzed and applied to a real healthcare dataset which is used as a benchmark dataset in most of the available literature. The classification algorithms were compared using the evaluation parametres as mentioned in Sect. 4. Different classification algorithms perform best in different conditions of datasets and the best method for classifying the patient's condition in the dataset was analyzed using the evaluation parameters.

4 Performance Assessment

The classifier algorithms are assessed on how exactly it has categorized the classes of the test dataset. The performance assessment is, therefore, done on the testing dataset. Validation and evaluation measures used are discussed in this section.

Cross-validation is a method where the available dataset is divided into k subsets where $(k-1)$ subsets are used for training of the supervised classifier and 1 subset is used for testing. This is repeated k times considering each subset once for testing and then the average of the cross-validation error is used as a performance assessment indicator. Cross-validation using 5 folds is used to assess the performance of the classifiers used in terms of performance.

The possible perception for the classified output of a classifier can be categorised as follows [1]:

1. True Positive (TP)- An actually positive class classified as positive.
2. True Negative (TN)- An actually negative class classified as negative.
3. False Negative (FN)- An actually positive class misclassified as negative.
4. False Positive (FP)- An actually negative class misclassified as positive.

In case of a 2-class or binary problem, this is represented by a 2×2 dimensional matrix and as shown in Fig. 1. This 2×2 matrix is known as confusion matrix and the predicted true classes are compared using this matrix. This can be easily expanded to $n \times n$ dimensional matrix for a n class classification and prediction problem as Fig. 2. Accuracy is an evaluation measures that describe how correctly a classifier can classify the different classes or how good it is performing. Mathematically, it can be represented as Eq. 4.

$$\text{Accuracy} = \frac{\text{Number of correct predictions}}{\text{Total number of predictions}} \tag{4}$$

In confusion matrix terms, accuracy is given as Eq. 5 below.

$$\text{Accuracy} = \frac{TP + TN}{TP + TN + FN + FP} \tag{5}$$

Precision and recall are metrics that can be used to compare the result. Precision or positive predictive value gives the measure of how much positively identified classes are actually correct. Recall also known as sensitivity gives the measure of the portion of the correctly identified actual positive classes. This can be represented by the Eqs. 6 and 7.

		Predicted class		
		p	**n**	**total**
True class	**p′**	True positive (TP)	False negative (FN)	P′ =TP+FN
	n′	False positive (FP)	True negative (TN)	N′ =FP+TN
	total	P=TP+FP	N=FN+TN	

Fig. 1 Confusion matrix of a 2-class problem

		Predicted class			
		$\mathbf{c_1}$	$\mathbf{c_2}$		$\mathbf{c_j}$
True class	$\mathbf{c'_1}$	M_{11}	M_{12}	...	M_{1j}
	$\mathbf{c'_2}$	M_{21}	M_{22}	...	M_{2j}
		....	...	....	....
	$\mathbf{c'_i}$	M_{i1}	M_{i2}	...	M_{ij}

Fig. 2 Confusion matrix of a multiclass classification

$$\text{Precision} = \frac{TP}{TP + FP} \tag{6}$$

$$\text{Recall} = \frac{TP}{TP + FN} \tag{7}$$

The same can be extended for a multiclass problem as the Eqs. 8, 9 and 10

$$\text{Accuracy} = \frac{M_{11} + M_{22} + \cdots M_{jj}}{\sum_{i=1}^{n} \sum_{i=j}^{n} M_{ij}} \tag{8}$$

$$\text{Precision of } i\text{th class} = \frac{M_{ii}}{\sum_{j=1}^{n} M_{ji}} \tag{9}$$

$$\text{Recall of } i\text{th class} = \frac{M_{ii}}{\sum_{j=1}^{n} M_{ij}} \tag{10}$$

5 Case Study

The application of supervised classifiers as mentioned in Sect. 2, were applied to a Parkinson's disease dataset [2] to verify its performance in the healthcare sector.

5.1 Dataset Description

Parkinson's disease is a disorder that occurs in the central nervous system and one of the early symptoms observed in such patients is voice change. Therefore, the data was collected from 252 persons both male and female and of age group 41 to 82. Among them, 64 were healthy persons and 188 persons had symptoms. The data of healthy persons were considered as the controlling data. A microphone set at 44.1 KHz was used to collect the voice data which recorded the phonation of /a/ vowel three times each for each subject.

5.2 Results

Table 2 shows the performance of the different algorithms using a 5 fold cross validation.

The performance parameters: accuracy, true positive (TP), true negative (TN), false negative (FN) and false positive (FP) are compared. The accuracy of all the supervised classifiers ia above 68.9% with a maximum of 91.8% for Fine K-Nearest Neighbour classifier. The KNN classifier distinguishes between the two classes with a medium speed. The TP gives the actually positive cases which are detected as positive and TN gives the actually negative cases detected as negative for the classifier which is 79% and 96%, respectively. The FN classifies the actually positive cases as negative cases which can be very dangerous in healthcare data classification. This value is 4% in case of the Fine KNN classifier. The other classifiers giving less FN value has comparatively less accuracy because of the less TP value.

For the Parkinson's dataset, the supervised classifier Fine KNN classifies the positive and negative cases with the best accuracy among the different supervised classifiers. This can vary for other datasets. Therefore, this proves that the supervised classifier application in healthcare dataset is possible.

Table 2 Performance of supervised classifiers in the Parkinson's disease dataset

	Accuracy (%)	TP (%)	TN (%)	FN (%)	FP (%)
Fine tree	77.6	55	85	45	15
Medium tree	79.0	52	88	48	12
Coarse tree	78.2	48	88	52	12
Linear SVM	86.1	54	97	46	3
Quadratic SVM	88.0	66	96	34	4
Cubic SVM	88.8	69	95	31	5
Fine Gaussian SVM	74.6	0	100	100	0
Medium Gaussian SVM	84.7	45	98	55	2
Coarse Gaussian SVM	78.0	17	99	83	1
Fine KNN	**91.8**	79	96	21	4
Medium KNN	82.8	41	97	59	3
Coarse KNN	76.9	11	99	89	1
Cosine KNN	84.5	56	94	44	6
Cubic KNN	80.8	32	98	68	2
Weighted KNN	84.5	46	98	54	2
Ensemble boosted trees	88.5	66	96	34	4
Ensemble bagged trees	86.9	60	96	40	4
Ensemble subspace discriminant	84.0	68	90	32	10
Ensemble subspace KNN	68.9	38	80	63	20
Ensemble RUSBoosted trees	85.4	79	88	21	12

6 Conclusion and Future Works

The paper discusses the popular supervised learning algorithms and have shown its application in a healthcare dataset. The evaluation parameters for supervised classifiers are also discussed and calculated for the same. Application of the supervised classifiers to the Parkinson's disease dataset shows that the performance of all the classifiers are comparable and can be applied to the healthcare sector to make decisions. The training datasets, however, must be correct and with information content so as to train the classifier properly before applying it for practical situations.

Healthcare areas where supervised machine learning can be effectively explored on future are: identifying and diagnosing diseases, early stage discovery of drug and its manufacturing, medical image diagnosis, decreased clinical trial time, predicting any epidemic outbreak, etc.

References

1. The Basics of Classifier Evaluation: Part 1 (2015). https://www.svds.com/the-basics-of-classifier-evaluation-part-1/
2. Parkinson's Disease Classification Data Set (2018). https://archive.ics.uci.edu/ml/datasets/Parkinson
3. Aggarwal CC (2014) Data classification: algorithms and applications. CRC Press
4. Aravind K. Raja P, Mukesh K, Aniirudh R, Ashiwin R, Szczepanski C (2018) Disease classification in maize crop using bag of features and multiclass support vector machine. In: 2018 2nd international conference on inventive systems and control (ICISC). IEEE (2018)
5. Bansal V, Poddar A, Ghosh-Roy R (2019) Identifying a medical department based on unstructured data: A big data application in healthcare. Information 10(1):25
6. Carnevale L, Celesti A, Fiumara G, Galletta A, Villari M (2020) Investigating classification supervised learning approaches for the identification of critical patients' posts in a healthcare social network. Appl Soft Comput 90:106155
7. Dua P, Bais S (2014) Supervised learning methods for fraud detection in healthcare insurance. In: Machine learning in healthcare informatics. Springer, pp 261–285
8. Dua S, Acharya UR, Dua P (2014) Machine learning in healthcare informatics, vol 56. Springer
9. Esteva A, Robicquet A, Ramsundar B, Kuleshov V, DePristo M, Chou K, Cui C, Corrado G, Thrun S, Dean J (2019) A guide to deep learning in healthcare. Nat Med 25(1):24–29
10. Glaser JI, Benjamin AS, Farhoodi R, Kording KP (2019) The roles of supervised machine learning in systems neuroscience. Progress in neurobiology
11. Hazarika J, Kant P, Dasgupta R, Laskar SH (2018) Neural modulation in action video game players during inhibitory control function: an eeg study using discrete wavelet transform. Biomed Signal Process Control 45:144–150
12. Jiang F, Jiang Y, Zhi H, Dong Y, Li H, Ma S, Wang Y, Dong Q, Shen H, Wang Y (2017) Artificial intelligence in healthcare: past, present and future. Stroke Vasc Neurol 2(4):230–243
13. Miotto R, Wang F, Wang S, Jiang X, Dudley JT (2017) Deep learning for healthcare: review, opportunities and challenges. Brief Bioinform 19(6):1236–1246
14. Soleimanian F, Mohammadi P, Hakimi P (2012) Application of decision tree algorithm for data mining in healthcare operations: A case study. Int J Comput Appl 52:21–26
15. Subasi A, Khateeb K, Brahimi T, Sarirete A (2020) Human activity recognition using machine learning methods in a smart healthcare environment. In: Innovation in Health Informatics. Elsevier, pp. 123–144
16. Yadav CS, Sharan A (2020) Feature learning using random forest and binary logistic regression for ATDS. In: Applications of Machine Learning. Springer, Singapore, pp 341–352
17. Yadav M, Verma VK, Yadav CS, Verma JK (2020) MLPGI: multilayer perceptron-based gender identification over voice samples in supervised machine learning. In: Applications of Machine Learning. Springer, Singapore, pp 353–364
18. Zhang G, Ou SX, Huang YH, Wang CR (2015) Semi-supervised learning methods for large scale healthcare data analysis. Int J Comput Healthcare 2(2):98–110

Gridding and Supervised Segmentation Method for DNA Microarray Images

Bolem Sai Chandana, Jonnadula Harikiran, Battula Srinivasa Rao, and T. Subbareddy

Abstract This paper mainly focusses on the image gridding and segmentation methods of microarray analysis. The process of gridding is to divide the image into sub-array of spots (sub-gridding) and sub-arrays are again divided into spot areas (spot detection). Most of the existing methods depend on parameters such as number of rows/columns, spots count in each row/column, and size of sub-array. In this paper, a gridding algorithm is presented without any human intervention removing any parameter initializations. In the segmentation step, first the pixels are classified as spot/background using Support Vector Machine (SVM). This classification result is used for segmentation of spot area in gridded image block. The results show the proposed algorithms that perfectly grids the microarray image and perfectly segments the spot area from background. The log-ratio values calculated for each spot determines the transcription abundance of each gene.

Keywords Microarray image · Image segmentation · Mathematical morphology · Support vector machines

1 Introduction

Microarray technology is used for parallel analysis of gene expression values for thousands of spots [1]. The output of microarray experiment is an image extracted from hybridized microarray slide using a sensor with two different wavelengths Cy3 and Cy5. The analysis of this microarray image is carried out in three stages, Gridding, Segmentation, and Quantification. The process of gridding is carried out in two stages, first divide the image into sub-arrays which is called as sub-gridding and these sub-arrays are divided into gene spot areas which is called as spot detection

B. Sai Chandana (✉)
Department of CSE, Gudlavalleru Engineering College (GEC), Gudlavalleru, India
e-mail: bschandana@gmail.com

J. Harikiran · B. Srinivasa Rao · T. Subbareddy
School of CSE, Vellore Institute of Technology, VIT-AP University, Amaravathi, India

N. Chaki et al. (eds.), *Proceedings of International Conference on Computational Intelligence and Data Engineering*, Lecture Notes on Data Engineering and Communications Technologies 56, https://doi.org/10.1007/978-981-15-8767-2_8

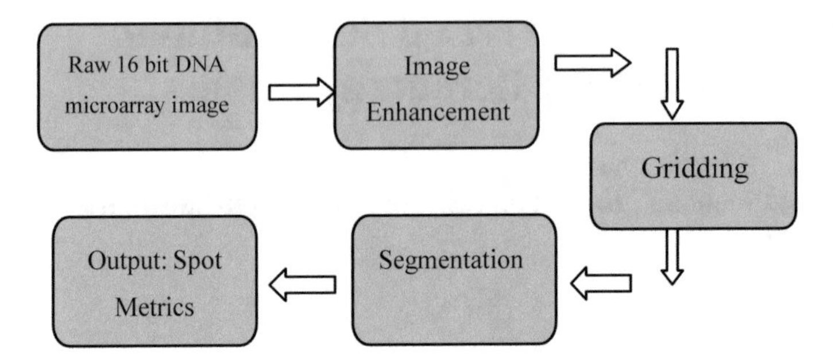

Fig. 1 Architecture for microarray analysis

[1]. Segmentation is the process of grouping spot pixels and background pixels in each spot area. Quantification is the process of finding the gene expression value which is equal to the log ratio of green to the red intensity values of spot [2].

The gene expression value depends on the intensity values of the foreground pixels (spot area) and background pixels. Most of the existing algorithms used in analysis of microarray image are semi-automatic which means that human intervention is required for initialization of parameters to execute the gridding algorithm [3]. In this paper, a fully automatic gridding algorithm is presented. This microarray technology is used in research on genetics, disease diagnosis, drug discovery, and many applications in pharmacology [4]. The microarray image analysis mechanism is shown in Fig. 1.

In this paper, Sect. 2 presents gridding procedure, Sect. 3 presents microarray segmentation, Sect. 4 presents results followed by conclusions.

2 Microarray Gridding

Accurate gridding improves the efficiency of segmentation and quantification steps in sequential analysis of microarray image. To capture each gene spot from microarray image, first the image is divided into sub-arrays, second these sub-arrays are divided into spot areas. The first step is called global gridding (sub-gridding) and second step is called local gridding (spot detection). The final output of gridding is an area with single spot and background. Existing algorithms for gridding are categorized into manual and semi-automatic procedures. In manual gridding, the user must specify all parameters such as number of sub-arrays, number of spots, and spot size, required for gridding. In semi-automatic procedures, users partially provide the parameters such as number of rows/columns and number of spots in each row/column. In this paper, an automatic gridding algorithm is presented using horizontal and vertical profiles of the microarray. The sub-gridding and spot detections algorithms are presented in Figs. 2 and 3.

Sub-gridding Algorithm:

Convert the original microarray image into grayscale image

Compute the standard deviation of pixel intensities of every row, producing the profile of horizontal standard deviation

$$S_i = \frac{1}{M-1}(I(i,j) - \tfrac{1}{N}\sum_{j=1}^{N} I(i,j))$$

for i=1,2,.....M and I(i,j) is the intensity of pixel located at row i and column j.

Compute the standard deviation of pixel intensities of every column, producing the profile of vertical standard deviation

$$S_j = \frac{1}{N-1}(I(i,j) - \tfrac{1}{M}\sum_{i=1}^{M} I(i,j))$$

for j=1,2........N and I(i,j) is the intensity of pixel located at row i and column j.

Find the minimum value in S_i and S_j.
MIN1= minimum (S_i)
MIN2= minimum (S_j)

These standard deviation profiles have to be thresholded for the estimation of row width and column width.
TS_i = 1, if S_i > MIN1+ Th * MIN1
= 0 otherwise
TS_j = 1, if S_j > MIN2+ Th * MIN2
= 0 otherwise

Calculation or row width and column width: The value in TS_i and TS_j looks like 1's followed by 0's again 1's and so on.

The procedure for row width calculation is as follows:
i. for i = 1 to M,
Count the number of 0's in TS_i (between the elements in TS_i having 1) = RW_{yp}, Where p = 1, 2, 3 ..., k.
Count the number of 1s in TS_i (between the elements in TS_i having 0) = RW_{xp}, Where p =1, 2, 3 ..., k.
ii. for p=1,2,...,k
RW_{zp}= RW_{xp} + (RW_{yp} + $RW_{y(p+1)}$)/2,
iii. Row width (RW^1) = median (RW_z).
v. Using the value RW^1 draw horizontal grid lines at positions of i, where i=1, 2 ..... M with step increment RW^1.
for i=1to M step RW^1
for j=1 to N
G^1[i,j] =1;

The procedure for column width calculation is as follows:
i. for j = 1 to N,
Count the number of 0's in TS_j (between the elements in TS_j having 1) = CW_{yp}, Where p = 1, 2, 3 ..., k.
Count the number of 1's in TS_j (between the elements in TS_j having 0) = CW_{xp}, Where p = 1, 2, 3 ..., k.
i. for p=1,2,...,k
CW_{zp}=CW_{xp} + (CW_{yp} + $CW_{y(p+1)}$)/2,
i. Column width (CW^1) = median (W_z).
v. Using the value CW^1 draw horizontal grid lines at positions of j, where j=1, 2 ..... N with step increment CW^1.
for j=1to N step CW^1
for i=1 to M
G^1[i,j] =1;

Map these grid matrix G^1 onto the microarray image producing sub-gridding.

Fig. 2 Sub-gridding algorithm

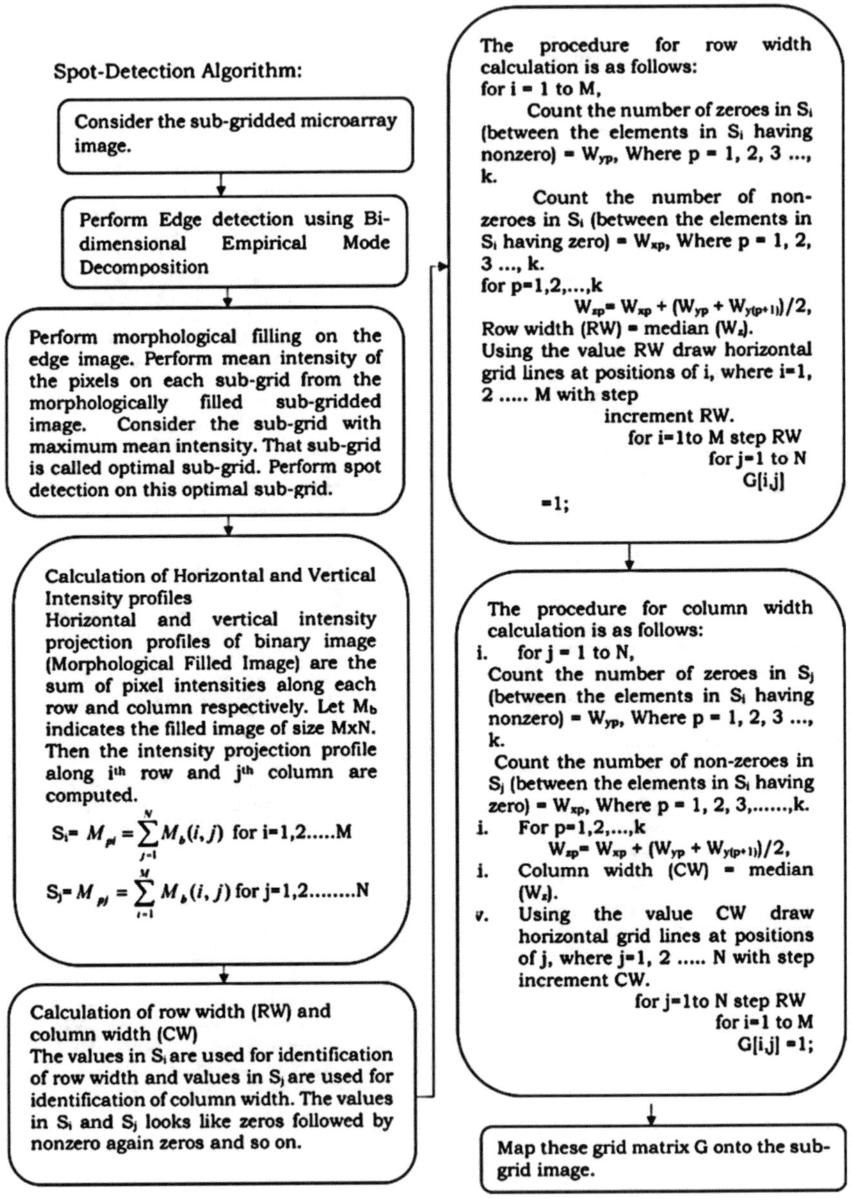

Spot-Detection Algorithm:

Consider the sub-gridded microarray image.

Perform Edge detection using Bi-dimensional Empirical Mode Decomposition

Perform morphological filling on the edge image. Perform mean intensity of the pixels on each sub-grid from the morphologically filled sub-gridded image. Consider the sub-grid with maximum mean intensity. That sub-grid is called optimal sub-grid. Perform spot detection on this optimal sub-grid.

Calculation of Horizontal and Vertical Intensity profiles
Horizontal and vertical intensity projection profiles of binary image (Morphological Filled Image) are the sum of pixel intensities along each row and column respectively. Let M_b indicates the filled image of size MxN. Then the intensity projection profile along i[th] row and j[th] column are computed.

$S_i = M_{pi} = \sum_{j=1}^{N} M_b(i,j)$ for i=1,2.....M

$S_j = M_{pj} = \sum_{i=1}^{M} M_b(i,j)$ for j=1,2........N

Calculation of row width (RW) and column width (CW)
The values in S_i are used for identification of row width and values in S_j are used for identification of column width. The values in S_i and S_j looks like zeros followed by nonzero again zeros and so on.

The procedure for row width calculation is as follows:
for i = 1 to M,
Count the number of zeroes in S_i (between the elements in S_i having nonzero) = W_{yp}, Where p = 1, 2, 3 ..., k.
Count the number of non-zeroes in S_i (between the elements in S_i having zero) = W_{xp}, Where p = 1, 2, 3 ..., k.
for p=1,2,...,k
$W_{zp} = W_{xp} + (W_{yp} + W_{y(p+1)})/2$,
Row width (RW) = median (W_z).
Using the value RW draw horizontal grid lines at positions of i, where i=1, 2 ..... M with step increment RW.
for i=1to M step RW
for j=1 to N
G[i,j] =1;

The procedure for column width calculation is as follows:

i. for j = 1 to N,
Count the number of zeroes in S_j (between the elements in S_i having nonzero) = W_{yp}, Where p = 1, 2, 3 ..., k.
Count the number of non-zeroes in S_j (between the elements in S_i having zero) = W_{xp}, Where p = 1, 2, 3,......,k.
i. For p=1,2,...,k
$W_{zp} = W_{xp} + (W_{yp} + W_{y(p+1)})/2$,
i. Column width (CW) = median (W_z).
v. Using the value CW draw horizontal grid lines at positions of j, where j=1, 2 ..... N with step increment CW.
for j=1to N step RW
for i=1 to M
G[i,j] =1;

Map these grid matrix G onto the sub-grid image.

Fig. 3 Spot detection algorithm

3 Microarray Segmentation

Image segmentation is the process of dividing image into regions, and with respect to microarray images, it is a process of dividing sub-grid into regions of foreground and background. This foreground region corresponds to spot area. Using this spot area of segmented image, the gene expression levels are estimated. This segmentation is a challenging task, as the intensities in this region are not uniform and spot sizes and shapes are also different from one another. Several statistical segmentation methods are proposed for segmentation of microarray image. Circular shape-based segmentation [5] in which a circular mask with some radius is used for identification of spots. But all spots are not of equal shape. Region growing method [6]: in this, the spot area and background area are separated by selecting seed pixels in both areas and regions which are extracted by using some predefined criteria. Selection of seed pixel is a difficult task. Thresholding method [5]: by using the histogram of the image, a suitable threshold is estimated, and the image is divided into two regions. This estimation is done using Mann Whitney test. Morphology based segmentation [7] uses the morphological operations such as hit or miss transforms to segment the image. Here the process depends on the selection of mask used for morphological operations. In this paper, supervised learning-based segmentation using support vector machine algorithm is used to segment the image. The features of each pixel are extracted and classified using SVM as a binary class classification problem. This classification result is used for segmentation of spot area in an image.

i. Feature Extraction:

The features extracted in each area are

(i) Intensity value of each pixel
(ii) Mean value of pixels in a surrounding neighborhood
(iii) Co-ordinate value of the pixels
(iv) Distance of pixel to the centroid of spot (Euclidean distance)

ii. Support Vector Machine (SVM): SVM is a discriminator and modeled by a discriminative hyperplane. It is a representation of data as points in space that are mapped, so that the points of different categories are separated by a gap as wide as possible. These hyperplanes are boundaries for classifying the data samples. Data points can be assigned to different classes irrespective of its position to the hyperplane. The number of features plays a crucial role in deciding dimension of the hyperplane. The hyperplane is just a line, if the number of independent features are two. The hyperplane becomes a two-dimensional plane, if the number of independent features are three.

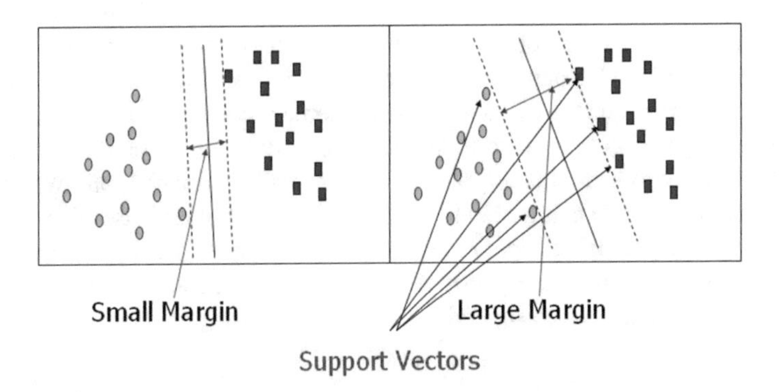

The data samples which are closer to the hyperplane are called as support vectors. These support vectors influence the orientation and position of the hyperplane. We maximize the margin of the classifier, by using these support vectors. After training SVM, if we get greater than 1 as output of SVM function, we label it as one class, otherwise we label it as another class in binary-class classification. Using the features of pixels in each gridded spot image, the SVM algorithm creates a maximal-margin hyperplane which divides the image pixels into two classes, spot area and background.

4 Experimental Results

The qualitative and quantitative analysis of proposed sub-gridding and spot detection algorithms are performed on two microarray images of breast cancer aCHG tumor tissue. The qualitive analysis of proposed gridding procedures is shown in Figs. 4 and 5.

The gridding accuracy is estimated by given formula [8]

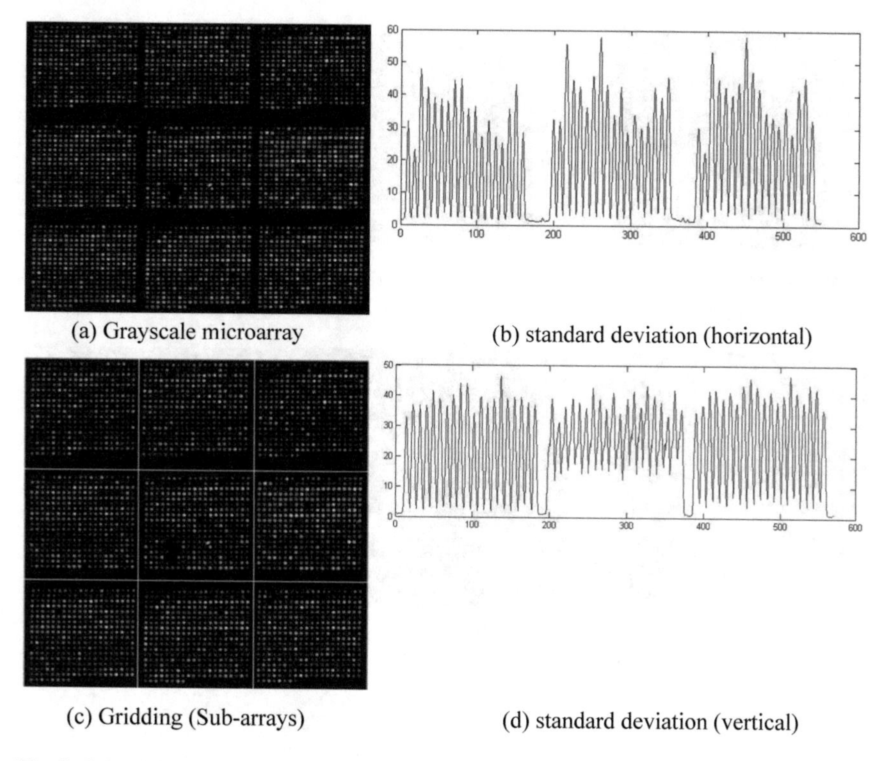

(a) Grayscale microarray

(b) standard deviation (horizontal)

(c) Gridding (Sub-arrays)

(d) standard deviation (vertical)

Fig. 4 Sub-gridding

$$\text{Percentage accuracy} = \frac{\text{Number of spots perfectly gridded}}{\text{Total number of spots}} * 100$$

The gridding procedure presented in this paper got 96% accuracy. After sub-gridding and spot detection algorithms, individual spots are extracted. These individual spots are segmented using SVM algorithm. The segmented step on a single spot is shown in Fig. 6.

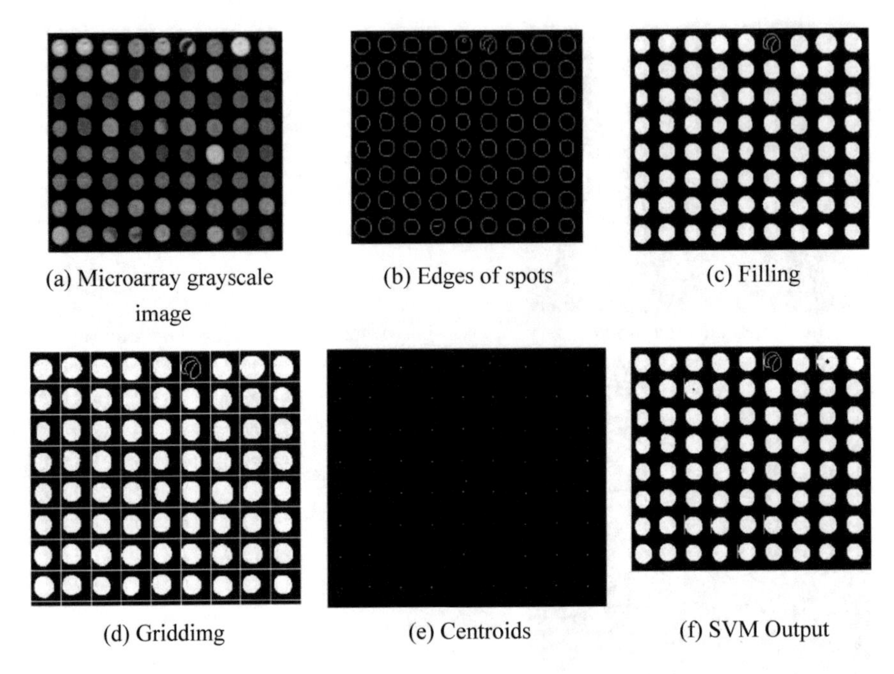
(a) Microarray grayscale image (b) Edges of spots (c) Filling

(d) Griddimg (e) Centroids (f) SVM Output

Fig. 5 Spot detection

5 Conclusions

The microarray image analysis is a sequential process with gridding, segmentation, and quantification process. Any error in the gridding and segmentation stages, the gene expression value will be affected. This paper presents an automatic gridding algorithm and estimates the parameters required for clustering in segmentation process. The experimental results show that the proposed algorithm grids the image with 96% accuracy and decreases the number of iterations by estimating the required parameters for segmenting the image by clustering algorithm.

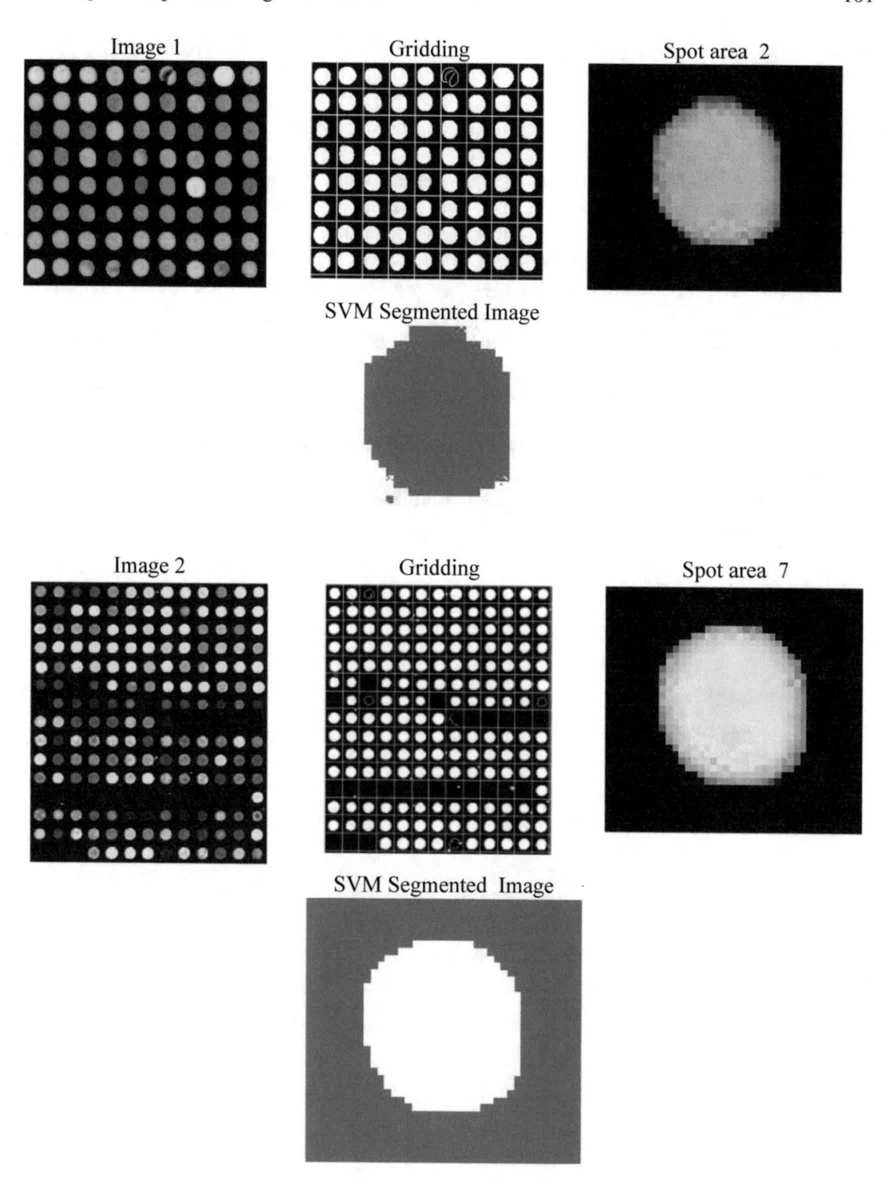

Fig. 6 Segmented result

References

1. Harikiran J et al (2014) Fast clustering algorithms for segmentation of microarray images. IJSER 5(10):569–574
2. Harikiran J et al (2017) Noise removal in microarray images using variational mode decomposition technique. TELKOMNICA Telecommun Comput Electron Control 15(4):1750–1757
3. Harikiran J et al (2012) Spot edge detection in microarray images using bi-dimensional empirical mode decomposition. In: Proceedings of C3IT-2012, vol 4, pp 19–25
4. Shao G et al (2019) Automatic microarray image segmentation with clustering based algorithms. Plus One Res
5. Chen Y et al (2009) Microarray image analysis and gene expression ratio statistics. In: CSAG, pp 1–19
6. Farouk RM et al (2019) Microarray spot segmentation algorithm based on integro-differential operator. Egypt Inform J 173–178
7. Katsigiannis S et al. MIGS-GPU: microarray image gridding and segmentation on the GPU. IEEE JBHI 21(3):867–874
8. Ni SH et al (2009) Spotted cDNA microarray image segmentation using ACWE. RJIST 12(2):249–263

Predicting Breast Cancer Using Machine Learning

Ch. V. Narayana, P. Manasa, M. Preethi, A. Mounika, and A. Bharadwaja

Abstract Malignancy is a type of sickness which occurs due to the change in the growth of cells in the body and increment past typical development and control. Bosom or breast malignancy is one of the continuous sorts of disease. The anticipation of bosom malignancy repeat is profoundly required to rise the endurance pace of patient experiencing bosom disease. With the headway of innovation and AI methods, the malignancy analysis and recognition exactness have improved. AI (ML) procedures offer different probabilistic and factual strategies that permit savvy frameworks to gain recurring past encounters to recognize and distinguish designs from a dataset. The exploration work exhibited a review of the AI procedures in malignancy sickness by applying learning calculations on bosom disease by using the dataset from the Wisconsin diagnostic breast cancer—support vector machine, random forest, K-nearest neighbor, and decision tree. The outcome result shows that Random Forest performs superior to different procedures.

Keywords Machine learning · Classification · Support vector machine (SVM) · Random forest · K-nearest neighbor · Decision tree

1 Introduction

Bosom malignant is the biggest sequential disease surrounded by women, affecting 2.1 million ladies each year, and also causing more passing among women related to the disease. In 2018, 627,000 females are reported to have kicked the bucket from bosom disease—approximately 15% of all malignant development among females. While the rates of bosom disease among women are higher in increasingly developed areas, rates are increasing comprehensively in about each area [1]. Bosom malignant growth is the most well-known disease among ladies around the world, with about 1.7 million different cases analyzed in 2012, speaking to around 25% of all

Ch. V. Narayana · P. Manasa (✉) · M. Preethi · A. Mounika · A. Bharadwaja
Department of CSE, Lakireddy Bali Reddy College of Engineering (Autonomous), L.B. Reddy Nagar, Mylavaram, Krishna (Dt.) 521230, AP, India
e-mail: pachavamanasa99@gmail.com

N. Chaki et al. (eds.), *Proceedings of International Conference on Computational Intelligence and Data Engineering*, Lecture Notes on Data Engineering and Communications Technologies 56, https://doi.org/10.1007/978-981-15-8767-2_9

tumors among ladies. Frequency rates all-inclusive shift generally from 27 for every 100,000 in North America. It is the most frequent element of malignancy in ladies, with an expected 522,000 passings (6.4% of aggregate). It is additionally the most regular reason for malignancy demise in ladies from locales with lower advancement or potentially pay rates (14.3% of passing's), and the second generally visits in areas with higher improvement as well as pay rates (15.4% of passing's), following lung disease. The danger of bosom malignant growth duplicates each prior decade menopause, after which the expansion eases back. Be that as it may, in the wake of accepting the post-nation rate inside a couple of ages, bosom disease is progressively normal. This shows the ecological components that are imperative to sickness advancement. By and large, endurance rates for bosom disease vary worldwide yet have expanded when all is said in done. This is on the grounds that in numerous countries access to clinical consideration is expanding and most instances of bosom disease are dealt with sooner and locally. On the other hand, upgraded activities and customized adjuvant consideration regiments are accessible. The 5-year endurance rate in numerous nations for ladies determined to have organized one/two (little tumors are kept to hubs under the arm) is 80–90%. In stages three/four (huge tumors spread further past the bosom or to removed organs), the pace of endurance tumbles to 24%. The pervasiveness of bosom malignant growth in ladies per 100,000 is 665 in Western Europe, 745 in North America, and 170 in East Asia [2]. The disease figures referred to in the third master report are from the database GLOBOCAN 2012. As indicated by the WHO, harm in the chest comprises 2,09 million cases and 627,000 travels around the world. It is the most generally perceived illness in Indian females and reports 14% of every single female tumor. It can occur at any age yet the occurrence rates in India start to ascend in the mid-30 s and top at ages 50–64 years [3].

2 Literature Survey

S. No	Techniques in machine learning	Results
1	Support vector machine (SVM) [4]	Breast cancer prognosis
2	Decision Tree [5]	Predicting whether there are cancerous tumours or not
3	Neural networks [6]	Medical screening
4	Bagging methods on medical repositories, using DT, NN and SVM [7]	Predicting the recurrence of malignant tumours
5	Artificial neural networks (ANNs), Logistic Regression (LR), and decision tree [8]	Prediction of the repetitiveness of breast cancer

3 Existing System

A mammogram is a breast X-ray image. It can be used for screening breast cancer in women who have no symptoms or signs of the disease. If you have a lump or some other symptom of breast cancer, it can also be used.

Steps to be followed while doing this process:

- Step-1: A molecular pathologist chooses a list of important genetic variations which he/she would like to study.
- Step-2: Within medical literature, the molecular pathologist is searching for evidence that somehow is important to the genetic variations of invest a tremendous amount of interest.
- Step-3: Eventually the amount of time the molecular pathologist analyzes the evidence relating to each of the variants to identify them.

Typically, we will get the results in a few weeks, though it depends on the facility. A radiologist will read the mammogram and then report the results within 30 days.
Disadvantages of Existing System:

- The accuracy of the procedure depends in part on the technique used and the experience and skill of the radiologist.
- It generally takes more time to predict the results.

4 Proposed System

4.1 Dataset

Dataset used in this paper is "Wisconsin Breast Cancer Data Set (WBCD)" [9]. We got this UCI—Repository data from UCI and the dataset contains 569 * 32, where 569 is the number of tests performed on women and 32 is the number of attributes which are used for detecting the cancerous cells.

The underneath referenced attributes are resolved for each mammogram image. For all these 10 features, the mean, normal error, and "most conspicuously bad" (mean of the three greatest features) are solved, which achieves more than 30 features. For example, field 2 is known as the Mean Radius, field 12 is known as the Radius SE, and field 22 is known as the Worst Radius. The qualities that are utilized are determination, mean of surface, border, zone, smoothness, conservativeness, concavity, curved focuses, balance, fractal measurement, sweep. Fig. 1 speaks to the stream chart of work in which Categorization performed on WBCD data Use AI approaches.

Dataset contains ten critical authentic regarded features which are figured for each cell center:

1. Radius (mean of separations among focus and focuses on the edge)
2. Perimeter

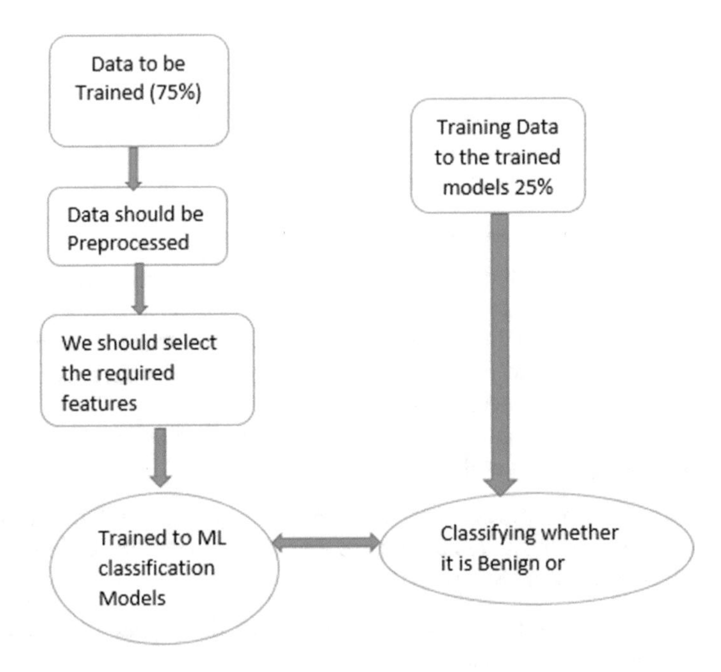

Fig. 1 Flow diagram

3. Texture (standard deviation of dark scale esteems)
4. Compactness (perimeter^2\zone − 1.0)
5. Area
6. Smoothness (neighborhood variety in lengths of range)
7. Concavity (severity of curved segments of the form)
8. Symmetry
9. Concave focuses (number of curved portions of the shape)
10. Fractal measurement ("coastline guess" − 1)

The workstream of the examination work does is as per the following: we divided the dataset as 75% for training and 25% for testing the models.

4.2 Methods

Bosom malignant growth is the most noticeable ailment in the region of clinical analysis which is expanding each year. A near examination of four generally utilized AI strategies are operated on Wisconsin Breast Cancer Database (WBCD) to anticipate bosom disease repeat:

- Support vector machine,
- (SVM), Random Forest,

- K-nearest neighbour (K-NN),
- Decision Tree.

Support Vector Machine:

Learning model is administered by Bolster Vector Machine. It is preferred for its implementation of the structure. In the calculation of SVM, each data object is defined as organized in the n-measurement volume. Here, n is all out of the amount of characterization highlights used, and the directions of the information point speak to the estimation of each element. SVM includes a hyperplane of choice which is used to separate the knowledge purposes of different classes using the most extreme edge. The knowledge focuses on which untruth is called as support vectors near the hyperplane. This grouping technique establishes non-direct limits of choice and focuses details on orders that have no representation of vector space.

Random Forest:

Random forest is called as a supervised learning algorithm which is used for categorization as well as regression. By the by, it is utilized basically for arrangement issues. Since we know a forest comprises of trees, more trees mean a more grounded forest. Likewise, the random forest The algorithm builds data sample decision trees and then obtains the forecast from each and uses voting to choose the best solution. This is a group technique which is better than a single decision tree since it diminishes the overfitting by joining the result.

K-Nearest neighbor:

K-Nearest Neighbor Scheme is a non-parametric method that is used for regression and grouping analysis. The details include K, K is different classes in the data, and the whole number is a little and positive. Most of its neighbors organize and recognize some detail. The information is allocated to the higher class in its nearest K neighbor along these lines Right now, the estimation of $k = 2$, will do the information to a class of two closest neighbors at that point since this dataset contain mainly two classes. The courses are toxic to cancers and non- hazardous to tumors. The differentiation from the Euclidean is used to measure the difference between two points of information.

Decision tree:

Decision Tree (DT) is an administered Machine Learning (ML) technique utilized for arrangement and relapse examination. Choice tree depends on the partition and vanquishes procedure. It isolates the parcel by two strategies: Numerical segments: Typically, segments are shaped based on discrete qualities with certain conditions and Nominal segment: The allotments are framed based on ostensible characteristics. It drives parting of the tree contingent upon traits esteems. This procedure creates a tree that can have straight out just as discrete qualities. The precision of each generated norm is evaluated to decide the request where characteristic will be positioned in the decision tree.

The most widely recognized technique is to confine the normal error of each center to

$$I(S) = \sum_{j=0}^{n} Pj \cdot f\left(Pj, P^2 j, P^3 j, P^C j\right)$$

where n = number of allotments, PCj is the amount of Class C v components falling into hub j, Pj is the likelihood of falling into hub j and f is a polluting impact characteristic.

4.3 Methodology

The presentation assessment of machine learning techniques is the fundamental goal of the experiment. The display is determined to be reliable, precise, interpretation, and Flscore, in order to improve early bosom malignancy detection. Measures to implement are characterized as follows:

- Precision: It is quantifying what number of right matches are found among all informational collections.
- Recall: It is the association of matches and correspondences with the opposite. The effect is the breaking of essential models.
- Accuracy: It is used to express an estimate's closeness to its true value.

5 Results

Much testing was carried out using Jupyter notebook. To determine bosom malignant growth, four techniques (RF, KNN, SVM, and Decision Tree) are done. The relative investigation among these AI systems is performed regarding Accuracy, Precision, and Recall is referenced in Table 1. Here, precision and recall performed on all properties.

As indicated by Table 1.

Table 1 ML techniques performance measures

Classification	K-NN	SVM	DT	RF
Accuracy	95.1	95.1	93.7	95.8
Precision	95	95	94	96
F1 Score	95	95	94	96
Recall	95	95	94	96

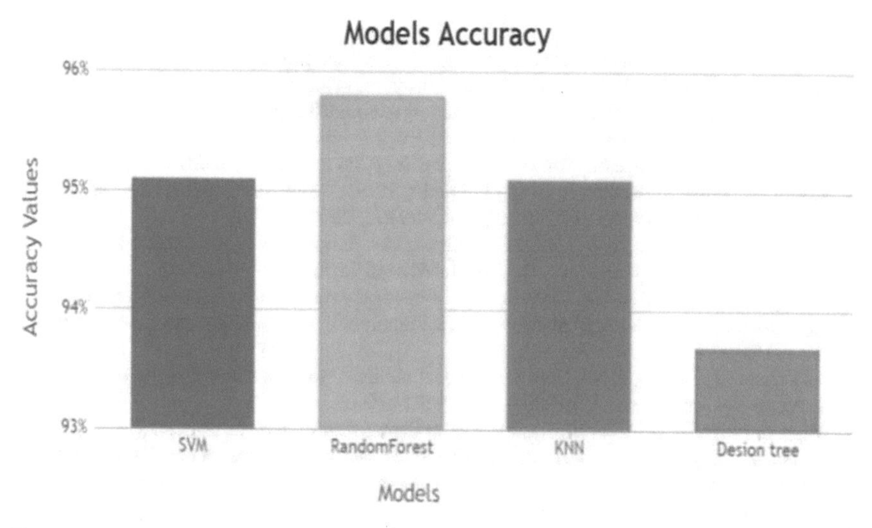

Fig. 2 Graph showing the results

- Random Forest works better than other model groupings when all measures are used in the planning, concealment of layers
- The accuracy of Random Forest (RF) stands highest in terms of accuracy followed by K- NN, SVM those followed by DT.
- When the top five parameters are implemented, the Random Forest model works better than other characterization models.
- Random Forest works better than the other K-NN, SVM, and DT classifications.
- The RF is the maximum in terms of accuracy, followed by K-NN, SVM, and then DT (Fig. 2).

6 Conclusion

Fundamental focal point of the examination work directed is to improve the expectation of bosom malignant growth so as to expand the exactness of finding. The greater part of the examinations is exhibited which have been proposed in quite a while and accentuation on the advancement of prescient models for bosom malignant growth finding/visualization utilizing AI strategies and characterization. The similar investigation of four generally utilized AI methods: Random Forest (RF), Support vector machine (SVM), K-Nearest Neighbor (KNN), and Decision Tree (DT) are performed. Precision display, RF is stronger when compared to different methods. RF approach also performs better than different procedures when measures of cross-validation are used in the expectation of malignant growth.

References

1. https://www.who.int/cancer/prevention/diagnosis-screening/breast-cancer/en/
2. https://www.wcrf.org/dietandcancer/breast-cancer
3. https://www.nhp.gov.in/breast-cancer-awareness-month2019_pg
4. A gene signature for breast cancer prognosis using support vector machine. In: 2012 5th international conference on biomedical engineering and informatics (BMEI). IEEE
5. Delen D, Walker G, Kadam A (2005) Predicting breast cancer survivability: a comparison of three data mining methods. Artif Intell Med 34(2):113127
6. Shaikhina T, Khovanova NA (2017) Handling limited datasets with neural networks in medical applications: a small-data approach. Artif Intell Med 75:51–63
7. Tsirogiannis GL et al (2004) Classification of medical data with a robust multi-level combination scheme. In: Proceedings of 2004 IEEE international joint conference on neural networks, vol 3. IEEE
8. Chaurasia V, Pal S (2014) Data mining techniques: to predict and resolve breast cancer survivability. Int J Comput Sci Mob Comput 3(1):10–22
9. https://archive.ics.uci.edu/ml/datasets/Breast+Cancer+Wisconsin+(Diagnostic)
10. Gupta M, Gupta B (2018) A comparative study of breast cancer diagnosis using supervised machine learning techniques. In: 2018 second international conference on computing methodologies and communication (ICCMC)
11. https://link.springer.com/chapter/10.1007%2F978-981-13-3185-5_3

Gender Identification Over Voice Sample Using Machine Learning

Meenu Yadav, Chandra Shekhar yadav, Rakesh Kumar, and Prem Shankar Yadav

Abstract Point of this paper is to style a gender reorganization system that distinguishes the gender of the agent. Gender classification is associate rising space of analysis for the accomplishment of economical interaction between human and machine victimization speech files. Numerous ways in which are planned for the gender classification within the past. Speech recognition is a chief approach for the identification of the supply. In this work different feature for the gender classification are gait of person, lips shape, automatic face recognition, iris code, etc. There aretwo experiments performed, the 1st experiment deals with the gender classification of ten completely different languages, every language consists of fifteen audio files and also the best accuracy achieved by the machine learning technique that's straight forward logistical (87.33%). The second experiment deals with the formation of language identification system within which Russian and Korean has the simplest combination of language with the a hundred accuracy by multilayer perceptron.

Keywords Machine learning · Gender classification · J48 · REP tree · Random forest · SMO · Speech

M. Yadav
Sant Longowal Institute of Engineering and Technology,, Longowal, Punjab, India
e-mail: meenu_yadav99@rediffmail.com

C. S. yadav (✉) · R. Kumar · P. S. Yadav
School of Computer and System Sciences, Jawaharlal Nehru University, Delhi, India
e-mail: chandrtech15@gmail.com

R. Kumar
e-mail: rakesh.kmr2509@gmail.com

P. S. Yadav
e-mail: premit2007@gmail.com

C. S. yadav
Standardisation Testing and Quality Certification, MeitY, New Delhi, India

N. Chaki et al. (eds.), *Proceedings of International Conference on Computational Intelligence and Data Engineering*, Lecture Notes on Data Engineering and Communications Technologies 56, https://doi.org/10.1007/978-981-15-8767-2_10

1 Introduction

A person will identify the opposite person's voice or can distinguish between voices of number of individuals by recognizing the pitch, frequency, tone of the voice. Even with the voice, a person will guess the age and gender of the voice. A person can even generate an unfamiliar voice. It is known that, for someone solely, two words are enough of his shut friend to acknowledge his voice. The voice options are referred to as voice-prints, from which a person will acknowledge the speaker's age, gender, and emotion. We can utilize the appliance speech (sound) process in numerous areas like voice dialing, phone voice communication, doing route identification of decision, dominant domestic gizmo, conversion of speech (sound) to text and the other way around, lip vogue identification and synchronization, automation systems, etc. [1].

2 Gender Classification Related Work

"WEKA is taken into account as a landmark system in data processing and machine learning" [2]. This paper provided Associate in Nursing introduction of WEKA work table and delineated, however, WEKA has become thus common in academe and business fields. Bales et al. used a highly-instrumented Goodwin Hall good building for the activity of the vibrations created by peoples on the walking surface. They used Bagged call Trees, Boosted call Trees, Support Vector Machines (SVMs), and Neural Networks because the machine learning techniques for his or her ability so as to gender classification [3]. Brain Computer Interfaces (BCI) may be an affiliation between the brain and a tool that takes the signals from the brain to direct the external activity [4]. During this paper, they need recorded cerebral activities of sixty subjects (males & females) keeping their position in relax mode and with closed eyes. According to [5], once recognizers got up sure-handed with acoustic (audial) options at the side of body part features the accuracy of recognition get improved. Nivedita et al. [6] planned a contemporary analysis methodology that is single frequency filtering (SFF) in speech signals. In this paper [7], they investigated totally different approaches of auxiliary options with the assistance of Dynamic Bayesian Network or hybrid HMM/ANNs. The approaches tell that auxiliary options don't seem to be emitted directly from the HMM, however, they carry a higher level of data that's connected with customary features. The second experiment gained higher end in terms of average accuracy than the primary one [8]. This paper [9], aimed toward coming up with of gender system exploitation supervised machine learning techniques. This technique takes the speech files as associate input and offers output as a classification of gender.

3 Proposed Work

The section includes the reason concerning the dataset, preprocessing and also the planned design of the machine learning model that contains the multiple classifiers like Random Forest, J48, Bagging, REP Tree, Multilayer perceptron, simple logistic, logistic, and SMO for classifying the gender by analyzing the speech signals. A comparison of various machine learning classifiers has conjointly been provided. The results are compared with alternative previous add this domain. The framework is bestowed in Fig. 1.

3.1 Used Dataset

The dataset contains the speech signals of 1500 people with 680 males and 820 females their age is between 15–50 as a result of it contains the voice files of kids, adults, and senior voters. This dataset has 10 languages primarily based speech files like English, German, Japanese, French, Spanish, Mandarin, Korean, Russian, Italian, and Hindi.

3.2 Voice PreProcessing

Windowing could be a method to create the subsets of the larger set for analysis and processing. Like here windowing is performed on larger speech signal which is able to be distributed among a variety of little subsets of speech and these also are known as frames. In this paper, frame size is 20 ms and overlapping is of 10 ms, whereas the one hundred fifty speech files are of random length and it contains files of 10 languages within which every language has fifteen files. Rectangular window merely truncates the dataset before and once windowing , however, doesn't modify the contents of the window the least bit. The rectangular window may be defined by [1]

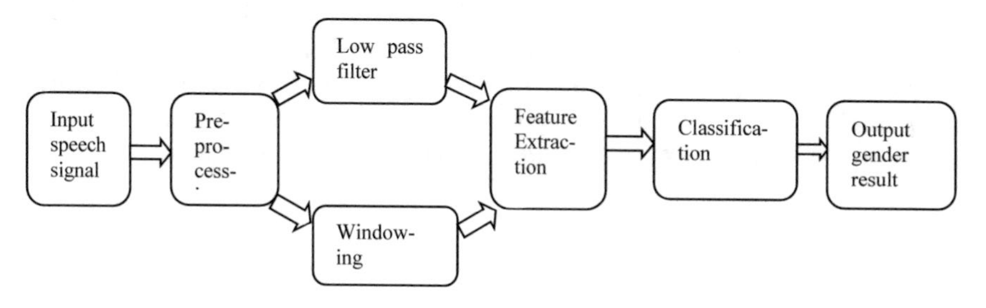

Fig. 1 Gender Classification System

$$w_R(n) \triangleq \begin{cases} 1, -\frac{M-1}{2} \leq n \leq \frac{M-1}{2} \\ 0, otherwise \end{cases} (1)$$

where w_R(n) is a window operate, M denotes window order and n may be a variety of voice samples.

3.3 Voice Features Extraction

The Fundamental frequency represents an all-time low frequency of the speech signal. it's correlative with a pitch. The variations in fold tension and sub speech organ gas pressure with the vocal fold vibration offer this frequency because of the output. Each category of a person includes a completely different fundamental frequency, for a man it varies between eighty-five to one hundred fifty-five cycle per second and for female frequency has a range from one hundred sixty-five to two hundred fifty-five Hz. The formula of F_0 (Fundamental frequency) is given in equation-2. Here τ_{ms} represents pitch length.

$$F_0 = \frac{1000}{\tau_{ms}} (2)$$

3.4 Classification

The term classification indicates to classify the varied observations into the various teams like classifying objects of same aspects into categories and marking or distinctive the gender of the speaker. The extraction of voice options is that the key task within the speech recognition system which means of feature extraction is that bring out those options of speech that facilitate the system in distinguishing the language of speaker and gender of the speaker. These options of sound (speech) are referred to as patterns. These patterns embrace the coaching set and that they are accustomed to apply the classification algorithmic rule. The advantage of this formula in distinguishing pattern of claiming of a specific gender is feature matching in conjunction with the extraction of options. The pattern recognization ought to have 2 properties, first is extracting patterns and objects even once part hidden or totally hidden and second is classifying unknown objects [1].

3.5 Used Models for Classification

There are several experiments that performed victimization variable techniques of machine learning. Those machine learning techniques are Random Forest, J48, Bagging, SMO, Multilayer perceptron, logistic, straightforward logistical, and REP Tree.

Random Forest: Random Forest is additionally known as random call trees that are used for classification, regression, and different tasks that operates by constructing the choice trees of solutions for the individual drawback at the coaching time of knowledge and provides output within the kind of categories (classification) [10, 11]. The posterior probability that x may be a purpose belongs to category c (c = 1, 2,…,n) be denoted by $\boldsymbol{P}\left(\boldsymbol{c}|\boldsymbol{v}_j(\boldsymbol{x})\right)$, showing in equation-3.

$$P\left(c|v_j(x)\right) = \frac{P\left(c, v_j(x)\right).}{\sum_{l=1}^{n} P\left(c_l, v_j(x)\right).} \tag{3}$$

J48: The C4.5 algorithmic program is employed for building the choice trees in rail tool as a classifier that is termed as J48. J48 encompasses a full name weka.classifiers.trees.J48. It's principally used for classification of instances.

Bagging: Bagging is employed to cut back the variance and retentive the bias. It happens once we reason the common of predictions in numerous areas of the input feature space [12]. The equation is defined in equation-4. L is a learning set, x is an input, φ_B denotes aggregation, $L^{(B)}$ is a repeated bootstrap sample.

$$\varphi_B(x) = av_B\varphi\left(\mathrm{x}, \mathrm{L}^{(\mathrm{B})}\right) \tag{4}$$

SMO: SMO is an associate algorithmic program that is employed to resolve the quadratic programming issues that occurred throughout the coaching amount of Support Vector Machine (SVM) [13]. SMO computes the minimum on the direction of the constraint by computing the second Lagrange multiplier factor shown in Eq. (5)

$$\alpha_2^{new} = \alpha_2 + \frac{y_2(E_1 - E_2)}{\eta} \tag{5}$$

REP Tree: REP Tree could be a quick call learner tree. It constructs a choice tree or regression with the assistance of the obtained data and cropped it with the help of Reduced Error Pruning with back fitting.

Multilayer Perceptron: The degree of error in associate output node j within the ordinal information (training example) by $e_j(n) = d_j(n) - y_j(n)$, wherever is that the target worth and is the value created by the perceptron [14]. The node weights will then be adjusted supported corrections that minimize the error within the entire output, given by Eq. (6)

$$\varepsilon(n) = \frac{1}{2}\sum_{j} e_j^2(n) \tag{6}$$

Using gradient descent, the amendment in every weight is showing by Eq. (7)

$$\Delta w_{ji}(n) = -\eta \frac{\partial \varepsilon(n)}{\partial v_j(n)} y_i(n) \tag{7}$$

where y_j is that the output of the previous somatic cell and is the learning rate, that is chosen to make sure that the weights quickly converge to a response, while not oscillations.

The spinoff to be calculated depends on the induced native field y_j, that itself varies. it's simple to prove that for associate degree output node this spinoff will be simplified to

$$-\frac{\partial \varepsilon(n)}{\partial v_j(n)} = e_j(n)\varnothing'\left(v_j(n)\right) \tag{8}$$

where Ø^' is the spinoff of the activation perform delineate higher than, that itself doesn't vary. The analysis is harder for the modification in weights to a hidden node, however, it will be shown that the relevant spinoff is

$$-\frac{\partial \varepsilon(n)}{\partial v_j(n)} = \varnothing'\left(v_j(n)\right)\sum_{k} -\frac{\partial \varepsilon(n)}{\partial v_k(n)} w_{kj}(n) \tag{9}$$

Simple Logistic: The condition for using Simple Logistic is that we must always have one nominal variable and one measure variable and that we wish to understand will the modification within the measure variable bring the modification in a nominal variable [15]. A multi-class LLR score fusion model is expressed in equation-10. $\boldsymbol{\alpha}_i$ and $\boldsymbol{\beta}_l$ are the regression coefficients and biases, respectively.

$$\overline{s}_{nl} = \sum_{i=1}^{C} a^i s_{nl}^{(i)} + \beta_l = \alpha^T s_{nl} + \beta_l \tag{10}$$

Logistic regression: Logistic regression may be a supervised machine learning classification algorithmic program. During this operate input values (x) are joined with the assistance of weights and also the coefficients values to induce the output values (y). It's showing by Eq. (11)

$$y = e^{\frac{(b0+b1\times x)}{\left(1+e^{(b0+b1\times x)}\right)}}$$

where y denotes the anticipated output, b0 represents bias or intercept term, and b1 denotes the constant of the one input price (x). There's an associated b coefficient (a constant real value) in our input file for each column that has got to be learned from our coaching data.

4 Experiment and Result

The experiment of the gender organization of speech files is introduced during this chapter. The info will be the recording of voice mistreatment any package in the system will use datasets that are utilized in previous papers of OGI, NIST, and decision Friend. However, during this paper, we've developed our own totally different dataset that may be an assortment of random speech files from the web or some from MSLT Corpus [16], and a few language info is downloaded from the kaggle web site. The whole hierarchical data structure of speech options is shown in Fig. 2.

The filtered, silence free, and noise-free speech files then enframed mistreatment the Rectangular Windowing method. It takes abundant time and power of RAM to access the long speech files therefore to create frames of those files this window operate is employed. during this section remaining all the options were extracted.

We have taken 1500 speech files of ten languages: English, German, Japanese, French, Spanish, Mandarin, Korean, Russian, Italian, and Hindi as employed

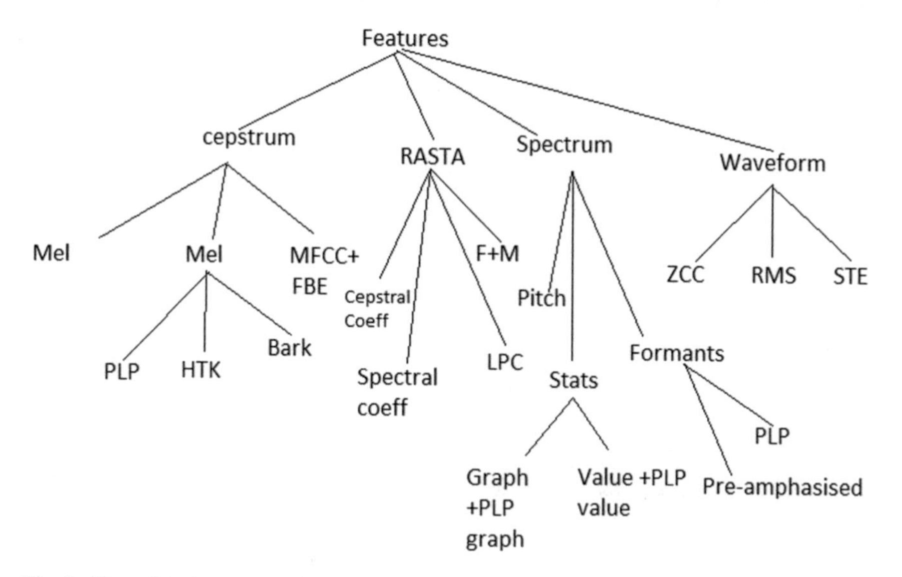

Fig. 2 Hierarchical structure of speech features

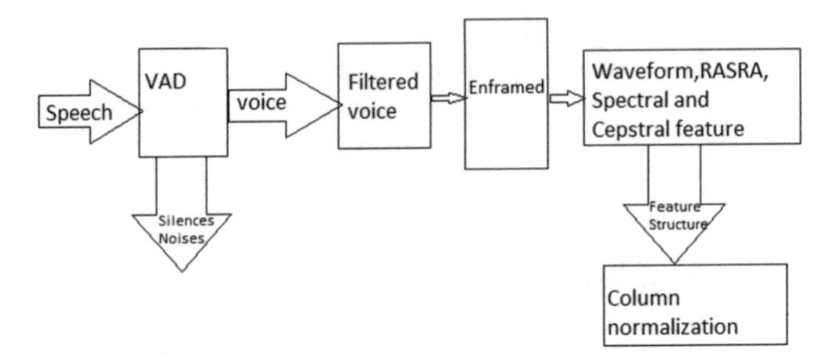

Fig. 3 The feature extraction process

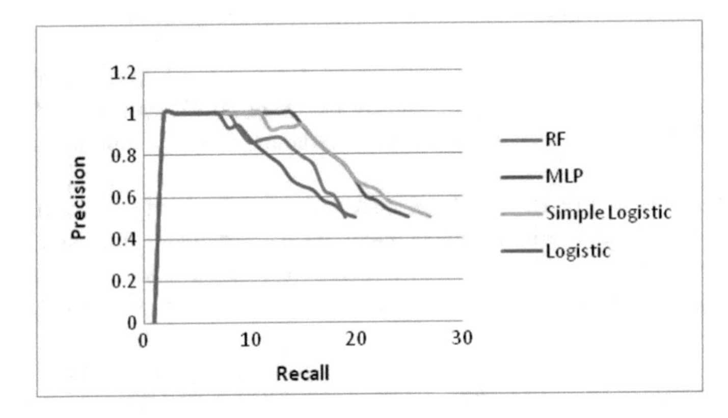

Fig. 4 PR curves of males for best four machine learning methods

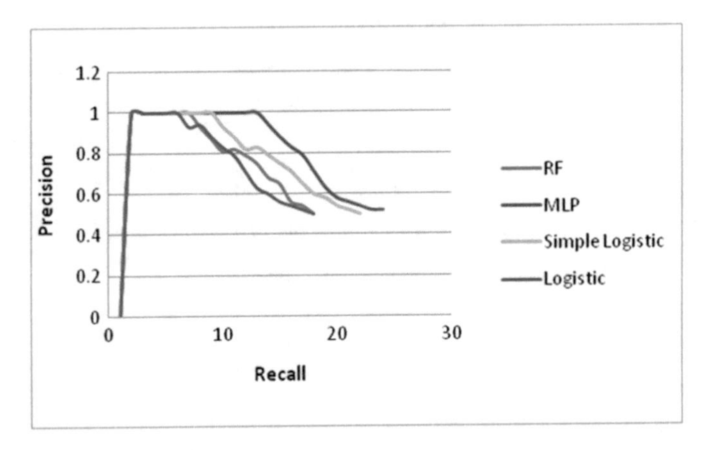

Fig. 5 PR curves of females for best four machine learning methods

Table 1 General information about the speech corpus

Language	Number of speech files	Number of male files	Number of females files	Size of files in sec (min < x < max)
English	150	50	100	2 < x < 60
German	150	60	90	30 < x < 60
Japanese	150	60	90	1 < x < 3
French	150	140	10	2 < x < 10
Spanish	150	100	50	2 < x < 10
Mandarin	150	80	70	40 < x < 60
Korean	150	50	100	1 < x < 4
Russian	150	100	50	5 < x < 9
Italian	150	70	80	59 < x < 90
Hindi	15	3	12	47 < x < 60

in [10], however, we have a tendency to use three completely different languages from those ten languages that are employed in [10] (Table 1).

Total fifteen hundred speech files experimented in ten languages that are Russian, Japanese, Mandarin, German, French, English, Hindi, Korean, Spanish, and Italian. It's a combination of males and feminizes speech files during which six hundred eighty are of male speech files and eight hundred twenty are of female speech files. Table a pair of shows the accuracies, F measure, ROC Area, PRC curve achieved by every machine learning methodology on the dataset. We've tested the dataset victimization tenfold cross validation methodology. We tend to get a stunning result that straightforward supply no heritable 87.33% that is highest among the eight machine learning strategies utilized in this paper in all aspects like accuracy, F measure, ROC Area, PRC curve. It's higher than the previous, add this space as a result of the last paper [10], regarding language and gender classification has achieved solely 76.29% accuracy victimization Random Forest. And results are clearly showing in below Table 2.

The eight Machine learning strategies that we've utilized in this work are the most effective methods for classification purpose. Figure four and Figure five shows the PR curves for Male and Females victimization best four accuracies of machine learning strategies.

5 Conclusion and Future Work

In this analysis work, we've given Associate in LID system which is completely different in several manners from the previous LIDs [1]. We have a tendency to computed four feature sets of speech files that are Cepstrum options, spectrum options, disciple options, and undulation features. In contrast to previous

Table 2 Results obtained using 6 machine learning methods

ML_Method	F_Measure	ROC_Area	PRC_Area	Accuracy
Random forest	86.7	94.5	94.7	86.7
J48	81.3	84.5	80.4	81.3
Bagging	84.6	90	88.4	84.7
SMO	77.4	77.6	71.7	77.3
REP Tree	79.3	82.4	79.6	79.3
Logistic	82.6	87.9	87	82.7
Multilayer	83.3	93.3	93.6	83.3
perceptron				
Simple	87.3	88	86.5	87.3
logistic				

add this space we've computed options directly from our speech files with none changes to intermediate features [2]. We have a tendency to achieved high results exploitation the 20 ms frame size that is five hundred times smaller than the frame size employed in previous LIDs experiments [3, 17]. In this work we have exploited eight machine learning strategies that aren't nonetheless employed in previous work as per our greatest information. In future this work can be used to achive the objectives of digitizations and e-govenrnance. Since, audio and vedio data is not directly readable by human therefore, to achive the given targets, first step is identification of gender, then equivalent text conversion and later followed by summarization approaches over the equivelent text [18, 19] other future directions for analysis are conducting experiments that modify comparison of performance as a perform of the frame size,and conducting experiments victimization a lot of speech files from more languages

References

1. Hong Z (2017) Speaker gender recognition system. Master's Thesis, D. Programme in Wireless, and C. Engineering, University of Oulu
2. Frank E, Holmes G, Reutemann BPP, Witten IH (1997) Random utility/multinomial logit model literature overview. Mar Policy 7(January 1996):13–21
3. Bales D et al (2016) Gender classification of walkers via underfloor accelerometer measurements. IEEE Internet Things J 3(6):1259–1266
4. Kaushik P, Gupta A, Roy PP, Dogra DP (2019) EEG-based age and gender prediction using deep BLSTM-LSTM network model. IEEE Sens J 19(7):2634–2641
5. Najnin S, Banerjee B (2019) Speech recognition using cepstral articulatory features. Speech Commun 107(February 2018):26–37
6. Chennupati N, Kadiri SR, Yegnanarayana B (2019) Spectral and temporal manipulations of SFF envelopes for enhancement of speech intelligibility in noise. Comput Speech Lang 54:86–105
7. Stephenson TA, Doss MM, Member S (2004) Speech recognition with auxiliary information 12(3):189–203

8. Ramdinmawii E, Mittal VK (2016) Gender identification from speech signal by examining the speech production characteristics, 244–249
9. Sengupta S, Yasmin G (2017) Classification of male and female speech using perceptual features
10. Kam HT Random decision forests 47:4–8
11. Yadav CS, Sharan A (2020) Feature learning using random forest and binary logistic regression for ATDS. In: Johri P, Verma J, Paul S (eds) applications of machine learning. Algorithms for intelligent systems. Springer, Singapore. https://doi.org/10.1007/978-981-15-3357-0_22
12. Breiman L (1994) Bagging predictors. Department of statistics university of california at berkeley 421
13. Platt JC (1998) Sequential minimal optimization: a fast algorithm for training support vector machines, 1–21
14. Rumelhart DE, et al (1985) Learning internal representations by error propagation. Institute for cognitive science university of California, San Diego La Jolla, California V
15. Sim KC, Lee K (2010) Adaptive score fusion using weighted logistic linear regression for spoken language recognition. In: Sim KC, Lee KA (eds) Agency for science, technology and research (A STAR), Singapore, 2010 IEEE international conference on acoustics, speech and signal processing, pp 5018–5021
16. Federmann C, Lewis WD (2016) Microsoft speech language translation (MSLT) corpus: the IWSLT 2016 release for English , French and German.
17. Yadav M, Verma VK, Yadav CS, Verma JK (2020) MLPGI: multilayer perceptron-based gender identification over voice samples in supervised machine learning. In: Johri P, Verma J, Paul S (eds) Applications of machine learning. Algorithms for intelligent systems. Springer, Singapore. http://doi-org-443.webvpn.fjmu.edu.cn/10.1007/978-981-15-3357-0_23
18. Yadav CS, Sharan A (2018) Automatic text document summarization using graph based centrality measures on lexical network. Int J Inf Retrieval Res (IJIRR) 8(3):14–32
19. Yadav CS, Sharan A (2015) Hybrid approach for single text document summarization using statistical and sentiment features. Int J Inf Retrieval Res (IJIRR) 5(4):46–70

Framework for Evaluation of Explainable Recommender System

Nupur Mukherjee and G. M. Karthik

Abstract Explainable recommendation system (ERS) in addition to recommending items to the user also explains why the recommendation is being made. Explanations improve user acceptance and system transparency. Since every recommender system (RS) has some strength and weakness, a combination of RS may be required to demonstrate both performance and explainability. In this paper, an innovative framework is proposed for systematic evaluation of different configuration of ERS compared with respect to performance and explainability of RS recommendations. Framework uses a novel approach to configure RS with different types of recommender models, hybridization of recommender models to create new models, well-defined metrics to compare performance and explainability of recommendation given by ERS. Simulation experiments show the efficacy of the framework in helping users gain insight into how various components of ERS effect explainability and recommendation quality.

Keywords Recommender systems · Explainability · Collaborative filtering · Metrics

1 Introduction

Recommender systems (RS) automatically recommend items to users when they carry out on-line purchase from websites. Websites offer overwhelming purchase options, leading to information overload on users. RS mitigates this problem by giving specific recommendations by identifying patterns of user activities using sophisticated algorithms. In the last couple of decades, RS has become an active research area for both industry and academia. Over a period of time number of RS have been developed for different domains, e.g., usenet articles (GroupLens [1]), music (Ringo [2]), jokes (Jester [3]). In the early 90s, different collaborative filtering (CF) techniques [4] were proposed to assist users in online purchase, e.g., item-based

N. Mukherjee (✉) · G. M. Karthik
SRM Institute of Science & Technology, Chennai, Tamil Nadu, India
e-mail: nupurmukherjee98@gmail.com

N. Chaki et al. (eds.), *Proceedings of International Conference on Computational Intelligence and Data Engineering*, Lecture Notes on Data Engineering and Communications Technologies 56, https://doi.org/10.1007/978-981-15-8767-2_11

collaborative filtering (IBCF) [5], and user-based collaborative filtering (UBCF) [1]. Low rank matrix factorization (LRMF) is another popular algorithm [6], based on latent features (where both items and users are characterized using a set of features).

Explainable recommendation system (ERS), apart from recommending items, also explain why the recommendation is being made making it more acceptable to the user [5]. Since every RS has some strength and weakness, a hybrid RS [7], that combines two or more basic RS may be required that demonstrate both performance and explainability. In this paper, an innovative framework is proposed to configure RS with different types of recommender models, hybridization of recommender models to create new models, well-defined metrics to compare performance and explainability of recommendation given by ERS. The paper has been organized as follows—in Sect. 2, literature survey on related work is presented. The proposed evaluation framework is described in Sect. 3. In Sect. 4, simulation experiments are described that show proposed approach effectiveness. In the last section, conclusion of the paper is given.

2 Related Work

Brief description of the literature survey of related work is given below. Study on ERS improving trust, user satisfaction, transparency, trustworthiness, and system debugging has been done [5, 8, 9]. Early research in ERS [10], demonstrated that recommendations can be explained using items that are familiar to the user, e.g., "product you are looking at is similar to these products you liked earlier". Initial approaches were usually collaborative filtering (CF)-based explanations [11], that leverage "wisdom of the crowds", e.g., "users that are similar to you loved this item" or "the item is similar to your previously loved items". On the other hand, explainability of recommendation by a latent factor-based LRMF is difficult as "latent factors" do not have intuitive meanings.

Evaluating ERS involves computing percentage of recommendations explainable by ERS model. Abdollahi and Nasraoui [12] defined explainability precision (EP) and explainability recall (ER) as a metric of explainability, defined in terms of recommendable items and explainable items. To improve the performance of RS, recommendation techniques are combined to develop hybrid RS. Survey of hybrid RS [7], has identified different types of approaches. In the present paper, weighted and switching approach has been used. In the weighted approach [13], RS outputs are numerically combined, whereas, in the switching approach [7], RS outputs are combined by selecting one output based on the confidence level of RS.

3 Proposed Work

In this paper, an evaluation framework has been proposed to identify the best ERS configuration in terms of performance and explainability. The framework (Fig. 1), has the facility to use different types of recommender models (e.g., UBCF, IBCF, LRMF), hybridization of recommender models to create new models, performance, and explainability metrics computation and item/user/tag-based explanations of recommendation given by configured RS.

3.1 RS Algorithm Module

The proposed framework implements basic RS algorithms of UBCF, IBCF, and LRMF. UBCF algorithm (see Eq. (1)), predicts rating of item i for user u based on ratings given to item i by the neighborhood of user u. Neighborhood of user u are users similar to u, where similarity is computed using Pearson correlation (see Eq. (2)). Equation (1) gives the formula for UBCF based prediction of item i for user u.

$$P_{ui} = r_{u,i} + \frac{\sum_{n \subset neighbors(u)} S_{u,n}^{user} \cdot \left(r_{n,i} - \bar{r_n}\right)}{\sum_{n \subset neighbors(u)} S_{u,n}^{user}} \tag{1}$$

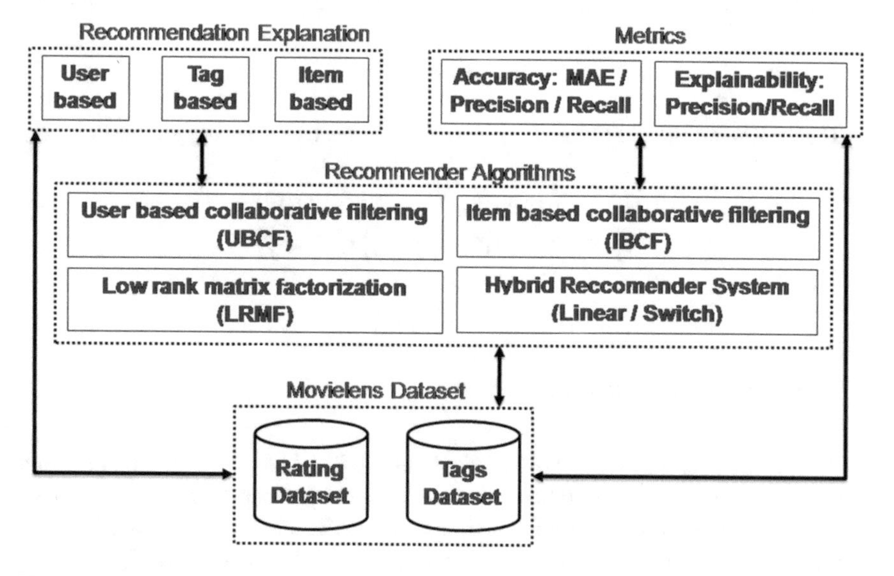

Fig. 1 Architecture

$$S_{u,n}^{user} = \frac{\sum_{i \subset CR_{u,n}} \left(r_{u,i} - \bar{r_u}\right)\left(r_{n,i} - \bar{r_n}\right)}{\sqrt{\sum_{i \subset CR_{u,n}} \left(r_{u,i} - \bar{r_u}\right)^2} \sqrt{\sum_{i \subset CR_{u,n}} \left(r_{n,i} - \bar{r_n}\right)^2}} \quad (2)$$

IBCF is similar to UBCF but based on item similarity rather than user similarity. LRMF predict user ratings by characterizing both users and items using a set of features deduced from the user ratings. Let X_i be feature vector of item i and θ_u be parameter vector of user u (both vectors are of dimensionality n). X_i indicates the degree to which features are possessed by item i. θ_u indicates the degree of user interest in corresponding features by user u. The resulting dot product (see Eq. (3)), gives a prediction of rating for item i by user u

$$\hat{r}_{u,i} = \theta_u^T X_i \quad (3)$$

Given a set of user ratings, feature vectors X_i and θ_u are learnt by RS by minimizing the cost function (see Eq. (4)), using stochastic gradient descent algorithm

$$J(X^1 \ldots X^{|I|}, \theta^1 \ldots \theta^{|U|}) = \min_{X^1 \ldots X^{|I|}, \theta^1 \ldots \theta^{|U|}} \frac{1}{2} \sum_{(i,j): r_{ij} \neq 0} \left(\theta_j^T X_i - r_{ij}\right)^2 + \frac{\lambda}{2} \sum_{i=1}^{|I|} \sum_{k=1}^{n} \left(X_i^k\right)^2 + \frac{\lambda}{2} \sum_{j=1}^{|U|} \sum_{k=1}^{n} \left(\theta_j^k\right)^2 \quad (4)$$

LRMF predicts $\hat{r}_{u,i}$, given each training case and the prediction error is computed

$$e_{ij} = \theta_j^T X_i - r_{ij} \quad (5)$$

The minimization of $J(X^1 \ldots X^{|I|}, \theta^1 \ldots \theta^{|U|})$ is done by following gradient descent

$$\begin{aligned} x_i^k &= x_i^k - \alpha\left(\left(\theta_j^T X_i - r_{ij}\right)\theta_j^k + \lambda x_i^k\right) \\ \theta_j^k &= \theta_j^k - \alpha\left(\left(\theta_j^T X_i - r_{ij}\right)x_i^k + \lambda \theta_j^k\right) \end{aligned} \quad (6)$$

3.2 Hybrid Mechanism Module

Each of basic algorithm viz. UBCF, IBCF, and LRMF has its own strengths and weaknesses. UBCF and IBCF are based on local neighborhood information while LRMF is based on global error information. IBCF has better computational efficiency and scalability compared to UBCF. Hybridization of RS algorithm [7], involves combining basic RS algorithms to get better aggregated performance than constituent

RS. The proposed framework has the facility to create hybrid RS by combining one or more basic RS either in a static or dynamic fashion. Top-n recommendation list suggested for a particular user by UBCF, IBCF, and LRMF are first combined into a single list and re-ranked by three RS to generate three ranked lists. For static numerical combination of three lists, the geometric mean of ranks of each item in the three lists gives new rank for the item in the hybrid recommendation list. In case of switching combination, maximum confidence value [14], of ranks of each item in the three lists (given by respective RS) gives new rank for the item.

3.3 Metric Computation Module

In the proposed framework, users can compute different type of metrics to compare ERS algorithms, e.g., performance metrics (mean absolute error, precision, recall, ranking metrics) and explainability metrics (explainability precision, explainability recall). To evaluate the accuracy of an RS algorithm, mean absolute error between predicted rating and true rating is computed. Precision and recall metrics determine the quality of top-n recommendation list depending upon how relevant are recommended item to the user. Ranking metrics measures how close is the ordering in recommendation list to the user preferred ordering of the items in the list.

To compute explainability of recommendation, explainability precision (EP) and explainability recall (ER) metrics are used (see Eqs. (8) and (9)), defined in terms of "explainable" recommended items to the user [12]. Explainable value of a recommended item j for user i as given by

$$Explainable\ Value\,(i,\ j) = \frac{\left|N_i^j\right|}{|N_i|} \tag{7}$$

where N_i is the set of nearest neighbors of user i, and N_i^j is the set of user i neighbors who have rated item j

$$Explainable\ Precision = \frac{|\{\text{Explanable items}\} \cap \{\text{Reccomended items}\}|}{|\{\text{Reccomended items}\}|} \tag{8}$$

$$Explainable\ Recall = \frac{|\{\text{Explanable items}\} \cap \{\text{Reccomended items}\}|}{|\{\text{Explanable items}\}|} \tag{9}$$

Here a set *Explainable items* has been taken as those items that have *Explainable value* greater than some threshold.

4 Result

The proposed framework has been used for RS evaluation experimentations using well-defined performance and explainability metrics as described earlier. Total of eleven RS algorithms have been evaluated—three basic algorithms (UBCF, IBCF, LRMF) and their eight possible hybridized combinations—H111 (UBCF + IBCF + LRMF), H110 (UBCF + IBCF), H101 (UBCF + LRMF), and H011 (IBCF + LRMF), each having numerical and switching hybridization version (indicated by suffix of N or S to hybrid algorithm name).

4.1 *Experiment 1: Performance of Basic Algorithms*

Performance of basic algorithms (UBCF, IBCF, LRMF) have been compared in terms of accuracy, precision, and recall metrics (see Fig. 2). As can be seen from the figure, LRMF has best MAE (0.4459) compared to UBCF (0.6054) and IBCF (0.6609). This is to be expected as LRMF is generally accepted to be state-of-the-art RS algorithm as far as accuracy is concerned. All MAE fall between 0.4 and 0.6 which is fairly good accuracy performance.

Quality of top-n recommendation have been measured using precision and recall metrics. Precision is the fraction of recommended items that are relevant to the user. As can be seen from the figure, again LRMF gives the best precision (0.9658) as compared to UBCF (0.9209) and IBCF (0.8750). Recall is the fraction of the relevant items that are recommended to the user. Here IBCF gives best performance (0.9493)

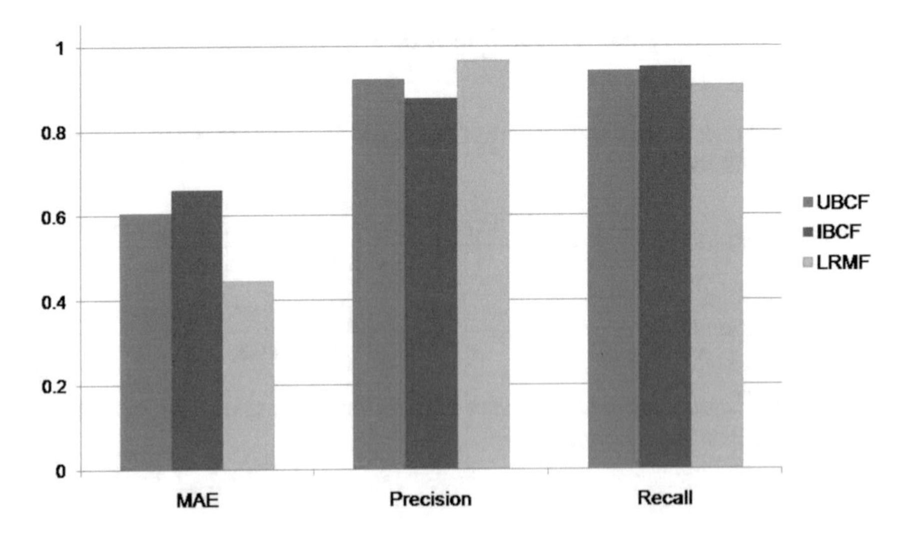

Fig. 2 Performance comparison of RS Algorithm

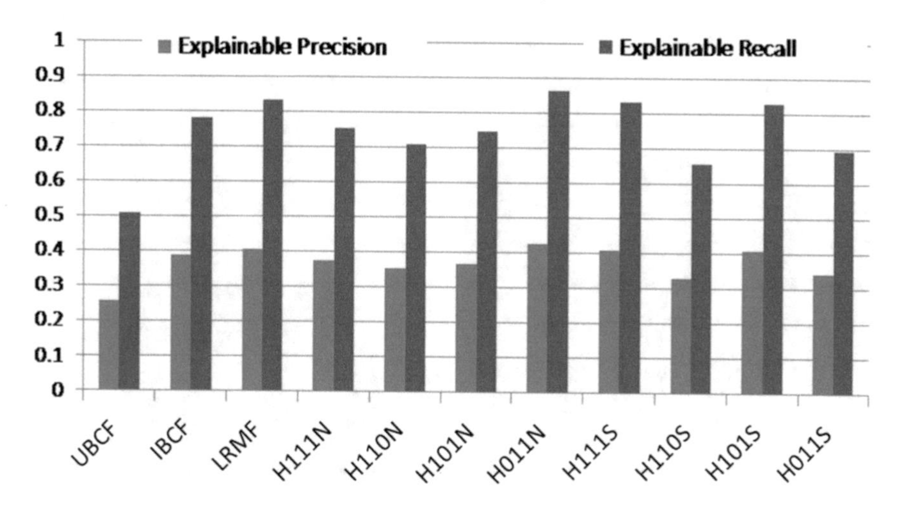

Fig. 3 Mean explainable precision and explainable recall

as compared to UBCF (0.9419) and LRMF (0.9077). Experimentation on different aspects of performance show that each of the basic algorithms does well in one aspect of performance.

4.2 Experiment 2: Evaluation of Recommendation Explainability

Evaluation of explainability has been done in terms of the percentage of recommendations that can be explained by the ERS model using EP and ER metrics. Hybrid algorithms (H011N, H111S, and H101S) show (Fig. 3), better explainability precision (greater than 0.4) and explainability recall (greater than 0.8) compared to other algorithms. This is because hybrid algorithms combine local neighborhood view of CF-based approach and global optimization view of LRMF. Best explainability precision (0.425) and recall (0.862) is given by H011N followed by H111S.

4.3 Experiment 3: Evaluation of Recommendation Ranking

Evaluation of recommendation ranking by an RS algorithm has been done in terms of the extent to which recommendation ranking agree with ranking done by the user. As can be seen from Fig. 4, LRMF (0.728) and UBCF (0.739), have comparable ranking performance although the worst ranking is demonstrated by IBCF (0.216). Switch based hybrid algorithms (H111S, H110S, H101S) demonstrate best ranking performance compared to other algorithms. This is due to the fact that switch based

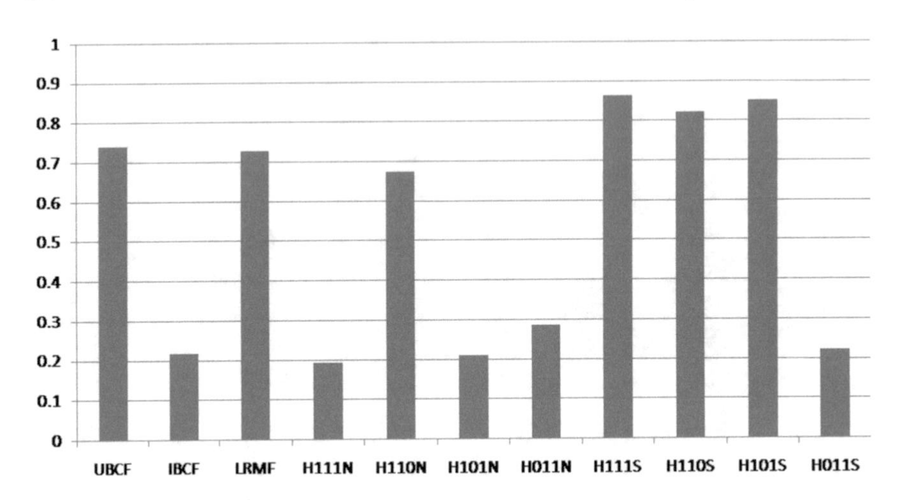

Fig. 4 Ranking metric

hybridization is dynamic in nature and adapts well to different user profile unlike numerical approach which is static in nature. Best ranking metrics are given by H111S (0.862) where the strengths of both CF-based and MF-based approach come into play.

Based on comparative experimentation carried out, it can be clearly seen that a switching hybrid approach (H111S) gives best overall results with respect to performance, explainability, and ranking metrics. The explainability of H111S recommendation is also one of the best due to the fact that H111S combines the benefits of local neighborhood view (of UBCF and IBCF) with global optimization view (of LRMF). In addition, H111S use context-based dynamic switching for hybridization that adapts well to different users unlike the static approach of numerical combination.

5 Conclusion

This paper presented details of the framework developed for systematic evaluation of different configuration of ERS with respect to performance and explainability of recommendation. The framework has the facility to configure RS with different types of recommender models, hybridization of recommender models to create new models, well-defined metrics to compare performance and explainability of recommendation given by ERS. Comparative evaluations were carried using performance metrics (MAE, Precision, Recall) and explainability metrics (EP, ER). Simulation results show that the proposed framework facilitates the systematic exploration of different ERS configuration for evaluation in terms of performance and explainability.

References

1. Resnick P, Iacovou N, Suchak M, Bergstrom P, Riedl J (1994) GroupLens: an open architecture for collaborative filtering of netnews. In: Proceedings of ACM CSCW '94. ACM, pp 175–186
2. Shardanand U, Maes P (1995) Social information filtering: algorithms for automating 'word of mouth. In: Proceedings of ACM CHI '95. ACM Press/Addison-Wesley Publishing Co., pp 210–217
3. Goldberg K, Roeder T, Gupta D, Perkins C (2001) Eigentaste: A constant time collaborative filtering algorithm. Inf Retr 4(2):133–151
4. Sarwar B, Karypis G, Konstan J, Riedl J (2001) Item-based collaborative filtering recommendation algorithms. In: Proceedings of the WWW conference
5. Herlocker JL, Konstan JA, Riedl J (2000) Explaining collaborative filtering recommendations. In: Proceedings of the 2000 ACM conference on computer supported cooperative work. ACM, p 241250
6. Ben Schafer J, Frankowski D, Herlocker J, Sen S (2007) Collaborative filtering recommender systems. In: Brusilovsky P, Kobsa A, Nejdl W (eds) The adaptive web. LNCS, vol 4321. Springer, Berlin, Heidelberg, pp 291–324
7. Burke R (2002) Hybrid recommender systems: Survey and experiments. User Model User Adap Inter 12(4):331370
8. Zhang Y, Chen X (2019) Explainable recommendation: a survey and new perspectives, vol XX, no XX, p 188
9. Sinha R, Swearingen K (2002) The role of transparency in recommender systems. In: CHI02 extended abstracts on human factors in computing systems. ACM, p 830831
10. Schafer JB, Konstan J, Riedl J (1999) Recommender systems in e-commerce. In: Proceedings of the 1st ACM conference on electronic commerce. ACM, p 158166
11. Ricci F, Rokach L, Shapira B (2011) Introduction to recommender systems handbook. In: Recommender systems handbook. Springer, p 135
12. Abdollahi B, Nasraoui O (2016) Explainable matrix factorization for collaborative filtering. In: Proceedings of the 25th international conference companion on world wide web. International World Wide Web Conferences Steering Committee, vol 5–6. International conference on knowledge discovery & data mining. ACM, pp 2060–2069
13. Bellogin A, Castells P, Cantador I (2011) Self-adjusting hybrid recommenders based on social network analysis. In: Proceedings of the 34th international ACM SIGIR conference on research and development in information retrieval, p 11471148
14. Van Setten M (2006) Supporting people in finding information: hybrid recommender systems and goal-based structuring. Report no. 016 (TI/FRS/016). Telematica Institut, Enschede, The Netherlands

Greedy-Based PSO with Clustering Technique for Cloud Task Scheduling

Y. Home Prasanna Raju and Nagaraju Devarakonda

Abstract Effective scheduling of cloud tasks is very much essential for a cloud computing environment. The cloud tasks are the user requests to be processed in a cloud environment. A number of cloud resources are consumed to process cloud tasks. Task scheduling optimizes the consumption of cloud resources and reduces the makespan time. The paper is aimed at reducing the makespan time in a cloud environment by introducing the new method Modified Greedy Particle Swarm Optimization with Clustered Approach (MGPSOC). The MGPSOC algorithm makes use of clustering with bio-inspired techniques. The proposed method showed good results when compared with the existing algorithm Greedy Particle Swarm Optimization Algorithm (G&PSO).

Keywords Cloud · Scheduling · Cluster · PSO · Greedy

1 Introduction

The Internet made things become easy for sharing data all over the world. The internet works with the distributed technology where the information is distributed globally to all the users. The speed of accessing data depends on the network bandwidth and the type of request made to the internet. The internet can be used not only for distributing data, but also for storing data into web applications or databases. It is very essential for any institution or organization to maintain their data or to provide services to

No academic titles or descriptions of academic positions should be included in the addresses. The affiliations should consist of the author's institution, town/city, and country.

Y. Home Prasanna Raju (✉)
Department of CSE, Acharya Nagarjuna University, Guntur, AP, India
e-mail: yhprasannaraju@gmail.com

N. Devarakonda
School of Computer Science and Engineering, VIT-AP University, Amaravati, AP, India
e-mail: dnagaraj_dnr@yahoo.co.in

N. Chaki et al. (eds.), *Proceedings of International Conference on Computational Intelligence and Data Engineering*, Lecture Notes on Data Engineering and Communications Technologies 56, https://doi.org/10.1007/978-981-15-8767-2_12

the users for less cost. Cloud computing can be a solution to the institutions, organizations or companies to maintain their information or run their businesses for less cost. Cloud computing [1] is a type of distributed technology which works with the internet. Cloud consists of resources like virtual machines, data Centers and network bandwidth. One can use cloud infrastructure without investing huge amounts to setup their infrastructure. Cloud works with the pay per use model. It means payment can be done to the use of cloud resources on demand. There are a variety of cloud services [2], available. They are platform as a service (Paas), infrastructure as a service (Iaas), and software as a service (Saas). Paas provides a platform to the users where they can build and run their applications without having to create a platform with their existing infrastructures for reducing complexity. Iaas provides infrastructure to the users in the form of physical servers and data centers. The users can reduce their expenses by using this service. Software as a service provides a required software through online to the users without having to install on their individual desktops. Like different types of cloud services, a variety of clouds [3], are available. They are private clouds, public cloud, and hybrid clouds. Each organization or institution can maintain their separate clouds which are called as private clouds. A cloud which can be available to everyone is called a public cloud. The integration of private and public clouds is called a hybrid cloud. Users can choose the type of cloud and type of service they want depending on their requirement. Cloud services should be reached to the end user when they are requested. It means response time should be minimized from the cloud environment when users initiate requests. One of the ways of reducing response time or tasks execution time is with the task scheduling techniques. Task scheduling schedules the user requests in the cloud and sees that the optimal assignment of tasks is done to virtual machines (VMs). Execution of cloud tasks is done by VMs. The overall time of execution process should be minimized to get the speedy results from the cloud environment. There are a variety of task scheduling algorithms [4]. They are priority-based scheduling, static and dynamic scheduling, heuristic scheduling, workflow scheduling. In case of priority-based scheduling, each cloud task has its own priority and these are executed by virtual machines depending upon their priorities. Sometimes they can be preemptive or non-preemptive scheduling techniques. In case of static scheduling, tasks are known to the scheduler in advance before they get scheduled. Dynamic schedulers schedule jobs dynamically as long as tasks come to them. In case of dynamic schedulers, complete tasks details are not known in advance to the schedulers. Sometimes cloud applications are of type NP-hard problems. Hence they need approximation solutions. In such cases, heuristic scheduling is necessary. These scheduling techniques come in a variety of ways. They are like ant colony optimization techniques [5–7], particle swarm optimization techniques (PSO), and genetic algorithms. The paper is aimed to address heuristic scheduling: particle swarm optimization technique with the clustering concept by making use of the greedy method to reduce makespan for the cloud environment. Greedy method is good at making better local optimal choices for global optimal results, whereas clustering can be used to group tasks according to their complexity levels.

The arrangement of paper is done as: Sect. 2 presents the work related to task scheduling in a cloud environment. Section 3 addresses the proposed methodology of the research paper. Experimental results of proposed and existing methodologies are discussed in the Sect. 4. Finally, conclusion is provided in the Sect. 5.

2 Related Work

Cloud task scheduling plays a major role to reduce makespan for the cloud environment. Related work is concentrated on cloud task scheduling with PSO techniques.

Abdi et al. [8] introduced a scheduling algorithm by modifying the existing PSO algorithm. The algorithm assigns the smallest cloud task to the fastest cloud processor as the initial random step. The paper also analyzed the performance with PSO and genetic algorithms and showed good results.

Awad et al. [9] presented a new technique LBMPSO for load balancing between cloud processors by taking execution time, transportation cost, task completion time, and time for round trip into account. The technique analyzed PSO and LCFP algorithms for showing better results.

Ali et al. [10] introduce an algorithm MDAPSO to better reduce the makespan. The algorithm is a mixture of cuckoo search and dynamic PSO algorithms. The MDAPSO showed better results when analyzed with PSO, dynamic PSO, and Cuckoo search algorithms.

Dordaie et al. [11] presented a hybrid PSO-hill climbing method for reducing the completion time. The technique assigns the particles in a random fashion with the PSO technique. Later particles are assigned to the processor using the HEFT technique. Finally, to optimize the solutions, hill climbing method was used.

Thanaa et al. [12] introduced an algorithm for load balancing called Binary load balancing-hybrid PSO with the gravitational search algorithm. The algorithm depends on the lengths of cloud task length and speeds of virtual machines. It showed good results when compared with the existing binary load balancing with hybrid PSO algorithm.

Sudheer et al. [13] presented a modified dynamic adaptive PSO technique for optimal utilization of cloud resources for cloud environment. The technique combined cuckoo search algorithm with dynamic adaptive PSO. It effectively optimizes resources when there is a heterogeneous workload.

Oqail Ahmad and Rafiqul Zaman Khan [14] presented an algorithm PSO-ALBA. The algorithm is the combination of existing PSO and adaptive load balancing algorithm. It effectively works when the cloud tasks are heterogeneous. The algorithm optimizes the makespan effectively.

Neha Miglani, Gaurav Sharma [15] presented a modified PSO algorithm for reducing the makespan. The algorithm categorizes the particles into two, namely communication intensive and computation intensive. The results are analyzed with PSO, min-min, and hybrid PSO techniques.

Zhong et al. [16] introduced a new technique G&PSO for better reducing the makespan. The algorithm is the combination of greedy method and PSO algorithm. The algorithm initializes the particles with the result of greedy output.

Mostly all the approaches did not make use of clustering concept for scheduling cloud tasks. The clustering concept can be used to separate the more likely complex tasks to groups before scheduling. Later the task groups can be assigned to cloud resources depending on their capacities which in turn reduce the makespan. Prasanna Raju and Nagaraju DevaraKonda [17] presented a new technique KPSOW for effectively reducing the makespan. The algorithm used the clustering approach for separating the complex tasks as the process of scheduling. It is the combination of PSO with the k-means algorithm. The same authors [18], also presented another technique KMPS to further reduce the makespan using clustering approach. This time the authors combined the PSO with the k-medoid approach. The later algorithm combined showed better results compared with the earlier one.

The proposed paper is aimed to reduce the makespan in the cloud environment by modifying the existing G&PSO algorithm with clustered approach. A greedy algorithm is very good at making local choices. Clustering was done using k-means approach to generate task groups.

3 Proposed Methodology

The proposed methodology is aimed to introduce Modified Greedy Particle Swarm Optimization with Clustered approach (MGPSOC) for reducing makespan. The MGPSOC algorithm is the modified version of the existing Greedy PSO approach. The existing algorithm computes the ETC (Expected time to compute) matrix for all the cloud task particle values to know the expected completion time. Let the matrix size be mxn. Each row specifies the cloud tasks and column specifies the cloud resources or VMs as represented in the Fig. 1. The entire row represents a cloud particle. Where i represents the cloud task which varies from 1 to m and j represents the cloud virtual machine which varies from 1 to n. The value $t(i_{m,}\ j_n)$ in ETC matrix represents the expected completion time of *i*th cloud task when assigned to nth virtual machine. The total completion time of each particle is calculated as the summation of all the first row expected completion time values which are shown in the Eq. 1. The total completion time is also called as fitness function. The fitness function must be minimized to get the best makespan in the cloud environment. The optimal objective function is the minimum of total completion time which is shown in the Eq. 2.

$$\text{Fitness function} = \sum_{j=1}^{n} t(i, j) \tag{1}$$

Fig. 1 ETC matrix

$$
\mathbf{ETC\ (i,\ j)} = \begin{bmatrix} t(i_1,j_1) & t(i_1,j_2) & \ldots\ldots & t(i_1,j_n) \\ t(i_2,j_1) & t(i_2,j_2) & \ldots\ldots & t(i_2,j_n) \\ . & . & & . \\ . & . & & . \\ . & . & & . \\ t(i_m,j_1) & t(i_m,j_2) & \ldots\ldots & t(i_m,j_n) \end{bmatrix}
$$

$$
\text{Objective function} = \min_{1<=i<=m}\left(\sum_{j=1}^{n} t(i, j)\right) \tag{2}
$$

In the original greedy particle swarm optimization algorithm, the initial allotment of cloud tasks to VMs and optimal solution as global best is considered for all particles based on a greedy algorithm. Later scheduling is done with PSO technique. The final result is considered as the best optimal solution. The allotment of cloud tasks to VMs is done as follows: as soon as the cloud task comes, it is allotted to a VM where it would get less execution time. If another cloud task is assigned to the same VM, then it would be allotted to the next VM with less execution time.

But the drawback of this approach is that there is a chance of assignment of small cloud task to the high capacity virtual machine. The desirable way is always to see that big cloud tasks are assigned to high capacity virtual machines which further reduces the makespan time. The proposed method MGPSOC effectively addresses the above said drawback by doing some modifications to the existing one. The architecture of the proposed method is represented in the Fig. 2.

The modification is done as follows: initially, the input cloud tasks are separated into groups using the k-means technique. The input k value 2 is considered for the k-means clustering technique. The technique separates the tasks into low level and high level cluster groups and virtual machines into low performance and high performance virtual machine groups. Low level cluster groups are allotted to low performance VMs and high level task groups are allotted to high performance VMs. The allotment is done as follows: for each group, tasks and VMs are organized in descending order according to their task size and performance, respectively. Later the greedy method is applied to each group as: first task is allotted to first VM and second cloud task is allotted to second VM and so on. When all virtual machines get equal tasks then the further tasks assignment is done from the beginning of the first virtual machine as said above. This process is repeated until there are no cloud tasks left for assignment. These are considered as initial task assignments to virtual machines. Later initial fitness functions are calculated for all particles in ETC matrix which in turn act as initial local bests for each particle and the best of all particles is considered as global best. These two local and global bests are taken as initial values for PSO algorithm and scheduling is done for each group. Finally, at the end of the

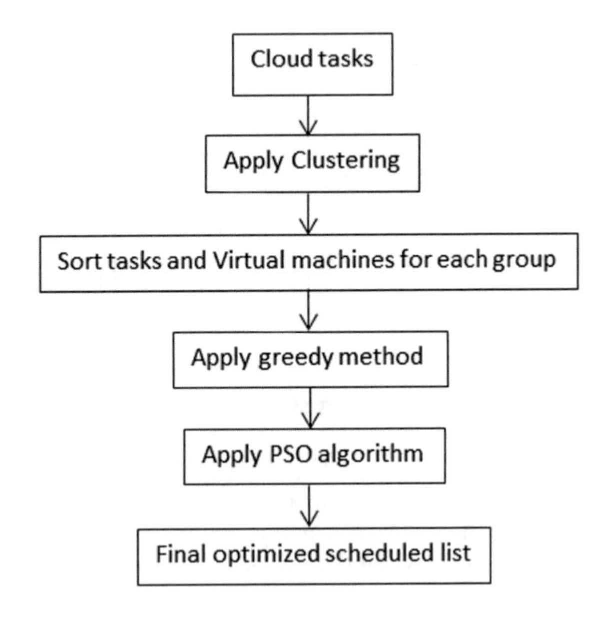

Fig. 2 Architecture of MGPSOC algorithm

scheduling process, the optimized global value is generated as a reduced makespan, and the equivalent task assignment to virtual machines is considered as an optimal task-scheduled list. The algorithm for the proposed MGPSOC is given below.

Algorithm MGPSOC(m,n,k)
{
Input: i) m number of cloud tasks with their task lengths.
ii) n number of virtual machines with their performance values.
iii) k value is 2 for k-means clustering technique.
Output: a list of final optimized (cloud task, virtual machine) mapping pairs.
Start:
Step1) Separate input m cloud tasks into 2 cluster groups using k-means.
Step2) Separate virtual machines into low and high performance VMs.
Step3) Arrange cloud tasks into non-increasing sequence in the two cloud cluster groups.
Step4) Virtual machines are ordered in decreasing fashion in the two low and high performance VM groups.
Step5) Apply greedy method to do initial mapping for already arranged above cloud tasks and virtual machines for each.
Step6) Apply PSO algorithm for each mapping group separately.
Step7) Collect optimized scheduled mapping pair list from each group.
Step8) Return the optimized mapping pair list.
Stop.
} // end algorithm

4 Experimental Results

The cloudsim tool was used to generate the experimental results of the proposed method. The number of cloud tasks is taken as 50 tasks. The lengths of cloud tasks were taken in between the range from 500 million instructions to 1000 MI. A total of 5 VMs was taken whose performances varied in the range from 500 to 900 MIPS. For PSO algorithm, the learning factor values for both c1 and c2 values were taken as 2. The population size or particle size was taken as 100 and the total number of iterations considered for the proposed method is 100. With these values, makespan parameter was tested with 100 iterations which are shown in the Fig. 3. The figure states that the makespan value taken for PSO algorithm is 17.92 s and for G&PSO algorithm is 17.80 s, whereas the proposed method took only 13.83 s for makespan, which is the best makespan value for the proposed method. It is also clearly stated that the proposed method exhibits an improvement of 12.89 percentage and 12.54 percentages when compared with the PSO method and G&PSO methods, respectively. The load of each virtual machine can be tested with the number of tasks assigned to them for execution. The load is calculated for all virtual machines after 100 iterations. The cloud tasks distributed to each virtual machine is shown in the Fig. 4.

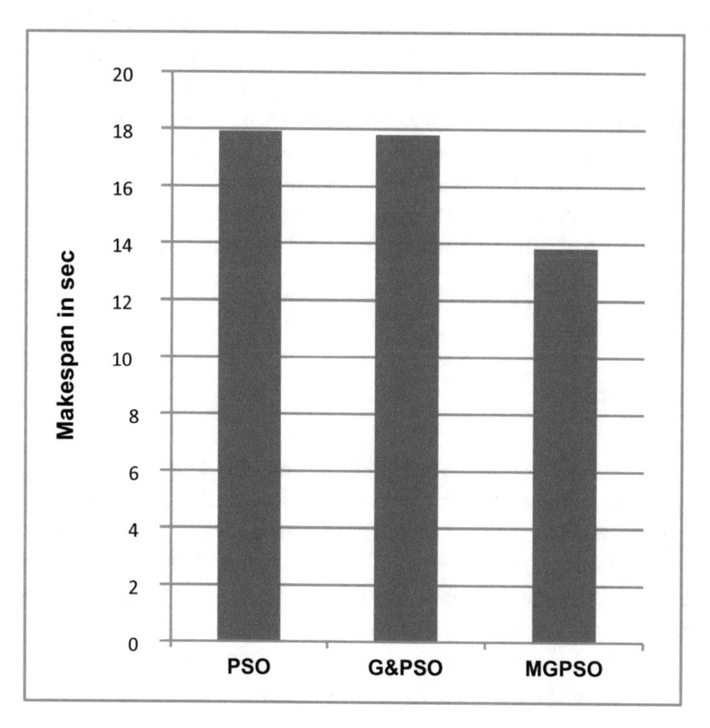

Fig. 3 Makespan comparison

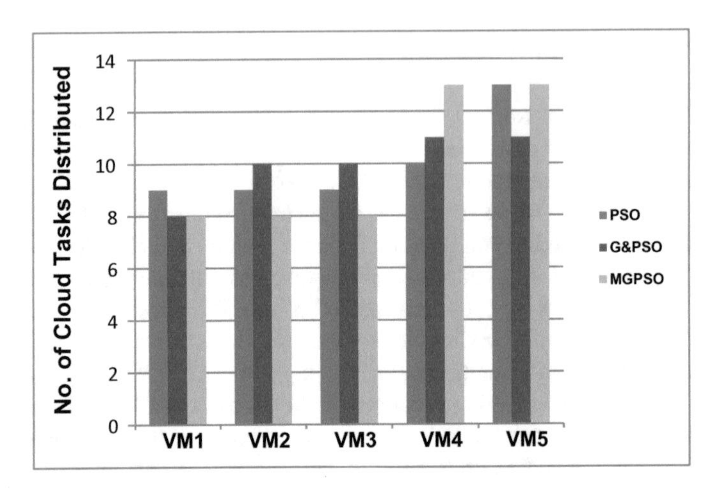

Fig. 4 Cloud tasks distribution comparison

The proposed method separated the 5 virtual machines into two groups. They are low performance virtual machines VM1, VM2, and VM3 and another one is high performance virtual machines which are VM4 and VM5. So the high performance VMs should get more tasks when compared to low performance VMs. The figure states that PSO and G&PSO techniques allocated more tasks to low performance virtual machines and less number of tasks to high performance virtual machines, whereas the proposed method MGPSO had allocated more number of tasks to VM4 and VM5 when compared to first three virtual machines. It states that heavy load was taken by high performance virtual machines compared to low performance virtual machines in the proposed method. Hence, the proposed MPGSO technique shows good results in load distribution when compared to other existing methods PSO and G&PSO.

5 Conclusion

Makespan represents the total completion time of scheduling tasks in cloud computing. Less makespan time in the cloud environment represents a better scheduling process. Hence, the order or assignment of cloud tasks to virtual machines decides the best scheduling process. The proposed MGPSOC algorithm showed good results when compared with the existing ones for reducing the makespan. The proposed algorithm used k-means clustering before task assignment to virtual machines and used the sorting process before applying the greedy method. Subsequently, PSO algorithm helped the scheduling process for giving better makespan results.

References

1. Hoefer CN, Karagiannis G (2010) Taxonomy of cloud computing services. In: IEEE globecom workshops, GC'10, pp 1345–1350
2. Akilandeswari P, Srimathi H (2016) Survey and analysis on task scheduling in cloud environment. Indian J Sci Technol Indian J Sci Technol 9(37). https://doi.org/10.17485/ijst/2016/v9i37/102058
3. Nagendra Babu P, Chaitanya Kumari M, Venkata Mohan S (2018) A literature survey on cloud computing. I-Manager's J Inf Technol 1(1):44–49
4. Sharma N, Tyagi S (2016) Task scheduling in cloud computing. Adv Comput Sci Eng 249–252
5. Li K, Xu G, Zhao G, Dong Y, Wang D (2011) Cloud task scheduling based on load balancing ant colony optimization. In: Chinagrid conference (ChinaGrid), 2011 Sixth annual, pp 3–9
6. Lu X, Gu Z (2011) A load-adapative cloud resource scheduling model based on ant colony algorithm. In: 2011 IEEE international conference on cloud computing and intelligence systems (CCIS), pp 296–300
7. Raju YHP, Devarakonda N (2018) Makespan efficient task scheduling in cloud computing. In: International conference on emerging technologies in data mining and information security. https://doi.org/10.1007/978-981-13-1951-8_26
8. Abdi S, Motamedi SA, Sharifian S (2014) Task scheduling using modified PSO algorithm in cloud computing environment. In: International conference on machine learning, electrical and mechanical engineering (ICMLEME'2014), 8–9 Jan 2014 Dubai (UAE)
9. Awad AI, El-Hefnawy NA, Abdel-Kader HM (2015) Enhanced particle swarm optimization for task scheduling in cloud computing environments. In: Procedia computer science. Elsevier Masson SAS, 65(Iccmit), pp 920–929. https://doi.org/10.1016/j.procs.2015.09.064
10. Al-maamari A, Omara FA (2015) Task scheduling using PSO algorithm in cloud computing environments. Int J Grid Distrib Comput 8(5):245–256. https://doi.org/10.14257/ijgdc.2015.8.5.24
11. Dordaie N, Navimipour NJ (2018) A hybrid particle swarm optimization and hill climbing algorithm for task scheduling in the cloud environments. ICT Express. Elsevier B.V., 4(4):199–202. https://doi.org/10.1016/j.icte.2017.08.001
12. Alnusairi TS, Shahin AA, Daadaa Y (2018) Binary PSOGSA for load balancing task scheduling in cloud environment. (IJACSA) Int J Adv Comput Sci Appl 9(5)
13. Sudheer MS, Vamsi Krishna M (2019) Dynamic PSO for task scheduling optimization in cloud computing. Int J Recent Technol Eng 8(2 Special Issue 11):332–338. https://doi.org/10.35940/ijrte.B1052.0982S1119
14. Md Oqail Ahmad and Rafiqul Zaman Khan (2019): Pso-Based Task Scheduling Algorithm Using Adaptive Load Balancing Approach For Cloud Computing Environment. In: International Journal Of Scientific & Technology Research Volume 8, Issue 11, November 2019.
15. Miglani N, Sharma G (2019) Modified Particle Swarm Optimization based upon Task categorization in Cloud Environment. Int J Eng Adv Technol (IJEAT) 8(4C). ISSN: 2249-8958
16. Zhong Z, Chen K, Zhai X, Zhou S (2016) Virtual machine-based task scheduling algorithm in a cloud computing environment. Tsinghua Sci Technol 21(6). ISSN ll1007-0214ll07/09llpp660-667
17. Raju YHP, Devarakonda N (2019) Cluster based hybrid approach to task scheduling in cloud environment. Int J Adv Comput Sci Appl 10(4):425–429. https://doi.org/10.14569/ijacsa.2019.0100452
18. Raju YHP, Devarakonda N (2020) A Cluster Medoid Approach For cloud Task Scheduling. KES J (Int J Knowl-Based Intell Eng Syst)

Brave Men and Emotional Women: Analyzing Gender Bias in Bollywood Songs

Ashish Gupta, Himanshu Ladia, Shefali Bansal, and Anshul Arora

Abstract Stereotypes exist in several sections of society including various means of popular entertainment. We believe that Bollywood songs are no exception as there has been a certain change in the characteristics of the songs' lyrics over the past few years. Hence, to computationally study Bollywood's songs lyrics from the Hindi movie industry, in this paper, we examine their style of writing and the presence of any biases. We analyze the changes in the songs' lyrics over time and quantitatively show the change in the pattern by evaluating the rank of certain sensitive words in the songs. We calculate embeddings for the lyrical vocabulary using Word2Vec, FastText, and GloVe algorithms and use the WEAT similarity score to show that Hindi songs indeed suffer from racial and gender bias. The metrics we obtain can be further used for more formal problems of music recommendation, lyrics generations and popularity prediction.

Keywords Natural language processing · Text mining · Document analysis

1 Introduction

The Hindi movie industry shows a glimpse of what we perceive about our society. The Indian Hindi movie industry (Bollywood) is the largest producer of films in the world. Bollywood movies are primarily musicals and have songs knitted throughout

---

A. Gupta · H. Ladia · S. Bansal (✉) · A. Arora
Department of Applied Mathematics, Delhi Technological University, New Delhi, India
e-mail: bansalshefali11@gmail.com

A. Gupta
e-mail: ashishgupta1350@gmail.com

H. Ladia
e-mail: hladia199811@gmail.com

A. Arora
e-mail: anshul15arora@gmail.com

N. Chaki et al. (eds.), *Proceedings of International Conference on Computational Intelligence and Data Engineering*, Lecture Notes on Data Engineering and Communications Technologies 56, https://doi.org/10.1007/978-981-15-8767-2_13

the script. According to a survey by the Indian Music Industry[1] a typical Indian spends 19.1 h a week listening to music, with their favorite genre being Bollywood. This is higher than the global average of 18 h. With the easy availability of the Internet via smartphones and tablets, people tend to browse for exact lyrics.

1.1 Motivation

People try to emotionally connect to songs via lyrics and this can affect the psych of a person in many ways.[2] Songs with degrading lyrics have an influence on the sexual behavior of teens [1]. As a motivation for the problem, we consider the following excerpt from a popular song Haseeno Ka Deewana from the Bollywood movie Kaabil:

Ladki ek dam right hai
Ladki dynamite hai
Ladki ke chakkar mein
Daily gali gali mein fight hai

Thoda hamein do time yaara
Love nahi hai crime yaara
Har din mujko better lage
Tu ladki hai ya wine yaara

Ladki (girl in English) is being objectified profusely in just a couple of paragraphs of this rap. She's been compared to a dynamite, a wine and is specified as a reason for fights in the neighborhood. We quantitatively show that such stereotypes exist throughout most of the Hindi songs in Bollywood.

1.2 Contribution

Prior work on song lyrics has been primarily manual and [2] used the recent advancements in natural language processing to perform the analysis on a large dataset. However, the analysis is only limited to English song lyrics. In this paper, we computationally study and highlight the degrading lyrics in Bollywood songs. The problem becomes two-fold as we deal with romanized English characters.

First, there's a need for an efficient and effective way of preprocessing such text. Transliteration of Hinglish text has been a popular choice among the research community [3]. However, such an approach induces a lot of errors. It misses transliteration

[1] https://indianmi.org/be/wp-content/uploads/2019/09/output.pdf.

[2] https://repository.upenn.edu/cgi/viewcontent.cgi?article=1094&context=mapp_capstone.

of many words because of spelling variations and the multilingual nature of Bollywood songs' lyrics. To avoid these issues, we build a custom stemming algorithm specifically for our purpose.

Second, we need to computationally process such a huge corpus of text to mine style and bias patterns. For the style analysis, we plot swear words and show how they change over the decade. For gender biases, we employ unsupervised word embedding algorithms and the WEAT metric to quantify the bias between two different target set of words.

1.3 Organization

The rest of the paper is organized as follows. We discuss the related work in Sect. 2. We explain the proposed methodology in detail in Sect. 3. Results of the proposed work are summarized in Sect. 4 and we conclude with future work directions in Sect. 5.

2 Related Work

The use of neural networks for generating lyrics has gathered increasing interest in recent years [4] have used LSTM models for English rap lyrics generation. The word embeddings used as an input to the LSTM models are often subject to various biases [5]. This can propagate through the LSTM models and reflect in a novel writing. Song emotion classification and popularity prediction are difficult but interesting business problems [6] have used Hindi sentiment lexicons, text stylistic features, and N-grams features for mood classification of Hindi songs. This involved transliteration of Hinglish text to UTF-8 characters using a dictionary[3] [7] have also suggested Hindi music recommendations based on lyrics. The approach avoids transliteration of Hinglish text and uses a custom stemming algorithm to deal with spelling variations in the text. Our approach contributes to the above approaches by analyzing style and bias in Hindi songs.

Madaan et al. [8] have examined stereotypes in the Hindi movie industry. Their work involved surveying movie scripts, posters, and the cast. However, songs too, have a major contribution to the entertainment industry [2] have computationally analyzed more than half a million English songs. In our work, we intend to perform an analysis of Hindi songs and study their style of writing. Instead of manually surveying the distributions [8], we computationally quantify the bias in sentence encodings using the word embedding association test [9] on Word2Vec [10], fastText [11] and GloVe [12] embeddings. The WEAT does so by computing the cosine

[3]https://tdil-dc.in/index.php?lang=en.

similarity between sets of words. We also use association rule mining [13] and study the support and confidence of thus obtained association rules.

3 Methodology

3.1 Dataset

There is no publicly available dataset for Bollywood songs' lyrics. So we built individual scrapers for several popular lyrics websites.[4,5,6] All the data we crawled was in romanized characters, ignoring lyrics in UTF-8 characters. We obtained a set of 7615 lyrics, which we use to perform our analysis.

Algorithm 1

```
stemlist ← [ 'ing','ana','yan','ane','eya','ies','aaa','aan','ake','on',
'ai','na','an','un','ey','en','se','aa','a','e','i','o','u','y','n','s' ]
Store all unique words in allvocab after sorting them alphabetically and
length wise
Create a dictionary dict
for each word_i in allvocab do
       dict[word_i] ← word_i
procedure RESTRICTED STEMMING
for each word_i in dict do
       if word_i can be broken into 2 parts where prefix is in allvocab and
suffix is in stemlist then dict[word_i] = first half of word_i
Define a list afterstem of all unique values in dictionary dict
Create a dictionary stemdict for each i in dict do
       stemdict[dict[i]] = dict[i]
procedure CORRECT SPELLING
spelldict ← {'aa': 'a', 'ee': 'i', 'w': 'v', 'j': 'z', 'oo': 'u'}
for each word_i in afterstem do
       for each key, value in spelldict do
              newword = word_i.replace(key, value)
              if newword in allvocab then
                     stemdict[word_i] = newword
for each i in dict do
       dict[i] = stemdict[dict[i]]
```

3.2 Preprocessing

The basic preprocessing involved removing irrelevant text such as HTML tags, URLs, punctuation, and repeat line marks (X2, X4, 2 times), other special characters. After

[4]https://www.metrolyrics.com/.

[5]https://www.lyricsia.com/.

[6]https://www.lyricsmint.com/.

tokenization, we obtained a dictionary of 31,402 unique words. We observed that there were many variations of the same word in the dictionary. For example, *bhool* (hindi word meaning mistake) had variations *bhoola, bhooli, bhoolna, bhoolun, bhule, bhuli, bhulna, bhulo, bhulon, bhulun, bhulana*, etc. It was important to preprocess these words and bring them to their root forms. Patra et al. [7] used a custom unsupervised stemmer for Hinglish. The major drawback of the approach is that words like *waari* turns to *waar*; *maasum, maana, maar* turns to *maa*, i.e., the root words are altered.

To avoid the above problem, we propose a two-step preprocessing method. The pseudo code is discussed in Algorithm 1. The first procedure of the algorithm performs restricted stemming and the words are brought to their root form. In the second procedure, frequent spelling variations in Hinglish language like *oo/u, ee/i* are taken into account and dealt with. After stemming, the dictionary is reduced to 24,409 unique words and after the second procedure, it is further reduced to 22,397.

3.3 *Style Analysis*

Songs before 2010 are relatively sparse. To solve this problem, we analyzed our data concluding over 4 years instead of 1 at the appropriate places for the songs before and upto 2010 given the nature of the data. For songs dated post 2010, the analysis on the songs has been done year on year due to a large number of songs for those years in the dataset.

We generated two-word clouds from the songs as shown in Fig. 1. One for songs before 2010 and one for after 2010. We can observe a major shift in the vocabulary over time. There is an influx of Punjabi, Harayanvi (regional languages of India) words which can be attributed to the rise in popularity of these songs over the past decade. The change in this pattern may in-turn explain the gender bias we aim to discover and quantify in the upcoming sections.

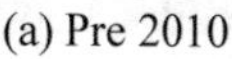
(a) Pre 2010

(b) Post 2010

Fig. 1 Word cloud

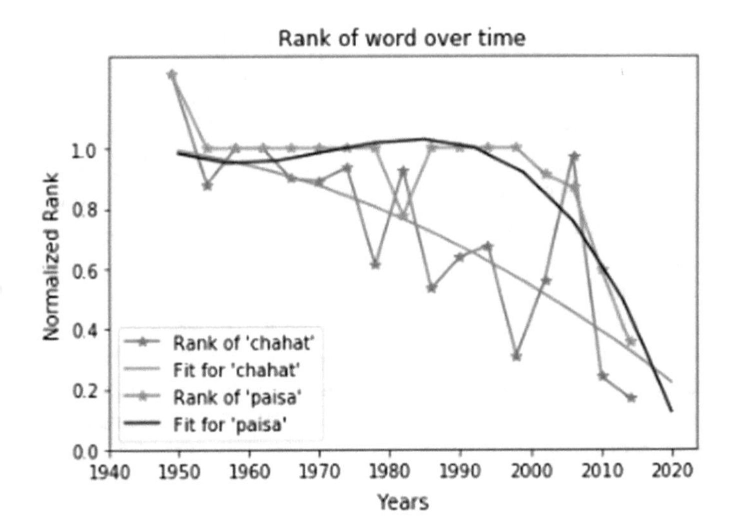

Fig. 2 Rank comparison of chahat and paisa

A trend analysis of various words over the years was also performed. We made a ranking system that ranks words on the basis of their frequency normalized by the length and number of songs over the years. This ranking comparison tool plots the relative ranks of words over the years. Lower rank means a more popular word in the dataset for that year. The results of songs are averaged by a factor of 4 years as seen in the Fig. 2.

This figure compares rank and hence the popularity of words chahat and paisa over the period from 1940 to 2019. For any given year Y, lower rank for a word X represent more frequent use of that word X in an Indian song lyrics in that year Y. We can observe that the two-word ranks are continuously dropping over the years denoting a shifting trend in the songs over the years. A three-degree polynomial curve is fitted to the curve to analyze the trend. We note that while except 2005–2007, our data shows a consistent drop in the rank of words over the years for chahat. The ranks close to 1 imply a rare or no use of the word.

We also compiled a list of mild and extreme swear words in Hinglish. Figure 4a, b shows a comparison between two words in song lyrics based on swear word usage over the period from 1940 to 2020. It is observed that Bollywood songs have steadily gained the swear word usage over time. There was little to no use of swear words before 2010. Generally, very abusive swear words occur extremely rarely in the song lyrics (as seen by the flat line in the curve for those songs in Fig. 4a) Hence such words give a flat line for songs before 2010.

Figure 3 shows a comparison between the top 50 words similar to ladki (hindi for girl) in years before 2010 and after 2010. The word embeddings show an interesting property of capturing the bias in the dataset they are trained upon. Word clouds generated from running a word2vec model on 2 datasets, one on Hindi songs before 2010 (refer Fig. 3a and the other on Hindi songs after 2010 (refer Fig. 3b) we see

(a) Pre 2010 (b) Post 2010

Fig. 3 Top 50 words similar to ladki

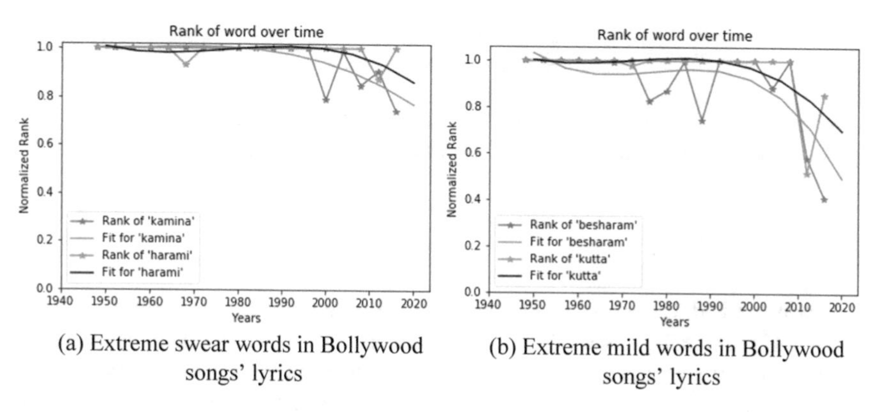

(a) Extreme swear words in Bollywood songs' lyrics (b) Extreme mild words in Bollywood songs' lyrics

Fig. 4 Ranks of swear words over the years

the inherent bias against women in Bollywood songs. Both clouds have words that describe women *as gudiya, nadan, bholi, pagli, gharwaali, hirni*, etc. (Fig. 4).

3.4 Stereotypes and Gender Bias

In this section, we aim to quantify the bias we hypothesized in 3.3 We create eight vector sets of words for soft attitude, strong attitude, color, cars, clothes, food, alcohol, and body looks.[7] We tried two different approaches to understand the bias between males/females and the words in the above vectors.[8]

[7]https://github.com/himanshuladia/bias-analysis/blob/master/attributes.

[8]https://github.com/himanshuladia/bias-analysis/blob/master/names.

Table 1 Word embedding association test results

Word2Vec	fastText	Glove
0.7381	0.8264	0.4458

3.4.1 Association Rule Mining

First, we use a probabilistic approach. This is a modification of association rule used to find interesting associations and patterns in text. For example, we take two sets, one to address females (example: *ladki, girl, chorri, lady*, etc.) and the other for different colors (example: *sanwali, saanwala, pink, red, laal, white*, etc.). To evaluate the association between the sets, two metrics are calculated: support and confidence. We define them as follows:

$$\text{Support} = \frac{(\text{number of documents that include at least one value from set 1 and one value from set 2}) * 100}{(\text{total number of documents})}$$

$$\text{Confidence} = \frac{(\text{number of documents that include at least one value from set 1 and one value from set 2}) * 100}{(\text{number of documents that include at least one value from set 1})}$$

The list of documents were created in three different ways. By taking entire song's lyrics, individual paragraphs, and individual lines as a single document.

3.4.2 WEAT Score

The word embedding association test [9] uses distributed representations of words like word2vec, fastText, and GloVe to evaluate bias between two sets of attribute and target words. It uses the cosine similarity metric to compute the similarity measure between sets of words. We use the WEAT score on our dataset to highlight the consistent presence of bias in Bollywood songs (Table 1).

4 Result and Discussion

Table 2 shows the results for support and confidence values for the discussed three cases. The results for entire paragraph as a single document is more dominant. This is due to the fact that in songs, we talk about an object in a single paragraph. The results are less dominant but still prevalent in rest of the cases. Support and confidence values for the color attribute is very high for females as compared to males. This explains that while talking about females in songs, colors are associated more often than males. They are usually used to define the color of their skin, eyes, lips, dress, etc. For *softAttitude* and *Bodylooks* attributes, we observe that even if support value may be slightly higher for males, the confidence is still higher for females. Which means although the number of documents with male names are more, the number of documents where females and attributes *softAttitude*, *Bodylooks* comes together is

Table 2 Association rule mining results

	Entire lyrics as doc				Each paragraph as doc				Each line as doc			
	Support		Confidence		Support		Confidence		Support		Confidence	
Attribute	F	M	F	M	F	M	F	M	F	M	F	M
Color	3.66	2.94	35.9	20.8	1.65	1.28	30.2	17.7	0.10	0.05	13.60	0.60
Soft	3.27	3.76	32.0	26.7	1.39	1.56	25.4	21.6	0.03	0.01	4.41	1.36
Strong	0.82	1.02	8.09	7.27	0.34	0.46	6.28	6.49	0.01	0.09	0.18	1.06
Cars	1.12	1.05	I 1.0	7.51	0.53	0.43	9.71	6.06	0.00	0.00	0.00	0.00
Clothes	1.18	1.51	11.6	10.7	0.53	0.64	9.71	8.87	0.04	0.05	0.55	0.60
Food	0.66	0.72	6.47	5.16	0.28	0.25	5.14	3.46	0.02	0.00	0.36	0.00
Alcohol	2.18	2.41	21.3	17.1	0.92	0.93	16.8	12.9	0.04	0.02	0.55	0.30
Looks	2.97	3.63	29.1	25.8	1.28	1.50	23.4	20.7	0.05	0.02	0.73	0.30

higher. For *strongAttitude*, the support and confidence values are higher for males. Considering results for support and confidence together for *cars, clothes, food,* and *alcohol* attributes, the values are higher for females as compared to males.

WEAT score is calculated between attribute sets *softAttitude*, *strongAttitude* and target sets *female_names*, *male_names* for distributed representations via word2Vec, fastText and GloVe. The results are shown in Table 1. A positive and close to 1 value indicates the presence of a high bias of *softAttitude* attributes toward *female_names*, and *strongAttribute* toward *male_names*. The results are consistent with that of association rule mining.

5 Conclusion and Future Work

We scraped Hinglish lyrics for Bollywood songs from popular websites and developed a preprocessing technique for reduction of the vocabulary size. Various experiments were performed on the dataset. Extensive qualitative and quantitative analysis in tandem highlight the presence of stereotypes and gender bias in Bollywood songs. We believe our work on bias in Bollywood will spark a change in the mindset of creative people.

In the future, we aim to study this bias on a multi-modal basis. YouTube is a popular source of video entertainment for this generation. We intend to use YouTube video frames, audio and lyrics collectively and develop a neural architecture to analyze the style, popularity and bias present in Bollywood songs.

References

1. Martino SC, Collins RL, Elliott MN, Strachman A, Kanouse DE, Berry SH (2006) Exposure to degrading versus non-degrading music lyrics and sexual behavior among youth. Pediatrics 118(2):e430–e441
2. Barman MP, Awekar A, Kothari S (2019) Decoding the style and bias of song lyrics. In: Proceedings of the 42Nd international ACM SIGIR conference on research and development in information retrieval, SIGIR '19. New York, NY, USA, ACM, pp 1165–1168
3. Kapoor R, Kumar Y, Rajput K, Shah RR, Kumaraguru P, Zimmermann R (2019) Mind your language: abuse and offense detection for code-switched languages. In: Proceedings of the AAAI conference on artificial intelligence, vol 33, pp 9951–9952
4. Potash P, Romanov A, Rumshisky A (2015) GhostWriter: using an LSTM for automatic rap lyric generation. In: Proceedings of the 2015 conference on empirical methods in natural language processing. Lisbon, Portugal, Association for Computational Linguistics, pp 1919–1924
5. Bordia S, Bowman SR (2019) Identifying and reducing gender bias in word-level language models. In: Proceedings of the 2019 conference of the North American chapter of the Association for Computational Linguistics: student research workshop. Minneapolis, Minnesota, Association for Computational Linguistics, pp 7–15
6. Patra BG, Das D, Bandyopadhyay S (2015) Mood classification of Hindi songs based on lyrics. In: Proceedings of the 12th international conference on natural language processing. Trivandrum, India, NLP Association of India, pp 261–267

7. Patra BG, Das D, Bandyopadhyay S (2017) Retrieving similar lyrics for music recommendation system. In: Proceedings of the 14th international conference on natural language processing (ICON-2017). Kolkata, India, NLP Association of India, pp 290–297
8. Madaan N, Mehta S, Agrawaal T, Malhotra V, Aggarwal A, Gupta Y, Saxena M (2018) Analyze, detect and remove gender stereotyping from Bollywood movies. In: Proceedings of the 1st conference on fairness, accountability and transparency, volume 81 of Proceedings of machine learning research. New York, NY, USA, PMLR, pp 92–105
9. May C, Wang A, Bordia S, Bowman SR, Rudinger R (2019) On measuring social biases in sentence encoders. In: Proceedings of the 2019 conference of the North American chapter of the Association for Computational Linguistics: human language technologies, vol 1 (Long and Short Papers). Minneapolis, Minnesota, Association for Computational Linguistics, pp 622–628
10. Mikolov T, Chen K, Corrado G, Dean J (2013) Efficient estimation of word representations in vector space. arXiv:1301.3781
11. Joulin A, Grave E, Bojanowski P, Mikolov T (2016) Bag of tricks for efficient text classification. arXiv:1607.01759
12. Pennington J, Socher R, Manning C (2014) Glove: global vectors for word representation. In: Proceedings of the 2014 conference on empirical methods in natural language processing (EMNLP). Doha, Qatar, Association for Computational Linguistics, pp 1532–1543
13. Velmurugan T (2013) A survey of association rule mining in text applications

Anomaly Detection in Crowded Scenes Using Motion Influence Map and Convolutional Autoencoder

Shilpi Agrawal and Ratnakar Dash

Abstract In this paper, we present a method to detect and localize unusual activity in crowded scenes. A large number of surveillance cameras are fixed at various places for security purposes. We propose an autoencoder-based deep learning framework to categorize abnormality. Optical flow is computed using motion influence map and fed to the convolutional autoencoder. Thus, the spatio-temporal features obtained from the output of the encoder are used for classification. K-means clustering has been utilized to classify the spatio-temporal features. Experiments were conducted on standard crowd datasets and it is observed that the proposed model achieves comparable accuracy measure with state-of-the-art techniques.

1 Introduction

CCTV cameras are used at various places like banks, shopping malls, ATM, streets, etc., to ensure public safety. A large amount of data are produced every day by these cameras. The amount of data containing suspicious behavior is very less. So, manual analysis of this huge data is not possible.

Studying video data is an important research issue nowadays. In video monitoring the main role is to track anomalous incidents such as collisions, suspicious behavior, and burglary. The goal is to identify the point of anomaly that diverges from normal patterns. The anomalous events are quite distinct. A person walking in a busy road may be considered as anomalous while the same person walking in a mall may be considered as normal. Therefore, the algorithm must not depend on prior information about these events.

S. Agrawal (✉) · R. Dash
Department of Computer Science and Engineering, NIT Rourkela, Rourkela, India
e-mail: 218CS2310@nitrkl.ac.in

R. Dash
e-mail: ratnakar@nitrkl.ac.in

N. Chaki et al. (eds.), *Proceedings of International Conference on Computational Intelligence and Data Engineering*, Lecture Notes on Data Engineering and Communications Technologies 56, https://doi.org/10.1007/978-981-15-8767-2_14

However, detection methods face a lot of hurdles in detecting objects in a vivid environment.

- **Illumination Changes**: It changes the appearance of the scene and causes divergence because of which chances of false alarm rates increases.
- **Camouflage**: It becomes difficult to distinguish when the object and the background are very similar to each other.
- **Uninteresting moving objects**: Every moving object such as a flowing river or moving leaves may not be an anomaly.
- **Shadows**: Objects may cast shadows because of changes in illumination which might also be classified as an anomaly.
- **Occlusion**: It becomes difficult for tracking algorithm to detect the exact position of an object when its view is blocked by another object.

This paper presents a model which extracts spatial features from Convolutional Autoencoder and temporal features from motion information algorithm at both pixel level and block level. K-means clustering has been used for classification which is based on the distance between the centre of the cluster and extracted spatio-temporal features.

The document is set out as follows. In Sect. 2, the recent development and research are outlined. The proposed method is discussed in Sect. 3. Section 4 describes the experimental setup and presents the result. The final section gives concluding remarks on some future research areas.

2 Related Work

In this area of study, a well-liked approach is to imbibe the usual video pattern and train the algorithm with this same pattern. Any deviation from this pattern would be considered anomalous.

A new structure was proposed by Xiang and Gong [1] using space and position context for anomaly detection. The behavior of object is presented using nuclear event, which contained swiftness, direction, and location of the object. An SF model was used by the Mehran et al. [2] for crowd behaviors [3] description. It did not involve any tracking methods. The disparity between the wanted and original speed we get from the interaction of the particle on the field of optical flow [4, 5] estimates the force of interaction. Mahadevan et al. [6] modeled information about appearance of normal function in crowded scenes with a blend of potent textures. It supposed both the features of space and time to detect anomalous activity in a congested scene. In video analysis and anomaly detection, trajectories have always been popular [7–10]. Trajectory based analysis is well known for the characteristic of deviation of nominal classes in the training period. And then in the testing phase against nominal classes compares the new test trajectories. An anomaly is indicated when it deviates from all classes. In the field of action recognition, there are many successful cases [11–14]. However, these methods work well with videos having precise labels without the involvement of blocked scenes. Pre-recorded video is unable to detect the unusual event whose occurrence is rare. Researchers have trained models using less or no supervision, including autoencoder [15], spatio-temporal features [16, 17], and dictionary learning [18]. Only unlabelled video data is required by them, which contains the event which is very unusual and also easily gets into the real-world application. Sultani et al. [19] utilized both high-quality and low-quality videos to learn the anomalies. The input names are at video level. Usual and unusual videos as bags and the sections of videos as occurrences in multiple instances learning (MIL) are considered by them in this method. Then their investigation of a model starts automatically which foretells a large score of irregularity for the fragments of the unusual video. Because of the usage of simply machine learning method, the classification precision is very less and needs improvement by using classifier and optimization techniques. The extraction of spatio-temporal energy measurements was performed by Huang et al. [20] having visual options whose level is low, motion map options, and energy options area unit. To illustrate the features of the model in the traditional patterns, three layers of convolution are trained. For the good representation of the crowd models, the theme of fusion is applied and for sensing the events of anomaly they applied the SVM model.

All the methods mostly focus either only on local activity or only on global activity. It is required to consider both of them with a single model. We propose one such model in which Convolutional Autoencoder learns the spatial features, and motion influence algorithm (Algorithm 1) learns the temporal features for both global and local unusual activities.

3 Proposed Method

The frame level unusual activity localization and detection of the pattern of the motion were studied by the proposed method and this paper considers both global and local activities. Local unusual activity refers to a non-human object occurring in a frame or a cart moving in pedestrian walking area. Global unusual activity means everyone in the frame starts to rush abruptly to getaway from the scene. Figure 1 shows the overview of the proposed method. Information of motion is evaluated first. Then, a motion influence map is created following which features are extracted from it. A feature vector is created and finally classification is done to detect anomaly.

For the extraction of spatio-temporal features the proposed scheme uses optical flow between different video frames. Convolutional Autoencoder places the output units of the encoder in less amount than the input to reduce the dimensionality during the extraction of the feature. Feature extraction from the input image is the main purpose of convolution and also a spatial relationship was preserved among the pixels. To extract the temporal features, motion influence map is computed from the motion values in each frame from information at block level. The motion influence map is divided into a consistent framework and for classification we used K-means clustering. The distance between each cluster and the computed spatio-temporal

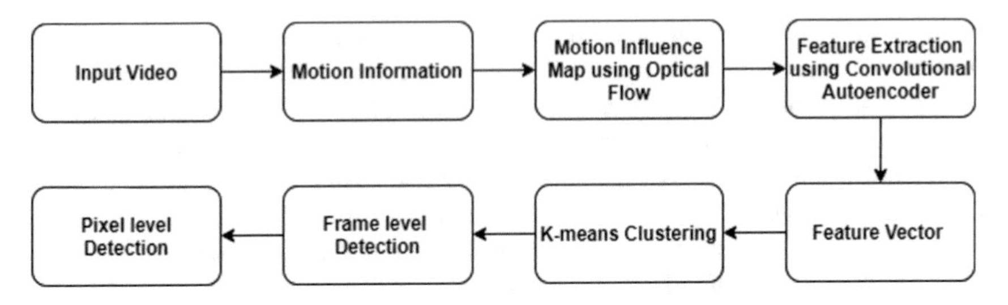

Fig. 1 Overview of proposed method

feature is used to classify whether the frame is unusual or not. Details of our proposed method is explained below. Our proposed method includes following steps:

- Motion descriptor calculation.
- Feature extraction.
- Unusual event detection and localization.

3.1 Motion Descriptor

The input RGB frames are converted to grayscale and applied to a motion descriptor. After preprocessing, FarneBack algorithm is used to calculate the optical flow for each pixel in the frame. Optical flow is the motion of objects between continuous frames of sequence. The frame is partitioned into uniform blocks of size M by N and then calculate optical flow for every block by averaging the optical flow of each pixel within the block as shown in Eq. 1.

$$b_m = \frac{1}{N} \sum f_m^n \tag{1}$$

where ith block optical flow was denoted by b_m, pixels number in a block was denoted by N and nth pixel of the optical flow of the mth block was denoted by f_m^n. The various factors which influence pedestrian motion are moving cart, other pedestrians, or an obstacle on the road. We assume that the two factors that affect the pedestrian are (1) Speed and (2) Direction.

Algorithm 1 is used to construct the motion influence map.

3.2 Feature Extraction

Now the step is to measure vector of motion influence of block j by considering all the blocks i affecting the motion of the block as Eq. 2,

Algorithm 1: Algorithm for Motion Influence Map

INPUT: M ← motion vector set, T ← block size, F ← a set of blocks in a frame
OUTPUT: MIP ← motion influence map
for $m \in K$ **do**
  $T_d = ||b_m|| \times T$
  $\frac{F_m}{2} = \angle bm + \frac{\pi}{2}$
  $\frac{-F_m}{2} = \angle bm - \frac{\pi}{2}$
  for $n \in F$ **do**
    if $m \neq n$ **then**
      Calculate the Euclidean distance $E(m, n)$ between b_m and b_n
      if $E(m, n) < T_d$ **then**
        Calculate the angle ϕ_{mn} between b_m and b_n
        if $\frac{-F_m}{2} < \phi_{mn} < \frac{F_m}{2}$ **then**
          $MIP^j(\angle b_m) = MIP^j(\angle b_m + exp(\frac{D(m,n)}{||b_m||})$
        end if
      end if
    end if
  end for
end for

$$H^j(i) = \sum w_{ij} \tag{2}$$

where $H^j(i)$ is motion vector of block j and the influence of block i on block j was indicated by w_{ij}. By using Convolutional Autoencoder, at each frame there is the addition of the motion vectors of the block for the extraction of the spatial features (Fig. 2) and concatenate motion vectors to get $8 \times t$-dimensional feature vector, where t is the count of frames. Frames are apportioned into blocks which are non-overlapping, every block of which is a blend by the influence of the value of motion. The motion influence value of a Megablock is the sum of motion influence values of all the tinier blocks constituting a larger block.

K-means clustering is performed for each mega block using spatial and temporal features. The centres are then set as codewords. For training, only normal video clip is used. So, codewords consist of patterns of normal activities only.

3.3 *Unusual Event Detection and Localization*

After extracting spatio-temporal features, in the testing phase, a minimum distance matrix is constructed for the mega blocks. Minimum Euclidean distance is used to define the value of an element between the mega block codewords and the current test frame feature vector as,

$$E(m, n) = \min_k ||f^{(m,n)} - w_k^{(m,n)}||^2 \tag{3}$$

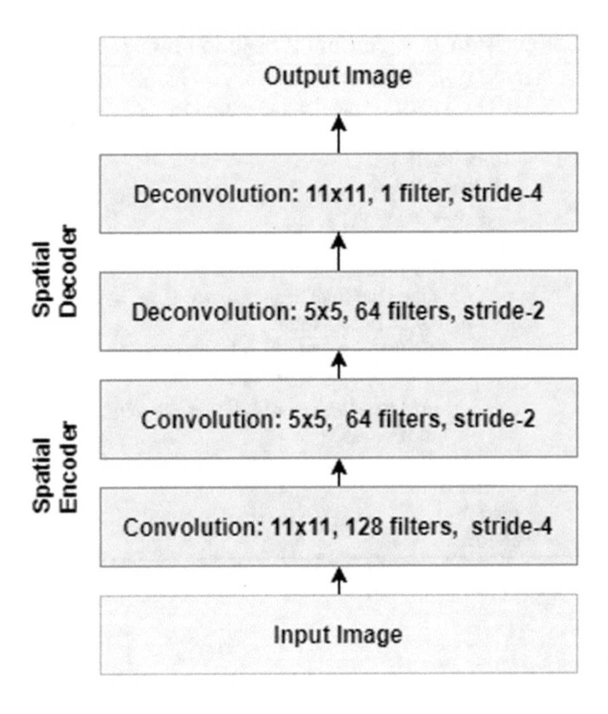

Fig. 2 Convolutional autoencoder

where $E(m, n)$ denotes (m, n)th element in E and $f^{(m,n)}$ denotes feature vector of (m, n)th mega block of the test frame.

If the value of Euclidean distance is small, then it is likely that it is not an unusual activity in the block. If the distance is large, it is likely that it is an unusual block and hence the unusual frame. If the value is higher than the threshold, then the classification of the frame by us as abnormal (frame level). The point of anomaly is also detected with the same approach and the same threshold value is used for every megablock to confine the activity which is unusual (pixel level).

4 Result and Analysis

Checking the efficiency of the approach proposed, we experiment on two publicly available dataset, UCSD [6] and UMN dataset, which contain local and global activities. We did frame and pixel level identification of local irregular events in the UCSD dataset and in UMN dataset, we detected only frame level global abnormal activities. To validate the proposed model, experiments were conducted using Python 3.7. We used AUC system to perform the study and equate it with state-of-the-art methodologies (Fig. 3).

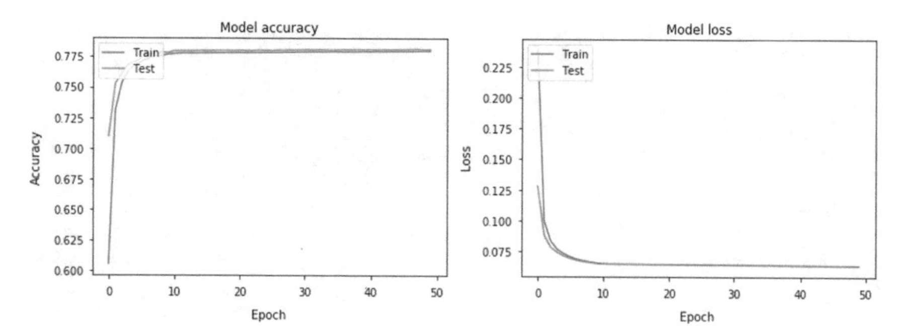

Fig. 3 Model accuracy and model loss

Fig. 4 Normal activity of UMN dataset

4.1 Dataset Description

There are 11 packed scenario video samples from three separate outdoor and indoor scenes in the UMN dataset. It contains 7740 frames, each 320 × 240 in dimension. In the video, people are seen walking around. It is considered a normal activity. It is considered a normal activity as shown in Figs. 4 and 5. It is called an unnatural behavior, when they make a sudden escape running action.

In the UCSD Dataset, there are two sets of videos recorded from different camera angle: (1) Ped1 and (2) Ped2. People walking in the pathway are considered as normal behavior. Ped1 provides 34 training videos and 36 testing videos. The size of the frame is 238 × 158. Ped2 has 16 videos for training and 12 videos for analysis. The dimensions of the frame are 360 × 240. The training video consists of the normal behavior of people walking around. In the test clips, there are some unusual activities like moving cart, bicycle, skating, etc.

4.2 Performance Analysis

The frame is partitioned into 8 × 8 blocks and threshold is set to maximum value of motion influence map of input images. In Table 1, it is observed that the model accuracy for different datasets was comparable to other models. Figure 3 shows model

Fig. 5 Normal activity of UCSD Ped1 and Ped2 dataset

Table 1 Accuracy report and comparison with other models

	AUC (%) in various datasets		
Method	UCSD Ped1 (%)	UCSD Ped2 (%)	UMN dataset (%)
Spatio-temporal autoencoder [21]	89.9	87.4	80.3
Motion influence map [22]	64.9	81.5	90.9
Proposed method	85.5	83.4	94.4

Fig. 6 Abnormal activity of UMN dataset

Fig. 7 Abnormal activity of UCSD Ped1 and Ped2 dataset

accuracy and model loss. While training the model, it was overfitting but after some time it performed well. Our model accuracy is less than spatio-temporal model [21], but the time taken to learn the model reduced drastically. However, it is more than the Motion Influence Map model [22]. The accuracy for Ped1 Dataset is 85.5%. For Ped2 dataset, it is 83.4% and for UMN dataset, it is 94.4%. Figures 6 and 7 display the results.

5 Conclusion

In this paper, we concentrated on irregular activity identification at both local and global rates. To detect and locate unusual activity in a crowded scene, we propose a deep learning system. The frames are graded as natural or anomalous based on the motion influence map and Convolutional Autoencoder for both time and space, respectively. The proposed model achieves better accuracy measure in standard datasets as compared to the state-of-the-art techniques.

The limitation of this paper is that it gives false results in case of occluded scenes. Also, the analysis is limited to a fixed viewpoint. We have not considered zoom or tilt functionality of the camera. It deals with only static camera datasets. By extending the proposed method, we can deal with these functionalities as future research work.

References

1. Xiang T, Gong S (2008) Video behavior profiling for anomaly detection. IEEE Trans Pattern Anal Mach Intell 30(5):893–908
2. Mehran R, Oyama A, Shah M (2009) Abnormal crowd behavior detection using social force model. In: 2009 IEEE conference on computer vision and pattern recognition. IEEE, pp 935–942
3. Helbing D, Molnar P (1995) Social force model for pedestrian dynamics. Phys Rev E 51(5):4282
4. Wu S, Moore BE, Shah M (2010) Chaotic invariants of Lagrangian particle trajectories for anomaly detection in crowded scenes. In: 2010 IEEE computer society conference on computer vision and pattern recognition. IEEE, pp 2054–2060
5. Ali S, Shah M (2007) A Lagrangian particle dynamics approach for crowd flow segmentation and stability analysis. In: 2007 IEEE conference on computer vision and pattern recognition. IEEE, pp 1–6
6. Mahadevan V, Li W, Bhalodia V, Vasconcelos N (2010) Anomaly detection in crowded scenes. In: 2010 IEEE computer society conference on computer vision and pattern recognition. IEEE, pp 1975–1981
7. Zhou S, Shen W, Zeng D, Zhang Z (2015) Unusual event detection in crowded scenes by trajectory analysis. In: 2015 IEEE international conference on acoustics, speech and signal processing (ICASSP). IEEE, pp 1300–1304
8. Li C, Han Z, Ye Q, Jiao J (2011) Abnormal behavior detection via sparse reconstruction analysis of trajectory. In: 2011 sixth international conference on image and graphics. IEEE, pp 807–810
9. Piciarelli C, Micheloni C, Foresti GL (2008) Trajectory-based anomalous event detection. IEEE Trans Circuits Syst Video Technol 18(11):1544–1554

10. Mo X, Monga V, Bala R, Fan Z (2013) Adaptive sparse representations for video anomaly detection. IEEE Trans Circuits Syst Video Technol 24(4):631–645
11. Tran D, Bourdev L, Fergus R, Torresani L, Paluri M (2015) Learning spatiotemporal features with 3d convolutional networks. In: Proceedings of the IEEE international conference on computer vision, pp 4489–4497
12. Ji S, Xu W, Yang M, Yu K (2012) 3D convolutional neural networks for human action recognition. IEEE Trans Pattern Anal Mach Intell 35(1):221–231
13. Karpathy A, Toderici G, Shetty S, Leung T, Sukthankar R, Fei-Fei L (2014) Large-scale video classification with convolutional neural networks. In: Proceedings of the IEEE conference on computer vision and pattern recognition, pp 1725–1732
14. Oneata D, Verbeek J, Schmid C (2013) Action and event recognition with fisher vectors on a compact feature set. In: Proceedings of the IEEE international conference on computer vision, pp 1817–1824
15. Sabokrou M, Fathy M, Hoseini M, Klette R (2015) Real-time anomaly detection and localization in crowded scenes. In: Proceedings of the IEEE conference on computer vision and pattern recognition workshops, pp 56–62
16. Lu C, Shi J, Jia J (2013) Abnormal event detection at 150 FPS in MATLAB. In: Proceedings of the IEEE international conference on computer vision, pp 2720–2727
17. Zhao B, Fei-Fei L, Xing EP (2011) Online detection of unusual events in videos via dynamic sparse coding. In: Proceedings of CVPR. IEEE, pp 3313–3320
18. Yen S-H, Wang C-H (2013) Abnormal event detection using HOSF. In: 2013 international conference on IT convergence and security (ICITCS). IEEE, pp 1–4
19. Sultani W, Chen C, Shah M (2018) Real-world anomaly detection in surveillance videos. In: Proceedings of the IEEE conference on computer vision and pattern recognition, pp 6479–6488
20. Huang S, Huang D, Zhou X (2018) Learning multimodal deep representations for crowd anomaly event detection. Math Probl Eng
21. Chong YS, Tay YH (2017) Abnormal event detection in videos using spatiotemporal autoencoder. In: International symposium on neural networks. Springer, pp 189–196
22. Lee D-G, Suk H-I, Park S-K, Lee S-W (2015) Motion influence map for unusual human activity detection and localization in crowded scenes. IEEE Trans Circuits Syst Video Technol 25(10):1612–1623

Learning-Based Image Registration Approach Using View Synthesis Framework (LIRVS)

B. Sirisha, B. Sandhya, J. Prasanna Kumar, and Chandra Sekhar Paidimarry

Abstract Image registration is a fundamental preprocessing step in varied image processing and computer vision applications, which rely on accurate spatial transformation between source and the reference image. Registration using iterative view synthesis algorithm is experimentally shown to solve a wide range of registration problems, albeit at the cost of additional memory and time to be spent on the generation of views and feature extraction across all the views. Hence, we have approached the problem by building a decision-maker model which could predetermine the possibility of registering the given input image pairs and also predict the iteration at which the image pair will be registered. The proposed approach incorporates a decision-maker (trained classifier model) into the registration pipeline. In order to ensure that the gain in time is considerable, the classifier model is designed using the registration parameters obtained from reference and source image. The trained classifier model can predetermine the possibility of registering the input image pairs and also the minimum number of synthetic views or iteration necessary to register the input image pair. Hence, for the images that have been registered, an additional time required for the proposed approach is tolerable. However, for the images that are not registered, the gain in time because of classifier model is extremely significant.

Keywords Image registration · View synthesis · Feature detection · Feature description · Classifier.

B. Sirisha (✉) · B. Sandhya · J. Prasanna Kumar
MVSR Engineering College, Hyderabad, India
e-mail: sirishavamsi@gmail.com

B. Sandhya
e-mail: sandhyab16@gmail.com

J. Prasanna Kumar
e-mail: pkumar62@gmail.com

C. S. Paidimarry
UCE, Osmania University, Hyderabad, India
e-mail: sekharpaidimarry@gmail.com

N. Chaki et al. (eds.), *Proceedings of International Conference on Computational Intelligence and Data Engineering*, Lecture Notes on Data Engineering and Communications Technologies 56, https://doi.org/10.1007/978-981-15-8767-2_15

1 Introduction

Image registration is concerned with the computation of a suitable transformation function [1, 2] between fully or partially overlapped images which vary due to different characteristics acquired from sensor, incidence angle or view angle, resolution, etc. Feature-based image registration can be achieved using standard and iterative approaches [3]. The standard approach for the registration includes the extraction of the local feature points, generation of matches, and their geometric validation with the computed homography (FIR). Standard registration approach is mainly suitable for simple registration problems, where the deformation complexity between the input image pair is very minimal [4, 5]. Iterative registration approach using view synthesis is widely implemented for registering complex registration problems, where the deformation complexity between the input image pair is extremely high. The concept of creating synthetic image views to enhance feature matching was originally explored by Lepetit and Fua [6]. Morel et al. [7] have integrated the view synthesis with DOG feature detector and SIFT matching. This approach is called Affine SIFT, effectively matches the challenging optical image pairs with orientation differences up to 80 degrees. Pang et al. [8] have substituted SIFT feature extraction [9] by SURF feature extraction [10] in the Affine SIFT approach to decrease the computational time. The resulting feature matching approach is called FAIR SURF (FSURF). Mishkin et al. [11, 12, 15] have proposed two-view feature matching approach, which integrates the view synthesis with Hessian Affine [13] and MSER [16] feature detectors and employs the Root SIFT [14, 17] feature matching. It is observed that the view synthesis can address a large range of deformations between the images effectively.

In this approach, synthetic view images are gradually increased as the iterations progress until the images are registered. The number of iterations required to register the image pair relies on the kind and amount of deformation between the input image pair. However, the time taken to register images increases with the increase in the iterations. Hence, the focus of the paper is to identify the minimum number of views needed or iteration required to register the input image pairs. In such a scenario, iterative approach is converted into a serial execution. However, the challenge lies in accurately predicting the optimal number of views because with less number of views images may not register and if more number of views are generated the gain in processing speed and storage overhead is nullified. The proposed algorithm incorporates a Decision -Maker (DM) (trained classifier model) into the pipeline of standard registration approach. The contribution made in this respect is twofold: [1.] Incorporation of knowledge into the feature-based registration algorithm by building a DM model that can be used for real-time registration of input image pairs. [2.] An effort has been made to identify the feature attributes to train and build a classifier model to predict the synthetic views required for registering the images. This paper is organized as follows: Sect. 2 describes the iterative registration algorithm using view synthesis and its technical challenges, Sects. 3 and 4 present the proposed approach, evaluation parameters, experimental results, and finally, the conclusion is provided in Sect. 5.

2 Iterative Registration Approach Using View Synthesis (IRVS)

IRVS generates the synthetic views of source and reference images and then applies the standard feature-based registration pipeline(FIR) on a set of images until the registration error is less than the threshold. In each iteration, the synthetic views of complementary angles are generated. IRVS has four major steps, in step:1—synthetic views of source and reference images are generated using Affine camera model to simulate various acquisition scenarios like view angle and scale. These synthesized views are used to find the feature points that are detected frequently under varied affine deformations. We have based the generation of synthetic views on affine camera model which uses three major parameters, i.e., latitude (θ), longitude (Φ), and scale (Ψ). In Step:2, feature points are detected from the synthesized source and the reference image views using SIFT extractor. In Step:3, the correspondence between the features detected in the source image and those detected in the reference image is established by the nearest neighbor ratio(NNR) matching technique using Bhattacharya distance. In Step:4, the true matches from the correspondences are generated using RANSAC algorithm. This algorithm eliminates the outliers and gives inliers, homography matrix as output. The source image is then transformed using the computed homography matrix. Registration error is computed between the transformed and reference images. If the error is above a predefined threshold, the process is iterated with the increasing the number of views.

Image registration using **Standard approach (FIR) and iterative registration using view synthesis (IRVS) approach is implemented on standard UK Bench dataset containing 7464 image pairs.** It can be observed that the **FIR approach could register only 3534 image pairs out of 7464 image pairs.** On the other hand, the **IRVS approach could register 5202 image pairs out of 7464 image pairs.** Hence, this method has been proved in finding more precise and correct feature correspondences over a dataset of 7464 input image pairs. Table 1 shows the total number of feature points detected and registration error obtained on a pair of images for iteration-1–6. The number of feature points detected increases as the number

Table 1 Table shows the total number of feature points detected and registration error obtained on a pair of images and Time (s) for iteration-1–6 I_1 to I_6

Detected feature points	FIR	Iteration-1	Iteration-2	Iteration-3	Iteration-4	Iteration-5	Iteration-6
Feature points	781	3047	3987	4896	5350	8879	18161
Registration error	FIR	Iteration-1	Iteration-2	Iteration-3	Iteration-4	Iteration-5	Iteration-6
RE	19.25	14.348	10.890	9.009	8.241	7.962	7.729
Time (s)		174.945	364.269	590.144	780.403	952.308	1068.039

of iterations (I1–I6) increases. Hence, by combining the feature points of all the synthetic views, we are improving the performance of feature detection, which further helps in feature matching and transformation estimation. From Table 1, important observation can be made, as the iteration increases from I1 to I6, the registration error decreases, which directly reflects the accuracy of registration.

3 Learning-Based Image Registration Approach Using View Synthesis Framework (LIRVS)

It is observed from the Sect. 2 that the view synthesis approach can address a large range of deformations between the images effectively. However, the time taken to register images increases with the increase in the iterations. Hence, we adapt the iterative view synthesis approach by employing DM, which could predetermine the possibility of registering the input image pairs before going through the iterations of registration and also predict the exact iteration required to register the input image pairs. The proposed LIRVS approach is decomposed into three stages, **Stage-1: Local Feature Extraction, Matching, and Transformation Estimation**, **Stage-2: Decision Maker Model (i.e., Trained Classifier Model)**, **Stage-3: Implementation of Registration using predicted views**. The figure shows the pictorial illustration of the proposed approach (Fig. 1).

Stage-1: Feature Extraction, Matching, and Transformation Stage:1 composed of feature extraction, matching, and transformation estimation. The feature points are detected from the source and the reference image using SIFT detector, where these detected features remain unaffected by scale and variations in view-

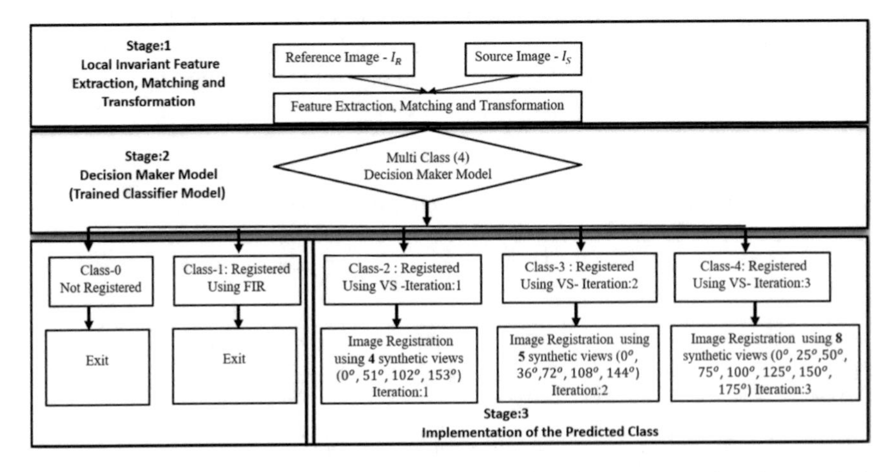

Fig. 1 Framework of proposed learning-based image registration approach using view synthesis framework (LIRVS

point. After feature detection is done, these features are described by using SIFT descriptor. The correspondence between the features detected in the source image and those detected in the reference image is established. SIFT descriptor are matched using two-way nearestNNR matching technique using Bhattacharya distance. If the ratio of the neighboring distances is greater than the threshold, then they are considered to be as the NNR matches. To increase the robustness of matches, matching is performed twice from source feature points to reference feature points and vice versa. A correspondence is considered only if it exists in both the scenarios. The true matches from the correspondences are generated using RANSAC algorithm [18]. This algorithm eliminates the outliers and gives inliers, homography matrix as output. The source image is then transformed using the computed homography matrix. Five vital parameters related to feature extraction, matching, and registration like

- Repeatability: Ratio of the number of matched points to the total number of key-points extracted from both the images
- Number of True matches
- Inlier Ratio(IR): Ratio of number of Inliers to the total number of matches.
- Registration Error(RE): Bhattacharya distance between matched points of source Image which are transformed with computed homography and target image matched points.
- Warp image error(WE): Distance between the transformed source image and the target image.

Stage-2: Decision-Maker Model The view synthesis approach can be improved by developing a DM (Trained Classifier Model) that has to decide whether a given pair of input images can be registered or not. If registered, it should predetermine the iteration required to register the input image pairs. DM connects Stage:1 and Stage:3. This stage includes the following activities: 1. Attributes Generation for the DM, 2. Attribute Selection, 3. Training Data Generation. **Attributes Generation for the Decision Maker** Hence, the output of Stage:1 is considered for attribute generation of DM. In order to ensure that the gain in time is considerable, the decision-making model is designed using the registration parameters obtained from reference and source images. Hence, an effort has been made to use image registration parameters as attributes for classifier. We have done experiments by building a classifier model using the registration parameters like repeatability score, number of matches, inlier ratio (IR), warp error, and registration error. The feature attributes are selected using information gain ratio; out of five attributes, inlier ratio (IR) and registration error are seleted as the key attributes for the DM Model.

Stage-3: Implementation of Registration using Predicted Views The output of DM is one of the five classes. Based on the output of the DM the registration approach exits successfully in the case of class:0, class:1, or continues the registration using view synthesis in all other classes.

4 Experimental Results

The following section describes the effectiveness of the Dm Model and time analysis of the proposed approach.

Effectiveness of Decision-Maker Model: In total, 7464 image pairs from UK-Bench dataset are used for generating the training data. Support vector machine classifier is trained on 7464 training records with five classes. Class distribution is divided as [class-0 = 2264, class-1 = 2534, class-2 = 1338, class-3 = 653, and class-4 = 675]. The performance of the support vector machine model is verified using tenfold cross-validation. The DM (trained SVM Classifier model's) effectiveness is assessed with help of accuracy measure. Table 5 shows the true positive rate, false positive rate, precision, recall, and ROC area of each class. It can be observed that ROC area is uniform across the classes. The DM model accuracy obtained with tenfold cross validation is about 95.682% (Table 2).

Time analysis of proposed approach: Time of proposed approach is compared with the standard and iterative image registration approaches. In order to obtain significant assessment, we have categorized the registration problem based on the geometric deformation between the image pairs as: 1. Simple registration problem—Image pairs which got registered in the standard approach. 2. Hard registration problem—Image pairs which got registered in the first iteration of view synthesis approach. 3. Very Hard registration problem—Image pairs which got registered in the second iteration of view synthesis approach. 4. Extremely hard registration problem—Image pairs which got registered in the third iteration of view synthesis approach. Figure 2 show the results of registering two extremely hard image pairs using iterative view synthesis (IRVS) and proposed (LIRVS) approach. As can be observed, in iterative view synthesis approach image pairs fails to register in first, second iterations and could be registered in the third iteration, whereas in proposed LIRVS approach, image pair could be registered in the third iteration directly because of the prediction from DM. Table 3 shows the time taken for two registration approaches for the varied registration problems. It can be observed for hard registration problem, the time required for registering is same for both the approaches, whereas for very hard and extremely hard registration problems, the proposed approach saves 79.303 s and 269.643 s, respectively. It is seen that for images which cannot be registered (class = 0) in the proposed approach, time saved is 1225.245 s as the algorithm quite without going through the iterations of view synthesis.

Table 2 Results for DM (7464 instances) Using SVM tenfold cross-validation

Classifier	Class	TP rate	FP rate	Precision	Recall	ROC area	Accuracy
SVM	−1	1	0	0.999	1	1	95.682%
	0	0.999	0.001	0.999	0.999	0.999	
	1	0.974	0.044	0.828	0.974	0.965	
	2	0.517	0.015	0.814	0.517	0.867	
	3	0.511	0.010	0.801	0.491	0.823	

Fig. 2 Proposed learning based image registration approach using view synthesis framework (LIRVS)

Table 3 Time taken for image registration in two registration approaches

	IRVS registration approach		LIRVS registration approach	
Sensed image	Iteration of registration	Time (s)	Predicted class	Time (s)
Hard	1	174	2	174
Very hard	2	364.53	3	285.227
Extremely hard	3	692.608	4	422.965
Cannot register	6	1323.374	0	98.25

5 Conclusion

In the proposed approach, we have adapted the view synthesis by incorporating a DM model which helps to predetermine the possibility of registering two images and also predict the iteration at which the image pair will be registered, without actually registering them. It is observed that the average time taken by the DM is significant compared to the time taken by any other process of image registration such as feature extraction, feature matching, and transformation estimation. Hence, for the images that have been registered, an additional time required for the proposed approach is tolerable. However, for the images that are not registered, the gain in time because of DM is extremely insignificant. The proposed approach is meant to improve the execution performance (time and space), unlike the iterative view synthesis approach. The proposed approach found to be efficient in addressing the extreme geometric deformations, which are not registrable with previous state of the art. The main drawback of iterative view synthesis approach is the algorithm is fast for simple registration problems and consumes space and time for extremely hard registration problems because The generation of synthetic views is done until a valid geometric estimate is achieved. For image pairs which cannot be registered, the iterative approach has to iterate through all the tilts and then conclude that the given image pair could not be registered. This setback is overcome by the proposed algorithm; in our approach, the algorithm exits dynamically when a valid geometric estimate is not attained.

References

1. Brown LG (1992) A survey of image registration techniques. ACM Comput Surv 24:325–376
2. Fonseca LMG, Manjunath BS (1996) Registration techniques for multisensor remotely sensed images. Photogramm Eng Remote Sens 1049–1056
3. Zitová B, Flusser J (2003) Image registration methods: a survey. Image Vis Comput 21(11):977–1000
4. Capel D, Zisserman A (2003) Computer vision applied to super resolution. IEEE Signal Process Mag 20(3):75-86. https://doi.org/10.1109/MSP.2003.1203211.
5. Eastman RD, Netanyahu NS, LeMoigne J (2011) Survey of image registration methods in image registration for remote sensing. Image registration for remote sensing
6. Lepetit V, Fua P (2006) Keypoint recognition using randomized trees. IEEE Trans Pattern Anal Mach Intell 28(9):1465–1479. https://doi.org/10.1109/TPAMI.2006.188
7. Morel J-M, Guoshen Y (2009) ASIFT: a new framework for fully affine invariant image comparison. SIAM J Imaging Sci 2(2):438–469
8. Pang Y, Li W, Yuan Y, Pan J (2012) Fully affine invariant surf for image matching. Neurocomputing 85:610. https://doi.org/10.1016/j.neucom.2011.12.006
9. Lowe DG (2004) Distinctive image features from scale-invariant keypoints. Int J Comput Vis 60:91–110
10. Bay H, Ess A, Tuytelaars T, Gool LV (2008) Speeded-up robust features (SURF). Comput Vis Image Underst 110:346–359
11. Mishkin D, Perdoch M, Matas J (2013) Two-view matching with view synthesis revisited. IVCNZ 2013:436–441
12. Mishkin D, Matas J, Perdoch M (2015) Mods: fast and robust method for two-view matching. CoRR abs/1503.02619
13. Mikolajczyk K, Schmid C (2004) Scale and affine invariant interest point detectors. Int J Comput Vis 60(1):63–86
14. Arandjelovic R, Zisserman A (2012) Three things everyone should know to improve object retrieval. In: Proceedings of CVPR
15. Mishkin D, Matas J, Perdoch M, Lenc K (2015) WxBS: wide baseline stereo generalizations. In: Proceedings of the British machine vision conference, BMVA
16. Matas J, Chum O, Urban M, Pajdla T (2002) Robust wide baseline stereo from maximally stable extrema regions. In: BMVC, pp 384–393
17. Kelman A, Sofka M, Stewart CV (2007) Keypoint descriptors for matching across multiple image modalities and non-linear intensity variations. In: Proceedings of CVPR
18. Lebeda K, Matas J, Chum O (2012) Fixing the locally optimized RANSAC. In: Proceedings of BMVC

A Walk Through Various Paradigms for Fake News Detection on Social Media

T. V. Divya and Barnali Gupta Banik

Abstract Around the globe, social media is serving as a significant source of news for millions of people because of its rapid dissemination, easy access and low cost. However, it has a significant risk in exposing fake news, which may mislead the readers, and it comes at the cost of dubious trustworthiness. Existing content-based analysis techniques are challenged by automatic detection of fake news. On social media, merits and demerits of different techniques of fake detection are studied in review work. For fake news detection, various techniques have been proposed in recent days. For given news, the precise statistical rating is not produced by existing works. Less variance is made by news category and input restrictions. Automatic fake news detection methods are studied in this review and concluded a method for detecting various news. Also, studied the ability of a technique in predicting fake news based on data sources.

Keywords Automatic detection and Classifiers · Fake news · Social media

1 Introduction

On social interactions, enormous impact is shown by internet spread and development of technology in recent days. For people, information is obtained using a social media as it becomes popular rapidly. Opinions, interest, activities of people are shared on various social media platforms.

Informations rapid spread, low cost, easy accessing of information are the major advantages of social media. Nowadays, people are using this social media for searching news rather than traditional sources of news like newspaper or television.

T. V. Divya (✉) · B. G. Banik
Department of Computer Science and Engineering, Koneru Lakshmaiah Education Foundation, Hyderabad 500075, Telangana, India
e-mail: tvdivya21@gmail.com

B. G. Banik
e-mail: barnali.guptabanik@ieee.org

N. Chaki et al. (eds.), *Proceedings of International Conference on Computational Intelligence and Data Engineering*, Lecture Notes on Data Engineering and Communications Technologies 56, https://doi.org/10.1007/978-981-15-8767-2_16

Classical news sources are rapidly replaced by social media in recent days [1]. With lot of advantages, online social media news is not qualified, when compared to classical sources of news. This is a major problem in it. In order to achieve various goals, social media contents are changed sometimes. In countries like United Kingdom, United States, Russia, Romania, Macedonia, these websites are available. Fake news are spread in a fast as well as broad manner because of this [2, 3].

In making user's decisions, social media news, opinions, reviews are playing a major role. On individual and societies opinions, negative effect is caused by low-quality news spread which is termed as fake news. Governments, businesses, society and individuals are affected because of this kind of fake news. Noticeable amount of damage may be caused to an organization because of fake news about that organization spread by malicious user or spam. So, most of the researchers are concentrating on detection of fake news [1, 4, 5]. Challenges are caused to available content-based analysis methods by automatic fake news detection. Major reason is that news interpretation is highly nuanced and need awareness about social context, common cause or political issue. Nowadays, knowledge about most advanced natural language processing algorithms are not able to gathered.

Bad actors are writing fake news intentionally, which may appear as a real news but has manipulated information or false information. Even well trained human experts are also not able to detect it [6, 7]. Less variance is made by news category and input restrictions. On social media, merits and demerits of different techniques of fake detection are studied in review work. For fake news detection, various techniques have been proposed in recent days. Automatic fake news detection methods are studied in this review and concluded a method for detecting various news. Also, studied about the ability of techniques in predicting fake news based on data sources [8, 9].

2 Literature Review

Different techniques of fake news detection are reviewed in this section.

Conroy et al. [10] provided two major classes of assessment methods which are emerged from typology of various varieties. They are network analysis technique and linguistic cue techniques. Network-based behavioural data is combined with linguistic cue and machine learning to form a hybrid approach for getting enhanced results. Fake news detector system design is not a straight forward problem and for feasible detection of fake detection system, operational guidelines are proposed.

Granik and Mesyura [11] used Naive Bayes classifier for designing a simple fake news detection system. This methodology was actualized as a product framework and tried against an informational collection of Facebook news posts. Accomplished characterization precision of around 74.

Long et al. [12] incorporated attention-based LSTM model with profiles of speaker for proposing a novel technique for detecting fake news. In two ways, contribution to this model is done by profile of speaker. They are inclusion in attention model and inclusion in additional data input. Profiles of speaker like credit and location history, title of speaker, party affiliation are added. Benchmark fake news detection dataset is used in experimentation and around 14.5.

Buntain and Golbeck [13] developed a system for automatically detecting Twitter fake news. In two credibility-focused Twitter datasets, accuracy assessments are predicted using learning. They are CREDBANK and PHEME. PHEME is a potential Twitter rumour dataset and it has those rumours about journalistic assessments. CREDBANK is a crowdsourced dataset, and it has Twitter events accuracy assessments.

For journalistic accuracy and crowdsourced assessments, features are identified in feature analysis and consistent results are obtained when compared with previous works. Credibility and accuracy discussions are closed and in Twitter, better performance of non-experts models compared with journalists in detecting fake news is also discussed.

Ruchansky et al. [14] combined all three characteristics of automated prediction and more accurate prediction for proposing a system. User and articles behaviours are incorporated as well as in addition to fake news propagating user's behaviour. Using these three characteristics, CSI model is proposed with three modules namely, Integrate, Score and Capture. Text and response forms the base for first model, and Recurrent Neural Network is used in this model for capturing user activities temporal pattern on a specified article. User behaviour is used in second model for learning source characteristics. An article is classified by integrating these two models. Exploratory investigation on certifiable information exhibits that CSI accomplishes higher precision than existing models, and concentrates important dormant portrayals of the two clients and articles.

Pan et al. [15] included B-TransE model for proposing a technique for fake news detection, which is based on content of news and uses knowledge graphs. Few technical challenges are addressed in these solutions. In order to cover required relations for detecting fake news are not done by computational-oriented fact checking. Validation triples about its correctness, which are extracted from news articles are more challenging. Kaggle's Getting Real about Fake News dataset is used for evaluating the techniques and few real articles in the main media stream are also used. Around 0.80 of F1 score is achieved by the proposed approach as shown in evaluations.

Della et al. [16] combined social context features and news content for proposing a novel ML fake news detection technique, where existing techniques are outperformed by this method and around 4.8.

Shu et al. [17] formed a real-world dataset, which represents trust level of users on fake news and experienced as well as naive users representative groups are selected. Fake news items are recognized as false by experienced group of representatives, and fake news items are recognized as true by naive group and representatives. Between these group of users, over implicit and explicit profile features, comparative analysis is performed and their effectiveness in differentiating fake news is revealed.

Gupta et al. [18] exploited an echo chambers presence for tackling fake news detection problem of social media. News articles latent representation in an informative as well as in effective way is obtained using an echo chambers presence in user's social network. Within the social network, echo chambers are modelled as a closely connected community for representing 3-mode tensor structure of new article. In a latent embedding space, news articles are encoded by proposing a method based on tensor factorization and to preserve structure of community. Two real-world datasets are used in experimentation for demonstrating effectiveness of the proposed method in detecting fake news. Collaborative News Recommendation and News Cohort Analysis are used for validating resulted in embeddings generalization. In detection and generalization, better performance is exhibited using this method.

Helmstetter and Paulheim [19] discussed weakly supervised technique, where hundreds of thousands of tweets are collected as a dataset automatically, but they are very noisy in nature. Based on source nature both untrustworthy or trustworthy source, tweets are labelled automatically in collection process, and this dataset is used for training the classifier. Then, non-fake and fake tweets are classified using this classifier and various classification targets are also classified using this classifier. In new classification target view, these labels are inaccurate and result in unclean inaccurate dataset. With 0.9 of F1 score, fake news can be detected.

Zhang et al. [20] introduced automatic fake news credibility inference technique termed as FAKE DETECTOR. Model of deep diffusive network is constructed in FAKEDETECTOR for learning subjects, creators and news articles representation, according to latent and explicit feature set extracted from textual information. Various traditional models are compared with FAKEDETECTOR, in experimentation using real-world dataset and proposed methods effectiveness is demonstrated using results of experimentation.

Xu et al. [21] used two perspectives for characterizing Facebook comments, reactions, shares on hundreds on popular real and fake news. Perspectives are content and websites. Diverse behaviour in registration and its timing is exhibited by real and fake news publisher's websites as shown in site analysis. After certain amount of time, from web, fake news tends to disappear.

In fake news detection, it is insufficient to apply latent Dirichlet allocation (LDA) and frequency-inverse document frequency (tf-idf) topic modelling as suggested by real and fake news content corpus characterization. Fake and real news can be predicted effectively using document similarity with word vectors and term as indicated in results.

Yang et al. [22] reviewed the unsupervised methods used for detecting fake news. Latent random variables are formed using news truth and credibility of users and for identifying opinions of users regarding news authenticity on social media, engagements of users are exploited.

Conditional dependencies between news truth, opinions of users and credibility of user are captured using Bayesian network model. Effective collapsed Gibbs sampling technique is proposed for solving inference problem for computing news truth and credibility of user without any data which are labelled. Two datasets are used in experimentation for demonstrating the effectiveness of the proposed technique while comparing with unsupervised techniques.

Shu et al. [23] implemented a tri-relationship embedding method TriFN which is used in the classification of fake news, where, interactions between user-news and relations between publisher-news are modelled simultaneously. Two real-world dataset is used in experimentation and results of experimentation shows that proposed method shown a better performance when compared with baseline methods in detecting fake news.

Hu et al. [24] acquired every news node representation by proposing multi-depth graph convolutional networks (M-GCNs) framework which utilizes graph embedding and captured neighbours multi-scale information using multi-depth GCN blocks and attention mechanism is used for combining them. LIAR is huge real-world public fake news dataset which is used in this experimentation, and M-GCN exhibited a better performance than recently implemented five techniques.

Table 1 gives a summary of all Inferences from Existing Work.

3 Inferences from Recent Work

Three major classes of techniques are available in existing techniques of detection of fake news. They are content-based, context-based and propagation-based techniques. In detection of fake news based on content-based techniques, linguistic (lexical and syntactical) features play a major role, and they are used for capturing writing styles and deceptive cues. Sufficiently sophisticated fake news are used for defining content-based technique, which is a major drawback of it and these news appear as fake immediately. There will be a language dependency of linguistic features, which makes the limitations of this technique. User demographics are included in social context features (such as age, gender, education and political affiliation, social network structure and user reactions (e.g. posts accompanying a news item or likes)). Most effective results can be obtained using propagation-based techniques. These methods study about the news proliferation process. It has been contended that the phony news scattering process is much the same as irresistible ailment spread and can be comprehended with organized scourges models. There is considerable observational proof that phony news proliferates uniquely in contrast to genuine news framing spreading designs that might be misused for programmed counterfeit news identification. By goodness of being content-rationalist, proliferation-based highlights are likely sums up across various dialects, districts and topographies, rather than content-based highlights that must be grown independently for every language. Deep learning has more complex procedures to compute features and to get accurate results so it needs some modifications to give reduced time complexity.

4 Solution

Informations rapid spread, low cost, easy accessing of information are the major advantages of social media. Nowadays, people are using this social media for searching news rather than traditional sources of news like newspaper or television. On other

Table 1 Inferences from existing work

S. No.	Authors name	Method	merits	Demerits
1	Conroy et al. [10]	Hybrid approach—combines linguistic cue and machine learning	High accuracy	Time consuming
2	Granik and Mesyura [11]	Naive Bayes classifier	Improved results	Cost expensive
3	Long et al. [12]	LSTM model	Provides valuable information for validating news articles credibility	Need to use other classifier
4	Buntain and Golbeck [13]	Predict accuracy assessments methods	Outperform	It does not using both crowd sourced and journalist assessors to evaluate the same data
5	Ruchansky et al. [14]	CSI	Achieves higher accuracy	Time consuming
6	Pan et al. [15]	B-TransE model	Fake news can be detected using imprecise knowledge graph	It does not use style-based approaches
7	Della et al. [16]	ML	Have over 0.80 F1-scores	It does not focusing on cases harder to classify
8	Shu et al. [23]	Measuring users trust level	Provides better performance	Need to use other methods for features can be aggregated
9	Gupta et al. [18]	Tensor factorization	Outperforms	Need to use Neural Network based methods for modelling echo-chambers
10	Helmstetter and Paulheim [19]	Weakly supervised approach	Yields very good results	Time consuming
11	Zhang et al. [20]	Deep diffusive network model	Very effective	Very expensive
12	Xu et al. [21]	Inverse document frequency (tf-idf) and latent Dirichlet allocation (LDA) topic modelling	Provides better results	Computational complexity is very high in this model

(continued)

Table 1 (continued)

S. No.	Authors name	Method	merits	Demerits
13	Yang et al. [22]	Collapsed Gibbs sampling approach	Significantly outperforms the compared unsupervised methods	Requires semi-supervised learning for enhancing unsupervised model performance
14	Shu et al. [23]	Tri-relationship embedding framework TriFN	Achieve good detection performance	Time consuming nature
15	Hu et al. [24]	M-GCN	Provides better performance	Computational complexity is very high

side, it gives a chance of spreading fake news and make society as well as individuals negative impact. To solve this issue, this review work studied about some methods for further enhancement and will introduce a features optimization method for reducing the time complexity of detection. And also will use some pre-processing for the input data before sending it to detection phase to improve fake news detection classification performance and improve accuracy. Recurrent neural network (RNN) may give improved results performances than other deep learning approaches, so it could also be used for detection of fake news.

4.1 Objectives

Major objective of this Survey is listed below

- To implement the detection of fake news framework that can predict social media fake news articles regardless of noises present in news.
- To implement a computerized model for verifying extracted Twitter news and for checking the same. For information accumulation, general answers are given by them and demonstrate the recognition of fake news.
- To assure enhancement of accuracy metric and time complexity would be reduced by using machine learning techniques to utilize progressively complex model.
- To implement a model for predicting fake and real news percentage with an input of news events and utilizing reviews of twitter and classification algorithms.
- To ensure the right news for the users, features-based detection of fake news system is introduced.
- To implement improved detection system by deploying recurrent neural network (RNN).

5 Observation

This assesses different classifiers results which are discussed in review work. Experiments were conducted using LIAR dataset. It is a huge dataset real-world public news and most widely used in detection of fake news. There are 12,836 short statements which are labelled and having an average token of 17.9 and for truthfulness ratings, it has six fine-grained labels. They are true, mostly true, half-true, barely-true, false, pants-fire. Three sets are formed by splitting this dataset.

Three sets are formed by splitting this dataset. They are testing set with 10 percentage of dataset, validation set with 10 percentage of dataset and training set with 80 percentage of dataset. There will be w proper balancing between label distribution. Dataset is having huge amount of profiles of speakers. Profile includes credit history, topics, speech location, home state, job title, party affiliations and speaker.

In Fig. 1, fake news classification performance is shown in the figure in terms of recall. The recall result of presented M-GCN technique is 87.5, which is greater compared to the previous NBC and LSTM methods that produce only 85.5 and 86 accordingly.

In Fig. 2, Fake news classification performance is shown in the figure in terms of recall. The recall result of presented M-GCN technique is 87.5.

In Fig. 3, overall performance comparison result of precision metric is shown in Fig. 3 for fake news classification. The precision result of presented M-GCN technique is 92, which is greater compared to the previous GPBT and TME methods that produce only 86 and 87 accordingly.

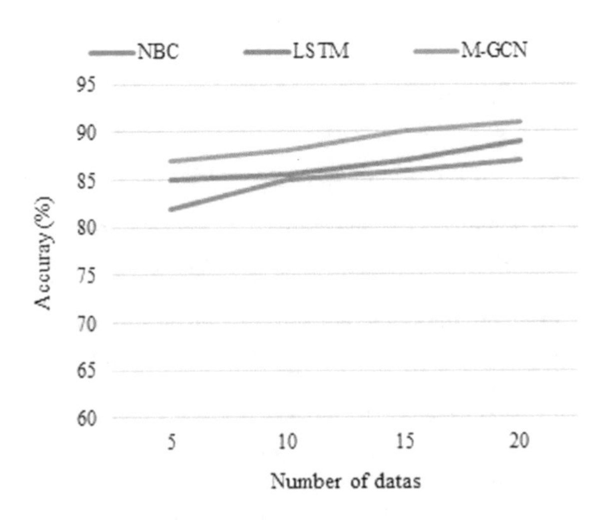

Fig. 1 Performance comparison of accuracy among different detection schemes

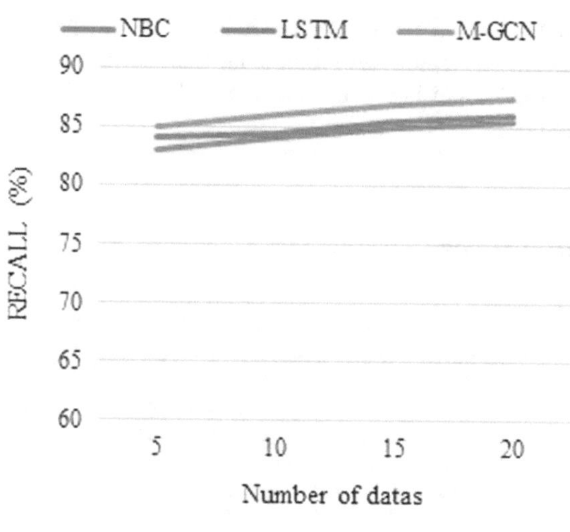

Fig. 2 Performance comparison of recall among different detection schemes

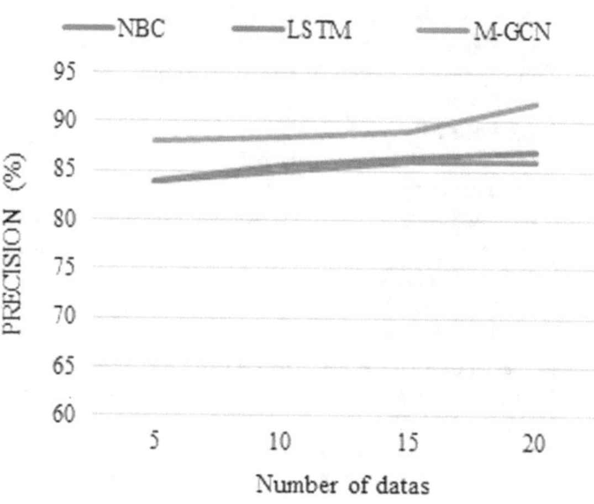

Fig. 3 Performance comparison of precision among different detection schemes

6 Conclusions

Before the emerging of internet, detection of fake news techniques was in use and they are used for detecting misleading information. This review work studied about the details of merits and demerits of various methods that are used for detecting social media fake news. Recently content-based approaches and propagation-based approaches are proposed for detecting fake news. For a specified news, precise statistical rating is not obtained in existing techniques due to their inefficiency in fake news detection, and it is producing less variance because of its input and news category restrictions. Finally, these reviews studied and concluded multi-depth graph convolutional networks (M-GCNs) that will be better and overcome the above-mentioned issues for automating fake news detection for various news. And also studied about how well this selected M-GCN is able to predict fake news in data sources.

References

1. Shu K, Bernard HR, Liu H (2019) Studying fake news via network analysis: detection and mitigation. In: Emerging research challenges and opportunities in computational social network analysis and mining. Springer, Cham, pp 43–65
2. Ahmed H, Traore I, Saad S (2017) Detection of online fake news using N-gram analysis and machine learning techniques. In: International conference on intelligent, secure, and dependable systems in distributed and cloud environments. Springer, Cham, pp 127–138
3. Shu K, Sliva A, Wang S, Tang J, Liu H (2017) Fake news detection on social media: a data mining perspective. ACM SIGKDD Explor Newslett 19(1):22–36
4. Tacchini E, Ballarin G, Della Vedova ML, Moret S, de Alfaro L (2017) Some like it hoax: automated fake news detection in social networks. arXiv:1704.07506
5. Tschiatschek S, Singla A, Gomez Rodriguez M, Merchant A, Krause A (2018) Fake news detection in social networks via crowd signals. In: Companion proceedings of the the web conference 2018, pp 517–524
6. Wang Y, Ma F, Jin Z, Yuan Y, Xun G, Jha K, Su L, Gao J (2018) Eann: event adversarial neural networks for multi-modal fake news detection. In: Proceedings of the 24th international conference on knowledge discovery & data mining, pp 849–857
7. Bourgonje P, Schneider JM, Rehm G (2017) From clickbait to fake news detection: an approach based on detecting the stance of headlines to articles. In: Proceedings of the 2017 EMNLP workshop: natural language processing meets journalism, pp 84–89
8. Monteiro RA, Santos RL, Pardo TA, de Almeida TA, Ruiz EE, Vale OA (2018) Contributions to the study of fake news in Portuguese: New corpus and automatic detection results. In: International conference on computational processing of the Portuguese Language. Springer, Cham, pp 324–334
9. Zhou X, Zafarani R (2018) Fake news: a survey of research, detection methods, and opportunities. arXiv:1812.00315
10. Conroy NJ, Rubin VL, Chen Y (2015) Automatic deception detection: methods for finding fake news. Proc Assoc Inf Sci Technol 52(1):1–4
11. Granik M, Mesyura V (2017) Fake news detection using naive Bayes classifier. In: 2017 IEEE First Ukraine conference on electrical and computer engineering (UKRCON). IEEE, pp 900–903
12. Long Y, Lu Q, Xiang R, Li M, Huang CR (2017) Fake news detection through multi-perspective speaker profiles. In: Proceedings of the eighth international joint conference on natural language processing (Volume 2: Short papers), pp 252–256
13. Buntain C, Golbeck J (2017) Automatically identifying fake news in popular Twitter threads. In: 2017 IEEE international conference on smart Cloud (SmartCloud). IEEE, pp 208–215
14. Ruchansky N, Seo S, Liu Y (2017) Csi: a hybrid deep model for fake news detection. In: Proceedings of the 2017 ACM on conference on information and knowledge management, pp 797–806
15. Pan JZ, Pavlova S, Li C, Li N, Li Y, Liu J (2018) Content based fake news detection using knowledge graphs. In: International semantic web conference. Springer, Cham, pp 669–683
16. Della Vedova ML, Tacchini E, Moret S, Ballarin G, DiPierro M, de Alfaro L (2018) Automatic online fake news detection combining content and social signals. In: 2018 22nd conference of open innovations association (FRUCT). IEEE, pp 272–279

17. Shu K, Wang S, Liu H (2018) Understanding user profiles on social media for fake news detection. In: 2018 IEEE conference on multimedia information processing and retrieval (MIPR). IEEE, pp 430–435
18. Gupta S, Thirukovalluru R, Sinha M, Mannarswamy S (2018) CIMTDetect: a community infused matrix-tensor coupled factorization based method for fake news detection. In: 2018 IEEE/ACM international conference on advances in social networks analysis and mining (ASONAM). IEEE, pp 278–281
19. Helmstetter S, Paulheim H (2018) Weakly supervised learning for fake news detection on Twitter. In: 2018 IEEE/ACM international conference on advances in social networks analysis and mining (ASONAM). IEEE, pp 274–277
20. Zhang J, Cui L, Fu Y, Gouza FB (2018) Fake news detection with deep diffusive network model. arXiv:1805.08751
21. Xu K, Wang F, Wang H, Yang B (2018) A first step towards combating fake news over online social media. In: International conference on wireless algorithms, systems, and applications. Springer, Cham, pp 521–531
22. Yang S, Shu K, Wang S, Gu R, Wu F, Liu H (2019) Unsupervised fake news detection on social media: a generative approach. In: Proceedings of the AAAI conference on artificial intelligence, vol 33, pp 5644–5651
23. Shu K, Wang S, Liu H (2019) Beyond news contents: the role of social context for fake news detection. In: Proceedings of the twelfth ACM international conference on web search and data mining, pp 312–320
24. Hu G, Ding Y, Qi S, Wang X, Liao Q (2019) Multi-depth graph convolutional networks for fake news detection. In: CCF international conference on natural language processing and Chinese computing. Springer, Cham, pp 698–710

Human Gait Classification Using Deep Learning Approaches

Abhishek Tarun and Anup Nandy

Abstract Human gait is one of the useful biometric traits which determines human's identity by the manner of their walk in the video surveillance. It is considered as behavioral biometric cues which depend on walking pattern of the people rather than the look of the people. Various factors affect the performance of gait such as carrying condition changes, clothing condition changes, and viewing angle variations. We develop a model for gait classification using deep learning methods. Separate gait databases are used in our experiment. The gait signatures are extracted from gait energy image (GEI) using convolutional neural network (CNN). The classifiers we used for classification of human gait are support vector machine (SVM), random forest, and long short term memory (LSTM). These models are tested on standard gait dataset CASIA A and our gait database is created using Microsoft Kinect device. The experimental results are very promising on both the datasets.

1 Introduction

Various image-based biometrics are available for person identification on which the more popular biometrics are fingerprints, face, iris, or gait. But it is found that iris and fingerprint biometric produce better performance as compared to the face and gait biometric. The biometrics like fingerprints, face, or iris require a cooperative subject, views from fixed aspects, and physical contact or proximity. For individual recognition who is non-cooperating from distance in the real world under different environmental conditions, this method is useful. The condition like clothing, shoes,

A. Tarun (✉) · A. Nandy
Department of Computer Science and Engineering, National Institute of Technology Rourkela, Machine Intelligence and Biomotion Research Lab, Rourkela, India
e-mail: abhishek23tarun@gmail.com

A. Nandy
e-mail: nandy.anup@gmail.com

N. Chaki et al. (eds.), *Proceedings of International Conference on Computational Intelligence and Data Engineering*, Lecture Notes on Data Engineering and Communications Technologies 56, https://doi.org/10.1007/978-981-15-8767-2_17

or environmental context degrade the performance of gait recognition. It is also affected by the physical injury which can change the walking pattern of the person. The gait recognition method can be categorized into two categories, i.e., model-based and model-free. The model-based [1–3] approaches extract the stride parameters of the subject that describes the gait of the human body. Images with high-resolution property are used by model-based methods and this makes it computationally expensive. Also, it is difficult to estimate. Model-free based methods consider the action or process of moving of the human body and it extracts the features of the gait from silhouette images. Subject detection, extraction of the silhouette, and classification are basic ideas behind the model-free approach. Our proposed work is also based on model-free based approaches. Silhouettes extraction is the primary step of the gait recognition algorithm. The dynamic physical appearance of a human is being considered by gait rather than the static physical appearance of a human, and due to this property, it is also called behavioral biometric. Gait only considers the walking pattern of the people rather than the look of the people. So most of the parts of physical appearance features are removed from the image during silhouette extraction. There is a challenge in gait identification to develop a better technique which is unvarying to many variate condition. This variate condition includes subject wearing or carrying conditions. The changes in viewing angle also affect its performance. The gait identification problem considers these varying conditions due to their frequent occurrence and also affects the performance of gait identification. Gait energy image (GEI) is chosen for effective gait recognition performance. It is a spatiotemporal gait representation and it is formed over the silhouettes [4] sequences. GEI contains information about body patterns and walking patterns of the humans and also it is expressed as a single image. It can give good competitive results as compared to alternative representations and also less sensitive to noise. The advantages of person identification using gait with GEI and CNN are discussed in this paper. We extract multiscale features of GEI with the help of CNN. We used pretrained CNN model, i.e., VGG16 which can do feature extraction with the help of filters. It can extract important features from any image. Our experiments provide confirmation that feature extraction done by the VGG network(deep CNN) are relevant for person identification. We applied three classifiers, i.e., SVM, LSTM, and Random Forest on the extracted features and the significant classification accuracy is achieved using these classifiers which demonstrates the relevancy of our method. The paper is divided into different sections and is as follows. Section 2 discusses about state-of-art work on gait recognition. Section 3 focuses on the proposed methodology for gait recognition using deep neural network. Section 4 addresses the result and analysis of the experiment. The conclusion and future direction of these work is presented in Sect. 5.

2 Related Work

For the effective working of any algorithm, the selection of data representation is very important. Liu and Sarkar [5] introduced a new metric for the representation of the sequence, i.e., an average of the silhouette. Han and Bhanu [4] proposed GEI which is equal to the mean of the silhouette images over one gait cycle. GEI is very simple and effective to use. Bashir et al. [6] introduced another gait representation template, i.e., Gait Entropy Image. Gait Entropy Image calculates the entropy of the silhouette images which represent one complete gait cycle. To solve the problem of changes in carrying and clothing condition Gait using Pal and Pal Entropy (GPPE) was proposed by Bashir et al. [7] as a strong gait representation. The performance evaluation of different gait representations using a large gait dataset by Iwama et al. [8] concluded that GEI works well in person identification. For reductions in dimensions, Principal Component Analysis (PCA) [9] was widely used in gait recognition. For the reduction of the dimensions in gait recognition, Linear Discriminant Analysis (LDA) [10] was also used. The projection of the representation was also done by LDA into space with a lower dimension with a separate class. PCA was used by Liang Wang et al. [11] for dimensionality reduction of the input feature and then feature extraction of gait was done. Han Bhanu [4] used both PCA and LDA for the feature extraction of the GEIs. PCA and LDA cannot take advantage of two-dimensional data were their drawbacks. Therefore, two-dimensional data is converted into one-dimensional data but this always not provides better recognition performance and sometimes may also result in poor recognition performance. One of the most advanced machine learning techniques is a deep convolutional neural network (CNN) because it can approximate non-linear functions. Alotaibi et al. [12] developed a specialized deep CNN architecture for person identification using gait and their architecture works better with common variations and occlusions.

3 Proposed Methodology

The proposed gait recognition method is discussed with the following diagram illustrated in Fig. 1.

3.1 Data Acquistion

For data acquisition, Microsoft Kinect v2.0 sensor is used which is placed at a distance of 3 m from the treadmill and the subject. We record Color Depth video at two different speed, i.e., 3 and 5 km/h and for this, and we used a system of 8 GB RAM and Intel i5 Core processor with Windows Operating System. In Kinect, there is no utility software to record depth video in any commonly readable formats so

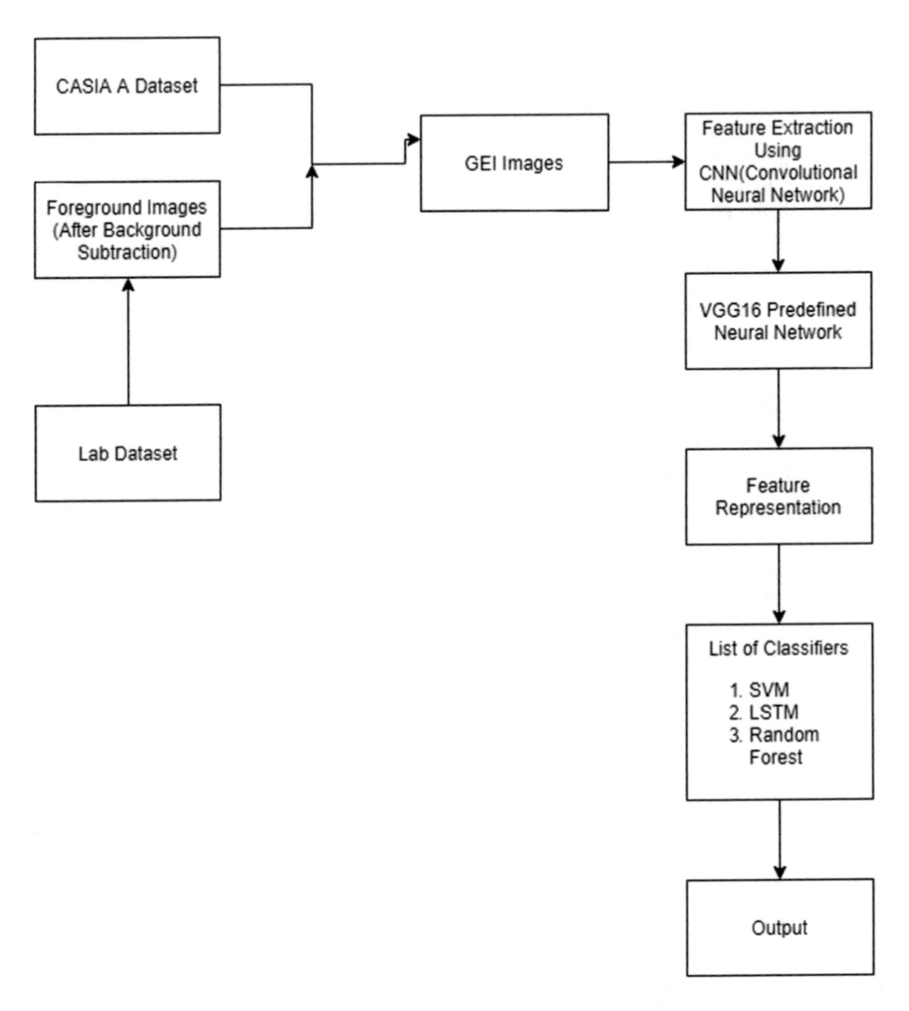

Fig. 1 Proposed approach

a screen recorder is used. All the videos are broken into frames. Figure 2 shows the experimental setup that we used for data acquisition in our lab. We also used a standard dataset for our research, i.e., CASIA A dataset [13]. The CASIA A dataset consists of a total of 20 subjects. Every subject consists of 12 image sequences, i.e., all the three directions 45°, 90°, and parallel) have four sequences. Each sequence length range varies from 37 to 127. A total of 19139 images are in this database.

Fig. 2 Experimental setup at our machine intelligence and biomotion research lab

3.2 Foreground Extraction

Firstly, we detect the person, i.e., our region of interest. This region of interest works as masks and indicates the regions of the image for both color and depth, which is generally a human object populates. The process of the region of interest extraction and Color Depth Image processing are done by detection and tracking of humans with the help of Kinect camera. It has three different color streams, which is in R, G, B (Red, Green, Blue). Color depth image assigns different colors and it depends on the distance of the subject from the camera. In our case, the subject is placed at a distance of 3 m, its color will lie in a specific range (yellow to green). So with the help of thresholding and binary masks, we can obtain the human as required.

The above procedure results in getting the area of the floor between the human on the treadmill. Also, the Kinect camera captures the area apart from the required object. So with the help of thresholding and binary masks, we can obtain the human as required. All work is done using HSV (Hue, Saturation, Value) format to ease the task of thresholding. We only focus on the human subject whose gait features are to be taken for research work and the human is specifically extracted. For this, we made some assumption that if the subject is standing with his back upright, the distance from the feet to the head will be the longest patch of pixels with almost the same

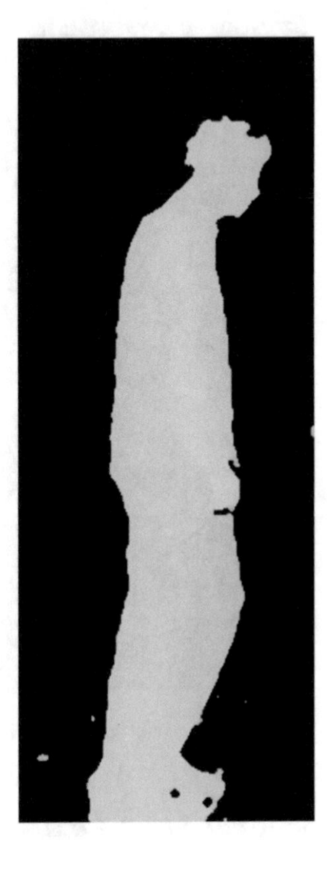

Fig. 3 Foreground extraction of the subject

depth (<30% variation in HSV). A semi-bounding box is fitted around the human subject to be able to quickly extract it. After background subtraction, the foreground image of the subject is shown in Fig. 3.

3.3 Gait Energy Image

The mean of the walking silhouette images is known as gait energy image, i.e., the summation of images of the walking silhouettes divided by the total number of the images. It cuts down the amount of gait period to a huge amount. Periodic sequences, phase transformation, and frequency information of the original image are also retained. The silhouette image was shown in Fig. 4 for CASIA A dataset and these images are of one complete gait cycle, GEI image of the gait cycle images are rightmost one and similarly Fig. 5 for our dataset. $B_t(x, y)$ is the preprocessed binary gait silhouette images at time t in a sequence is represented by Eq. (1), GEI is computed by

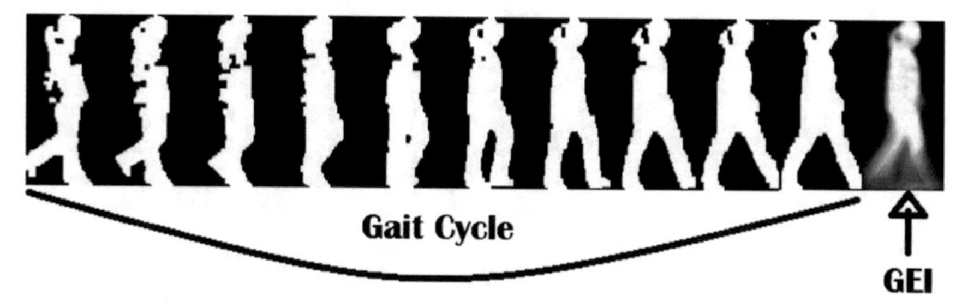

Fig. 4 GEI is mean of gait silhouettes over one gait cycle for CASIA A dataset

Fig. 5 GEI is mean of gait silhouettes over one gait cycle for our dataset

$$G(x, y) = 1/N \sum_{t=1}^{N} B_t(x, y) \tag{1}$$

In Eq. 1., the total length of the complete gait cycle is given by N, the image sequence is denoted by t, pixel coordinates of the images is denoted by (x,y).

3.4 Feature Extraction Using CNN

CNN was applied to GEI images for the multiscale features of the images. CNN can find out the significant features of the image by its multiple layers of neurons. CNN consists of the layers, i.e., convolutional, pooling, and fully connected (FC) layers. Initially, the raw image is given as an input to CNN and then the input for the next layer is the previous layer output. The convolution layer filter size has been decided in advance. The number of filters for the layer is equal to the output depth. The size of the stride is the amount each filter slides over the input at the time of convolutional operations. The reason behind this in local area image data tends to have a higher correlation rather than in a global area. The outstanding features of the receptive field were obtained by the convolution computation and it is the part of the input on which

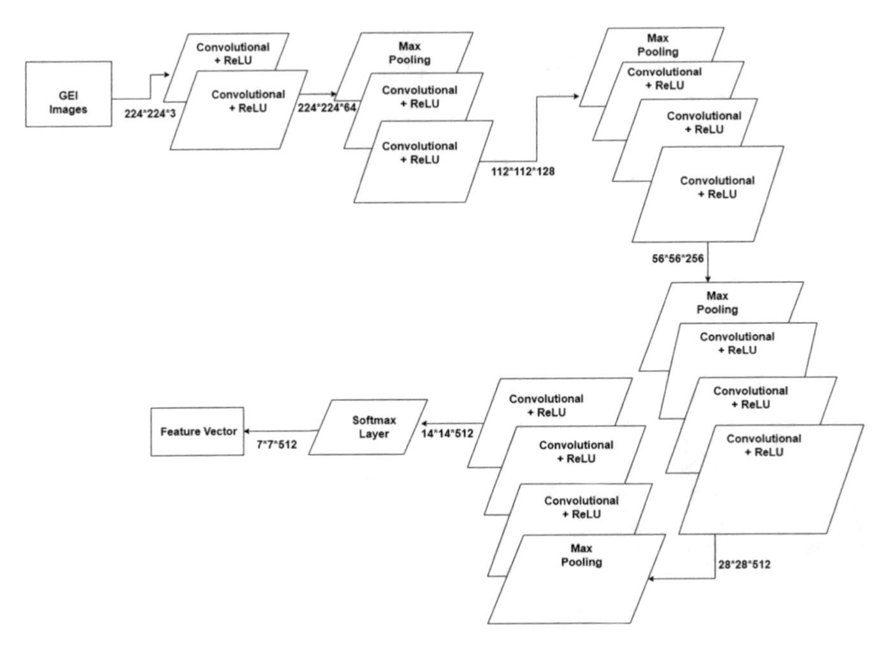

Fig. 6 VGG16 convolutional neural network model

filter is applied [14]. The output of the convolution layer can be treated as a feature that is extracted from each point of the input image known as the feature map [15]. The features of each point of the image is extracted from the convolutional layer, then the pooling layer combines the identical features. Then the network performance is constant to image distortion and its movement [14]. Dimensional reductions of the feature map are one of the characteristics of the pooling layer. The unit area of the feature map represents the input image region and its size was increased than before, and a new feature map that has a lower dimension that can show the features of the integrated input image was obtained by giving this feature map inputs to the next convolutional layer. The number of global feature extraction was increased as the action moves in a downward direction. The feature map used by our proposed method moves through every pretrained CNN layer as a feature for cross-domain image fetching. For classification of object, a CNN was trained with the ImageNet dataset which is a large image dataset [16] and also it works well for fetching of image of other datasets [17]. The training of CNN was done for classification that depends on a different dataset. The key feature extraction was done by the filters of the pretrained CNN. Our proposed methods used VGG16 network (Fig. 6) which is also a type of CNN. In this network, max pooling layers and multiple convolutional layers were repeated five times. At the end of the network, this network also contains three FC layers. FC layer's features do not have positional information. This is the reason that we do not use FC layer result as a feature. It is a 16 layer model which takes images of (224*224*3) size as input. It used only (3*3) convolutional layer and

(2*2) max-pooling layer throughout the network. We left the fully connected layers and took the output of the softmax layer just before the fully connected layers. So our output is (7*7*512) for each GEI image. That means we have a total of 25088 features for each GEI image.

3.5 Gait Recognition Using SVM (Support Vector Machine)

For the classification of the GEI test image, our proposed method used the SVM. For distinct classification of the data points in N-dimensional space (N –> the number of features), a hyperplane was found. On this particular objective, SVM works. There are many possible hyperplanes that are present. But we choose the hyperplane which has a maximum margin. Here maximum margin represents the maximum distance between data points of both classes.

3.6 Gait Recognition Using Random Forest

Random forest consists of a huge number of decision trees which can be seen as an ensemble. The prediction of the class was given by every tree and the class with the highest vote is our model's prediction. The model also gives predictions of ensemble which are not related and this makes them more efficient than any of the single predictions. The trees defend each other from their fault is the reason behind this effect but their fault is consistently not in the same direction. Most of the trees will be correct while few of the trees may be not correct but the group of the trees always moves in the right direction.

3.7 Gait Recognition Using LSTM (Long Short Term Memory)

The recurrent neural network has a type of network called the LSTM network that learns the dependency of order in the prediction of sequence problems. They regulate the information flow by learning which data in sequence to keep and which to leave with the help of gates. In this way, it can pass relevant information down the long chain of sequences to make predictions. Throughout the processing sequence, the cell state carries relevant information. Short-term memory effect would be lowered because the information from the earlier time steps can be transferred to later time steps. According to the relevancy of the information, it might be added or removed when the cell moves downward. This decision of relevancy of information was done by the gate during training which information is to be kept or forgotten.

4 Result Analysis and Discussion

The effectiveness of the proposed feature selection algorithms was evaluated on our dataset and the CASIA A Gait database [13].

The CASIA A dataset consists of a total of 20 subjects. Every subject consists of 12 image sequences, i.e., all the three directions (45°, 90°, and parallel) have four sequences. Each sequence length range varies from 37 to 127. A total of 19139 images are in this database with each image having height of 240 pixels and width of 352 pixels and we used all the images for our experiment. Therefore, we have a total of 240 GEI images. After applying VGG16 on all the 240 GEI images we got 240*25088 features. Similarly, we had also our dataset of 11 persons at two different speeds, i.e., 3 and 5 km/hr. We have 500 images for each person at each speed, so we have a total of 11000 images for our dataset and each image having height of 444 pixels and width of 180 pixels. After that we found GEI images for each person on both speeds. So we have a total of 22 GEI images. We do augmentation of each GEI image into 12 GEI images. So we have a total of 22*12 = 264 GEI images. After applying VGG16 on GEI images we got 264*25088 features. experiments which were performed by using GEI which is the real feature.

Table 1 Performance analysis metrics of SVM classifier on CASIA A dataset

Class label	Precision	Recall	f1-score
1.0	1.00	1.00	1.00
2.0	1.00	1.00	1.00
3.0	0.50	1.00	0.67
4.0	1.00	1.00	1.00
5.0	1.00	1.00	1.00
6.0	1.00	0.88	0.93
7.0	1.00	0.67	0.80
8.0	0.60	1.00	0.75
9.0	1.00	1.00	1.00
10.0	1.00	1.00	1.00
11.0	1.00	1.00	1.00
12.0	1.00	0.80	0.89
13.0	1.00	1.00	1.00
14.0	0.00	0.00	0.00
15.0	1.00	0.86	0.92
16.0	1.00	1.00	1.00
17.0	0.75	1.00	0.86
18.0	0.83	0.83	0.83
19.0	1.00	1.00	1.00
20.0	1.00	1.00	1.00

Table 2 Performance analysis metrics of LSTM classifier on CASIA A dataset

Class label	Precision	Recall	f1-score
1.0	1.00	1.00	1.00
2.0	1.00	1.00	1.00
3.0	1.00	1.00	1.00
4.0	1.00	1.00	1.00
5.0	1.00	0.50	0.67
6.0	1.00	1.00	1.00
7.0	1.00	1.00	1.00
8.0	1.00	1.00	1.00
9.0	1.00	1.00	1.00
10.0	1.00	1.00	1.00
11.0	1.00	0.86	0.92
12.0	0.64	1.00	0.78
13.0	1.00	1.00	1.00
14.0	1.00	1.00	1.00
15.0	1.00	1.00	1.00
16.0	1.00	1.00	1.00
17.0	1.00	0.67	0.80
18.0	1.00	0.67	0.80
19.0	0.88	1.00	0.93
20.0	1.00	1.00	1.00

CNN was used for feature extraction and it produces multiscale features that are important to perform the recognition. Here we had taken all the features extracted from CNN. We have not used any feature selection or feature reduction technique such as PCA or LDA because it may lead to the removal of some features which are important for gait recognition. Also without using any of these feature selection or feature reduction techniques, we are getting better results. The performance of our experiment shows that this CNN-based architecture improves the performance of person identification using gait. This CNN-based architecture helps find more features of the GEI which is good for recognition performance. We compare the recognition performance of CASIA A dataset and our dataset with SVM, LSTM, and Random Forest classifier.

The Table 1 shows the classification metrics of SVM classifier, Table 2 shows the classification metrics of LSTM classifier, and Table 3 shows the classification metrics of Random Forest classifier on CASIA A dataset. The Table 4 shows the classification metrics of SVM classifier, Table 5 shows the classification metrics of LSTM classifier, and Table 6 shows the classification metrics of Random Forest classifier on our dataset. The comparative analysis of Table 7 shows the comparison of the accuracy of our proposed model on CASIA A and our dataset. It shows that the

Table 3 Performance analysis metrics of random forest classifier on CASIA A dataset

Class label	Precision	Recall	f1-score
1.0	1.00	1.00	1.00
2.0	1.00	1.00	1.00
3.0	0.80	1.00	0.89
4.0	1.00	0.50	0.67
5.0	0.75	1.00	0.86
6.0	1.00	1.00	1.00
7.0	1.00	1.00	1.00
8.0	1.00	1.00	1.00
9.0	1.00	1.00	1.00
10.0	1.00	1.00	1.00
11.0	1.00	1.00	1.00
12.0	1.00	1.00	1.00
13.0	1.00	1.00	1.00
14.0	1.00	1.00	1.00
15.0	1.00	1.00	1.00
16.0	1.00	1.00	1.00
17.0	1.00	1.00	1.00
18.0	1.00	1.00	1.00
19.0	1.00	1.00	1.00
20.0	1.00	1.00	1.00

Table 4 Performance analysis metrics of SVM classifier on our dataset

Class label	Precision	Recall	f1-score
1.0	1.00	1.00	1.00
2.0	1.00	1.00	1.00
3.0	1.00	0.75	0.86
4.0	0.75	1.00	0.86
5.0	1.00	1.00	1.00
6.0	0.89	0.89	0.89
7.0	1.00	0.94	0.97
8.0	0.88	1.00	0.93
9.0	1.00	1.00	1.00
10.0	0.88	1.00	0.89
11.0	1.00	0.83	0.91

Table 5 Performance analysis metrics of LSTM classifier on our dataset

Class label	Precision	Recall	f1-score
1.0	1.0	1.0	1.0
2.0	1.0	1.0	1.0
3.0	0.80	1.0	0.89
4.0	1.0	0.91	0.95
5.0	0.90	1.00	0.95
6.0	1.00	0.83	0.91
7.0	1.00	0.89	0.94
8.0	1.0	1.0	1.0
9.0	1.0	1.0	1.0
10.0	1.00	0.75	0.86
11.0	0.82	1.00	0.90

Table 6 Performance analysis metrics of random forest classifier on our dataset

Class label	Precision	Recall	f1-score
1.0	1.00	1.00	1.00
2.0	1.00	1.00	1.00
3.0	0.88	1.00	0.93
4.0	1.00	0.88	0.93
5.0	1.00	1.00	1.00
6.0	1.00	1.00	1.00
7.0	1.00	1.00	1.00
8.0	1.00	1.00	1.00
9.0	1.00	1.00	1.00
10.0	1.00	1.00	1.00
11.0	1.00	1.00	1.00

Table 7 Comparative analysis of our model on CASIA A and our dataset

Classifier	CASIA A	Our dataset
SVM	91.54	93.58
LSTM	93.05	94.93
Random forest	97.18	98.71

performance of the proposed model on our dataset in comparison to CASIA A dataset is better. Random Forest classifier gives the best result out of all the three classifiers. The performance of LSTM is also good, and its performance might increase if we have a larger dataset. Here we do not have a large dataset of GEI in both CASIA A and in our dataset. We have 240 GEI from CASIA A dataset and 264 GEI from our dataset (Table 8).

Table 8 Comparison with other method

Methods	Accuracy (%)
DeepCNN [18]	97.58
Proposed approach (Random forest)	98.71

5 Conclusion and Future Work

The method of person identification using gait with the help of GEI and CNN proposed in this paper has advantages, i.e., with the help of CNN we extract the multiscale features of GEI. We used the pretrained CNN, i.e., VGG16 which can do feature extraction with the help of filters. It can extract important features from any image. Our experiments give confirmation that feature extraction done by deep neural network (the VGG network) is relevant for person identification. We used three classifiers, i.e., SVM, LSTM, and Random Forest and on these three classifiers we got high accuracy, and this shows the relevancy of our method. The future work includes exploring specific types of gait features (spatio-temporal, Kinematic, etc.) and applies different machine learning algorithms. The number of subjects in the experimental analysis was increased and it is also effective for gait recognition performance.

Acknowledgements This work is partially funded by Science and Engineering Research Board, Govt of India with project file number (ECR/2017/000408). We would like to extend our sincere gratitude to the students of Department of Computer Science and Engineering, NIT Rourkela for their uninterrupted co-operation and consented participation for data collection.

References

1. Yam C, Nixon MS, Carter JN (2004) Automated person recognition by walking and running via model-based approaches. Pattern Recognit 37(5):1057–1072
2. Tafazzoli F, Safabakhsh R (2010) Model-based human gait recognition using leg and arm movements. Eng Appl Artif Intell 23(8):1237–1246
3. Bobick AF, and Johnson AY (2001) Gait recognition using static, activity-specific parameters, In: Proceedings of the 2001 IEEE computer society conference on computer vision and pattern recognition. CVPR 2001, vol 1. IEEE, pp. I–I
4. Han J, Bhanu B (2005) Individual recognition using gait energy image. IEEE Trans Pattern Anal Mach Intell 28(2):316–322
5. Liu Z, Sarkar S (2004) Simplest representation yet for gait recognition: averaged silhouette. In: Proceedings of the 17th international conference on pattern recognition, ICPR 2004, vol 4. IEEE, pp 211–214
6. Bashir K, Xiang T, Gong S (2009) Gait recognition using gait entropy image
7. Jeevan M, Jain N, Hanmandlu M, Chetty G (2013) Gait recognition based on gait pal and pal entropy image. In: 2013 IEEE international conference on image processing. IEEE, pp 4195–4199

8. Iwama H, Okumura M, Makihara Y, Yagi Y (2012) The ou-isir gait database comprising the large population dataset and performance evaluation of gait recognition. IEEE Trans Inf Forensics Secur 7(5):1511–1521
9. Turk M, Pentland A (1991) Eigenfaces for recognition. J Cogn Neurosci 3(1):71–86
10. Belhumeur PN, Hespanha JP, Kriegman DJ (1997) Eigenfaces versus fisherfaces: recognition using class specific linear projection. IEEE Trans Pattern Anal Mach Intell 19(7):711–720
11. Wang L, Tan T, Ning H, Hu W (2003) Silhouette analysis-based gait recognition for human identification. IEEE Trans Pattern Anal Mach Intell 25(12):1505–1518
12. Alotaibi M, Mahmood A (2017) Improved gait recognition based on specialized deep convolutional neural network. Comput Vis Image Underst 164:103–110
13. Yu S, Tan D, Tan T (2006) A framework for evaluating the effect of view angle, clothing and carrying condition on gait recognition. In: 18th international conference on pattern recognition (ICPR'06), vol 4. IEEE, pp 441–444
14. LeCun Y, Bengio Y, Hinton G (2015) Deep learning. Nature 521(7553):436–444
15. Goodfellow I, Bengio Y, Courville A (2016) Deep learning. MIT press
16. Deng J, Dong W, Socher R, Li L-J, Li K, Fei-Fei L (2009) Imagenet: a large-scale hierarchical image database. In: IEEE conference on computer vision and pattern recognition. IEEE 2009:248–255
17. Razavian AS, Azizpour H, Sullivan J, Carlsson S (2014) CNN features off-the-shelf: an astounding baseline for recognition, vol 10. arXiv:1403.6382
18. Alotaibi M, Mahmood A (2015) Improved gait recognition based on specialized deep convolutional neural networks. In: IEEE applied imagery pattern recognition workshop (AIPR), pp 1–7

Generating Parking Area Patterns from Vehicle Positions in an Aerial Image Using Mask R-CNN

Manas Jyoti Das, Abhijit Boruah, Jyotirmoy Malakar, and Priyam Bora

Abstract Identification of visual patterns in cluster of similar objects in an aerial scene is a challenge for drone surveillance. Similar objects may form different patterns representing non-identical meanings in a scene. For example, a group of cars in a scene may form either traffic or a parking area, subjected to their position and arrangements. In this work, we proposed an approach to this issue by distinguishing between a parking area and any other arrangement of cars by generating a grouping pattern. Initial object detection in the aerial images is carried out by Mask-RCNN and then an algorithm is developed to form the arrangement pattern of the vehicles. A comparative analysis between parking spaces and non-parking spaces is provided in this work and distinct variations were reported between the generated patterns. The pattern generated showed formation of distinct geometric shapes in case of cars in a parking plot whereas cars in a traffic or non-parked condition displayed uneven shapes where many edges intersected at various points. However, the pattern generated will depend on the elevation of the camera from which the scene is captured and the ability of the object detection algorithm to detect cars. This approach can be used for understanding parking area organization, traffic managements, and direct cluster identification using deep learning methods for aerial surveillance applications in the near future.

Keywords Aerial image · Parking pattern · Mask RCNN

1 Introduction

Pattern generation and its usage through computer vision have a significant role in modern robotic applications. Tasks such as facial expression [5] and gesture recognition [9] have witnessed the development of various pattern recognition methods.

M. J. Das (✉) · A. Boruah · J. Malakar · P. Bora
Department of CSE, DUIET, Dibrugarh University, Dibrugarh, India
e-mail: mdas882@gmail.com

N. Chaki et al. (eds.), *Proceedings of International Conference on Computational Intelligence and Data Engineering*, Lecture Notes on Data Engineering and Communications Technologies 56, https://doi.org/10.1007/978-981-15-8767-2_18

Fig. 1 **a** Image of cars in traffic. **b** Image of cars in parking area

Many of these methods have also been used at different platforms to solve issues in real time which would otherwise be considered both labor intensive and time consuming. For example, threat detection in public places such as airports and railway stations using visual detection methods implement pattern analysis at a deeper level [12].

Visual patterns from a cluster of similar objects can be used to determine the co-relation between them. But this task becomes quite challenging when the objects are arranged in different order in various scenes. For example, it is highly unlikely that cars in two parking spaces will be arranged in the same order. But it is usually obvious that cars in parking spaces are arranged in a fixed order when compared to random arrangement of cars on the road. This paper describes a pattern generation method from the position of cars in different background scenes.

In this work, we have used the VisDrone 2019 dataset [15] which consists of aerial images collected from a drone on various landscapes. In scope of this work, we have considered only the images containing cars from two different backgrounds that are parked and non-parked as shown in Fig. 1. Here we have trained the model to obtain the bounding boxes on the intended object of interest and for that purpose we have employed Mask R-CNN [8] which is an object detection algorithm. It is a Faster R-CNN algorithm with an addition of a mask branch to generate the masks. The Faster R-CNN part of the algorithm perform objection and generate the bounding boxes whereas the mask branch generate masks over the detected objects. The coordinates of the bounding boxes are then extracted and used for calculating the centroids. This centroid calculation is achieved for each and every detected car. Unique patterns are generated by connecting the centroids. Pattern generation is done by calculating the Euclidean distance between the centroids and then connecting only those pair of centroids where the distance is lowest between the candidates of the pairs.

This paper is organized as follows. The previous works in line to the objective is discussed in Sect. 2. Details of the methodology along with the algorithm to generate pattern is discussed in Sect. 3. Results are elaborated in Sect. 4 and future direction of the work is discussed in Sect. 5.

2 Related Works

Substantial works have been carried out till date in relevance to identification of patterns in aerial scenes. Authors in [2] provided a framework for learning patterns of motion and sizes of objects in static camera surveillance. The proposed framework provides a means of performing higher level analysis to augment the traditional surveillance pipeline.

Ohn-Bar et al. [10] explored efficient means in dealing with intra-category diversity in object detection. Strategies for occlusion and orientation handling are explored by learning an ensemble of detection models from visual and geometrical clusters of object instances. The paper studies the role of learning appearance patterns of vehicles for detection and orientation estimation. Authors in [4] proposed a novel object detection algorithm based on shape matching using a single sketch of an object. The paper presented a practical circular arc extraction algorithm that uses the edge detection result of the image. Broken and damaged object boundaries are successfully reconstructed by the bottom-up processing of circular arc extension, which can be executed without the prior knowledge of the object. Authors in [13] have tried to present past and current research work in computer vision, pattern recognition, and image processing applied to LV modeling and analysis where he has tried to cover almost all cardiac imaging modalities, such as X-ray, CT, MR, PET, and SPECT. Their work concluded that learning techniques like neural nets and training are the future of global shape extraction for accurate boundary estimation.

These various approaches of identification of patterns provided solutions to the problem of parking space recognition. Wu [14] proposed a method to detect parking space where they first pre-processes the image and convert it into patches of 3 spaces and then applied multi Support Vector Machine with probabilistic output in order to recognize the patches and find their relationship with their surrounding patches. There may arise conflicts between two neighboring patches, so Markov Random Field (MRF) is used in order to improve the recognition.

One of the major issues is to detect parking occupancy in real time. Amato [1] used a CNN classifier which runs on a camera. The advantages of this approach are that the classifier works well irrespective of the different viewpoint of the test images. They created a dataset CNRPark which contains images of several real parking lot which were taken by cameras. Cazamias [3] explored two approached for classification of parking space: a binary classifier which labels an individual as parking space as empty or not and a multinomial classifier which returns the number of empty space for parking from a given image. The multinomial classifier shows lower accuracy in terms of correct prediction than the binary classifier. The binary classifier shows an overall accuracy up to 99%. Another work in [6] tries to make a comparison between classic supervised learning method and semi-supervised learning method which exploits the problem of empty versus non-empty parking plots from an image. Their results shows that supervised method outperformed the semi-supervised method.

Taking all the previous works into consideration, we proposed a simple algorithm for generating the vehicle arrangement patterns without utilizing deep learning modules, in order to reduce time and space complexity.

3 Methodology

3.1 Object Detection

For generating parking area patterns, the initial task is to detect cars in scene. For this, we used Mask RCNN on keras framework with tensorflow at the backend. Implementation is done on the Google© Colaboratory notebook with GPU configuration. The images are trained using single GPU which is a NVIDIA© tesla k80, and VGG image annotator [7] was used for the purpose of annotation.

Since the time required for training deep learning models is usually too high, we have employed transfer learning for better training and also to reduce the time for training. Initially, images are first fed into a block of convolutional neural network which is also the backbone network. The backbone network starts generating feature maps at various levels. The level of features increases as the image passes through different layers of the backbone network. The final result of the backbone network is a 2-D matrix. The 2-D matrix is then used by the region proposal network or RPN layer which scans the image and generate anchors depending on the position of the object in the image. These anchors signify the possible regions in the image that may hold an object.

The region of interest or ROI proposed by the RPN layer is then fed into a classifier which is a much denser network of fully connected layers. The classifier compares the region of interest proposed by the RPN layer with the feature maps generated by the backbone network. This allows the layer to classify the ROI to the specific classes and draw positive anchors around the detected objects.

After successful detection of the objects, masks are generated around the detected objects. The mask branch is a collection of CNN that considers the positive regions taken by ROI classifier and generates a binary mask around them as shown in Fig. 2.

3.2 Pattern Generation

The identified objects with bounding boxes are obtained after successful implementation of the object detection phase. The coordinates of the bounding boxes are extracted, and centroids for every bounding box are calculated, as shown in Fig. 3.

Then, distance between every centroid is calculated using Euclidean distance formula and the shortest distance for every centroid is considered, according to Eq. 1.

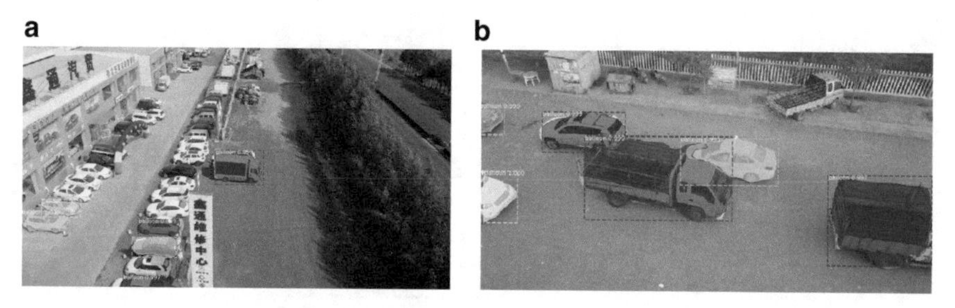

Fig. 2 **a** Image of detected cars in a parking plot. **b** Image of detected cars in traffic

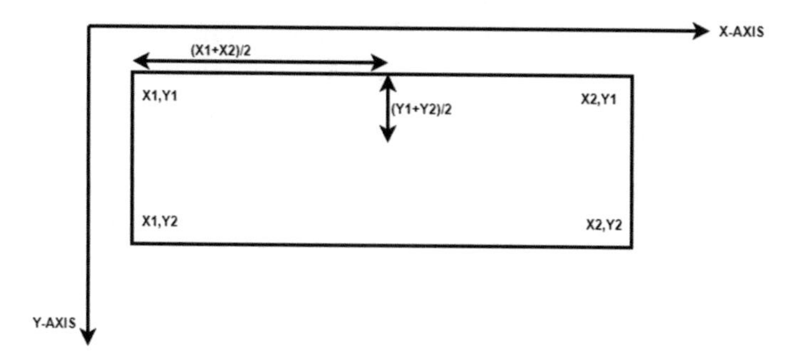

Fig. 3 Image representation for centroid calculation

$$d = \sqrt{(x_2 - x_1)^2 + (y_2 - y_1)^2} \quad (1)$$

where (x_1, y_1) is coordinate for centroid C_1 and (x_2, y_2) is coordinate for centroid C_2.

Following that centroids are sorted based on their closeness with one another using the traditional bubble sort algorithm. They are stored in a structure such that one centroid always forms a closest pair with its next neighbor. This ordered arrangement of centroids is then used for generating edges between every pair such that at the end we are left with pattern of the connections between the centroids. Flowchart of this complete process is depicted in Fig. 4.

4 Results

The pattern generation algorithm produces patterns with distinct visual differences in case of parked cars (Fig. 5a) and cars on the road (Fig. 5b) or other non-parking areas (Fig. 5c). In case of parked cars, the patterns form a specific geometric shape as seen in Fig. 6a, because the cars are arranged in a certain order according to the parking regulations. On the other hand, in a normal traffic or a non-parking area,

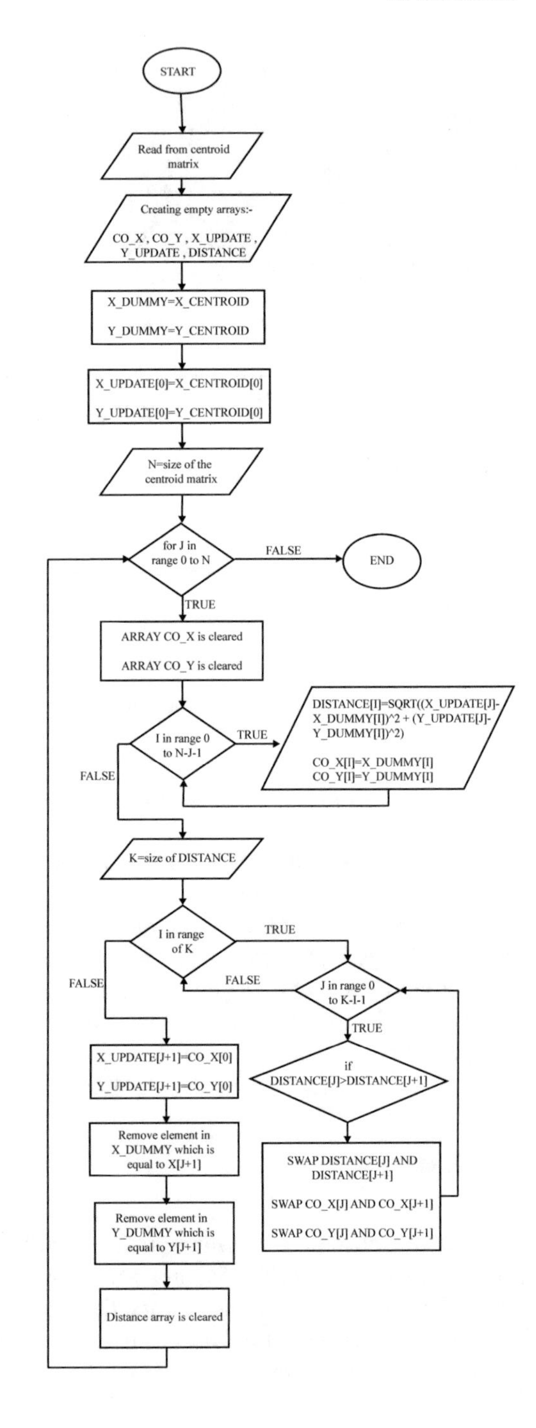

Fig. 4 Flowchart of the algorithm used for distance calculation

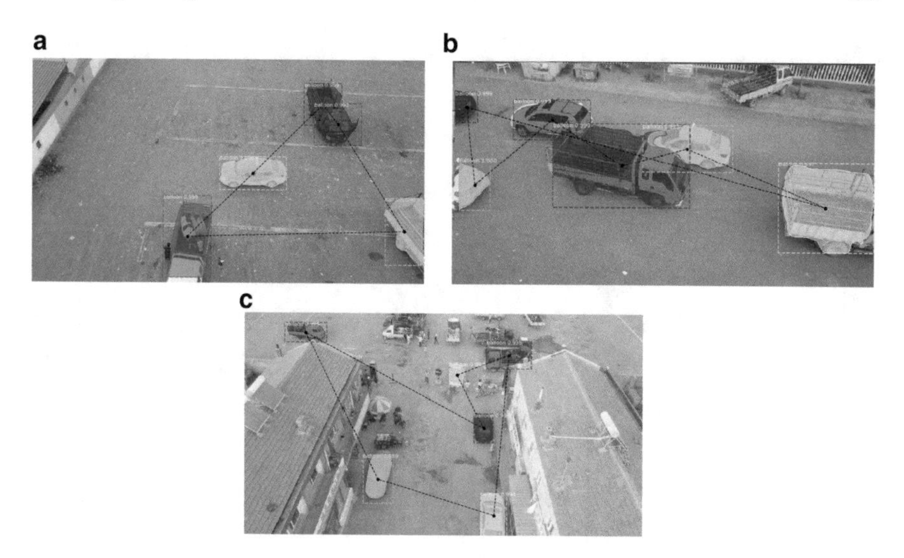

Fig. 5 **a** Pattern generated for parking area. **b** Pattern generated for on-road traffic. **c** Pattern generated for non-parking area

the patterns do not form any definite geometric shape (Fig. 6b, c), but will usually be any irregular polygon or any non-geometric shape, based on the number of cars detected.

As parking areas of different countries and cities follow different rules and regulations, generation of such patterns on a larger scale and their analysis can help in a better understanding of the features involved in aerial detection of parking areas. Moreover, these patterns generated over the same parking space at different period and at different parking conditions can be used to determine the amount of vacant spaces available at a given instant of time by identifying the missing centroids in edges and longer edges in a pattern. As seen in Fig. 6a, two edges with comparatively longer dimensions than the other edge reflects that there is parking space available on the edges of the captured image.

Moreover, we also witnessed the patterns formed by cars in traffic under different levels of congestion. These kinds of patterns can be used for studying the condition of traffic in real time and can be helpful in tackling traffic jams which have became a major issue globally. Study of these patterns also can help in deriving a statistical analysis over the existing traffic network topology [11] and finding those sections of the network which are responsible for lowering the performance of the overall network.

Speaking about the performance, the objection detection algorithm gave a training loss of about 0.21 whereas validation loss is about 0.39. However, better results can be achieved by training over a larger dataset and on a better hardware configuration. The pattern generation have an average complexity of $O(n^2)$.

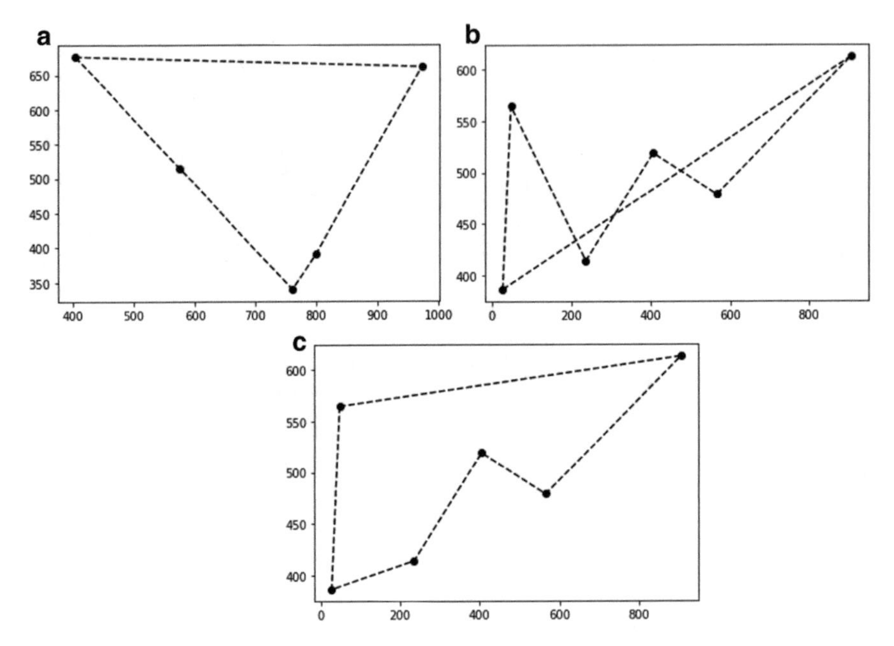

Fig. 6 **a** Extracted pattern of a parking area. **b** Extracted pattern of a on-road traffic. **c** Extracted pattern of a non-parking area

5 Conclusion and Future Scope

This work presents a simple approach to extract patterns from aerial images after objects are detected using Mask RCNN. The pattern generated from this technique can be used in determining the nature of the background and its condition based on how the objects are arranged in the scene. Any random variation in pattern when compared to usual patterns generated from the scene can be looked as a change in the condition of the background. For example, while analyzing the patterns obtained at different condition of traffic if a sudden change in pattern is observed then it may be an indication that an accident has occurred in the region. These patterns can be used as an input to another convolutional neural network for scene recognition based on background information, which is not included in this scope of work. This technique of pattern generations can also be implemented in other fields of study for analyzing the position of the object of interest. For example, in forestry, this pattern generation technique can be employed to check and analyze the growth of vegetation over a given period of time.

References

1. Amato G, Carrara F, Falchi F, Gennaro C, Vairo C (2016) Car parking occupancy detection using smart camera networks and deep learning. In: IEEE symposium on computers and communication (ISCC). IEEE, pp 1212–1217
2. Basharat A, Gritai A, Shah M (2008) Learning object motion patterns for anomaly detection and improved object detection. In: IEEE conference on computer vision and pattern recognition. IEEE, pp 1–8
3. Cazamias J, Marek M (2016) Parking space classification using convoluional neural networks. Technical report. Stanford University
4. Chang W, Lee SY (2013) Description of shape patterns using circular arcs for object detection. IET Comput Vision 7(2):90–104
5. Devi NS, Hemachandran K (2013) Automatic face recognition system using pattern recognition techniques a survey. Int J Comput Appl 975:8887
6. Di Mauro D, Battiato S, Patanè G, Leotta M, Maio D, Farinella GM (2016) Learning approaches for parking lots classification. In: International conference on advanced concepts for intelligent vision systems. Springer, pp 410–418
7. Dutta A, Zisserman A (2019) The vgg image annotator (via). arXiv:1904.10699
8. He K, Gkioxari G, Dollár P, Girshick R (2017) Mask r-cnn. In: Proceedings of the IEEE international conference on computer vision, pp 2961–2969
9. Noroozi F, Kaminska D, Corneanu C, Sapinski T, Escalera S, Anbarjafari G (2018) Survey on emotional body gesture recognition. IEEE Trans Affect Comput
10. Ohn-Bar E, Trivedi MM (2015) Learning to detect vehicles by clustering appearance patterns. IEEE Trans Intel Transport Syst 16(5):2511–2521
11. da Silva BC, Bazzan AL, Andriotti GK, Lopes F, de Oliveira D (2004) Itsumo: an intelligent transportation system for urban mobility. In: International workshop on innovative internet community systems. Springer, pp 224–235
12. Sindagi VA, Patel VM (2018) A survey of recent advances in cnn-based single image crowd counting and density estimation. Pattern Recognit Lett 107:3–16
13. Suri JS (2000) Computer vision, pattern recognition and image processing in left ventricle segmentation: the last 50 years. Pattern Anal Appl 3(3):209–242
14. Wu Q, Huang C, Wang SY, Chiu WC, Chen T (2007) Robust parking space detection considering inter-space correlation. In: IEEE international conference on multimedia and expo. IEEE, pp 659–662
15. Zhu P, Wen L, Bian X, Ling H, Hu Q (2018) Vision meets drones: a challenge. arXiv:1804.07437

Analysis of Inappropriate Usage of Cryptographic Primitives in Indian Mobile Financial Applications

Srinadh Swamy Majeti, Barnabas Janet, and N. P. Dhavale

Abstract Usage of cryptographic primitives while developing the Android applications must be employed in a precise way. But, developers are not doing this, yields to devastating the breaches of data security. The authors identified that there are 24 cryptographic weaknesses and categorized them into 4 groups. This motivates the authors to analyze the inappropriate usage of cryptographic primitives in Indian mobile financial applications. For this, the authors selected a total of 36 applications belongs to 3 different categories, namely mobile banking, mobile wallet, and e-commerce android applications and analyzed the weaknesses category-wise.

Keywords Mobile banking applications · Cryptographic primitives · Weaknesses

1 Introduction

Cryptography is the field of algorithms, functions, mechanisms, techniques, and procedures for providing privacy to protect our sensitive data. But advancements in technology, hackers are successful to break the security and steal our data. A lot of research is going on to improve the strength of security.

S. S. Majeti (✉)
NIT Trichy, Trichy, India
e-mail: msrinadhswamy@idrbt.ac.in

IDRBT, Hyderabad, India

B. Janet
Department of Computer Applications, National Institute of Technology (NIT) Trichy, Trichy, India
e-mail: janet@nitt.edu

N. P. Dhavale
Centre for Mobile Banking, Institute for Development and Research in Banking Technology (IDRBT), Hyderabad, India
e-mail: npdhavale@idrbt.ac.in

N. Chaki et al. (eds.), *Proceedings of International Conference on Computational Intelligence and Data Engineering*, Lecture Notes on Data Engineering and Communications Technologies 56, https://doi.org/10.1007/978-981-15-8767-2_19

Software development companies are recruiting developers, not cryptographers. Developers are not cryptographers. Cryptographers are responsible to use the appropriate cryptographic API while developing the security-related products. In this paper, the authors ask the developers/cryptographers that how efficiently they have incorporated the cryptographic primitives in the products? Nadi et al. [1] noticed that less than 50% of developers are only having cryptographic knowledge.

For instance, now all the banks had made a step ahead into the market to reach the services to the customers in an easy way via mobile banking. Mobile banking applications act as a vessel to store the sensitive information about the customers like the name of the customer, password, etc. Not only in mobile banking applications but also in mobile wallets, payment transaction based apps also belongs to this category. It indicates that a huge amount of sensitive content will be there in the apps. So, there is a need for providing security to the apps by using cryptographic API. But, due to the oversight of developers in developing the app, the cryptographic fundamentals are violated in developing the cryptographic primitives. Because most of the developers focused on the functionality, not on the strength of the security. Loopholes are forming due to this oversight and it leads a way to the hackers and results in the security violation.

In this work, the authors analyzed the inappropriate usage of cryptographic primitives in the Indian mobile applications containing money transactions. The authors identified mainly three categories in these applications. First, applications related to mobile banking. Second, applications related to mobile wallets. Third, e-commerce applications.

Category 1: Mobile banking applications

Different kinds of mobile banking applications available in the market based on the category of banks. Banks can be categorized as scheduled commercial banks and non-scheduled banks. Most of the mobile banking applications can be used by scheduled commercial banking applications. Within them, large public sector banks and new-generation private sector banks got a maximum share of business in terms of the number of users and number of transactions. It can also be said that these large public sector banking applications, new-generation private sector banks have the maximum deployment share in mobile banking app space.

Category 2: Mobile Wallets

A mobile wallet is a virtual wallet housed in an app on your mobile device which stores payment card information in it. Mobile wallets are a convenient way for users to make payment. Higher volumes of transactions are done in the retail market by using these mobile wallets. Simply, mobile wallets are digitalized version of our physical wallets. It can store personal identity cards like driving license number, credit card details, etc. Mobile wallets will reduce the payment time and waiting time too.

Category 3: E-Commerce Applications

In addition to mobile banking and mobile wallet applications, the authors analyzed the e-commerce applications. For example, one user wants to order the food through an online app (e.g., Swiggy—Online food delivery app in India). User will pay the necessary amount for food by using net banking/mobile wallets/other payment methods like credit card/debit card, etc., in the Swiggy app itself. Many of the mobile applications with different categories like entertainment, lifestyle, education, photography, business, sports, shopping, travel, medical, food and drink, etc., support financial services.

Static and dynamic analysis techniques were used in this work for analyzing. Static analysis is the process of analyzing the software artifacts to gain information about source code, binaries and configuration files without executing, i.e., analyzing the application "at rest." Dynamic analysis is the process of executing the artifacts at run time. It is the process of evaluating the application by using real-time data. The authors selected 36 mobile applications for analyzing the flaws in cryptographic primitives used in the above categories.

The remaining paper organized as, previous research work described in Sect. 2, 24 cryptographic inappropriate usages are listed in Sect. 3, methodology followed and results obtained are shown in Sect. 4.

2 Related Work

In this section, the authors described the previous works of researchers for detecting the flaws in cryptographic primitives.

Egele et al. [2] implemented a new system named as CryptoLint, analyzed the static program analysis. This system analyzed the cryptographic misuses quickly by disassembling the binary files. But, it works only on Dalvik bytecode, not on native code.

Mitchell et al. [3] implemented a new enumeration tool named as Murphi for detecting replay attacks, analyzing the cryptographic protocols.

Kruger et al. [4] implemented CrySL compiler. In this compiler, they defined the cryptographic parameters and algorithms as rules. CrySL automatically detects if an android app violating the CrySLencoded rules.

Das et al. [5] systematically compared the usage of cryptographic primitives in different programming languages.

Shuai et al. [6] implemented a new tool for detecting cryptographic vulnerabilities. But, it didn't detect the vulnerabilities which were not defined in the API.

Lazar et al. [7] detected 269 vulnerabilities from CVE database. Out of these, 223 vulnerabilities are caused due to oversight of developers while developing the product.

3 Identification of Inappropriate Usage of Cryptographic Primitives

In this section, the authors presented the inappropriate usage of cryptographic primitives by developers while implementing in the android applications. Twenty-four weaknesses were identified and divided into 4 categories shown in Fig. 1.

1. Weak Cryptography (WC)
2. Weak Implementations (WI)
3. Weak Keys (WK)
4. Weak cryptographic Parameters (WP)

3.1 Weak Cryptography (WC)

If cryptographic algorithms are implemented in an insecure manner, then weaknesses will arise. Suppose, for example, MD4, SHA1 are used in the application, then it is said to be weak algorithms because it can be easily broken. This will lead to WC1 weakness, i.e., weak cryptographic algorithm used. Inappropriate usages of Weak Cryptography are explained briefly in Table 1.

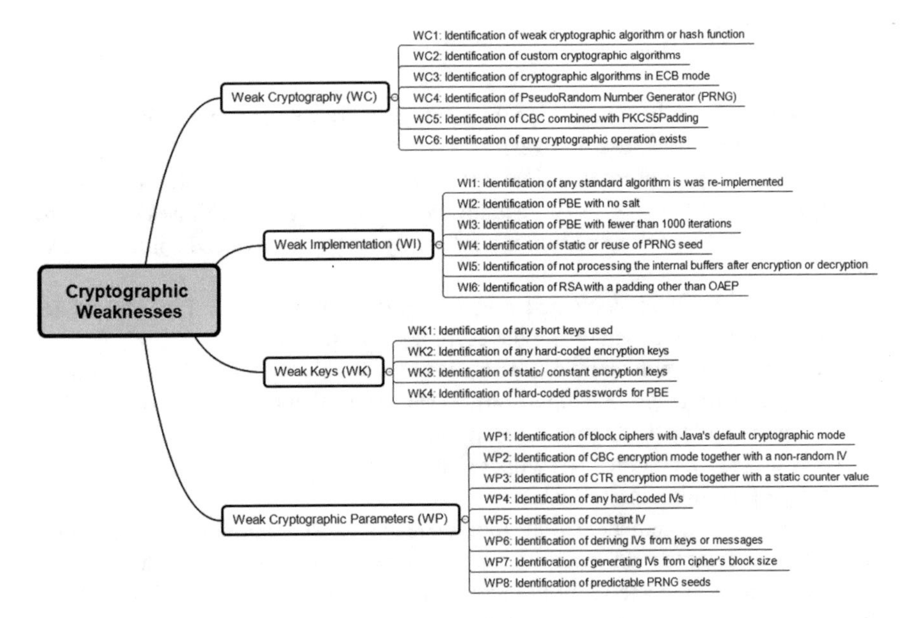

Fig. 1 24 inappropriate usages found in cryptographic primitives

Table 1 Brief explanation of weak cryptography (WC) misuses

Code	Name of the weakness	Explanation
WC1	Identification of weak cryptographic algorithm or hash function	For example, MD5, SHA1, DES are used in application, then it is a weakness
WC2	Identification of custom, cryptographic algorithms	If cryptographic algorithm implemented by own (not reviewed), then it maybe vulnerable to attacks
WC3	Identification of cryptographic algorithms in ECB mode	If algorithms were implemented in Electronic Code Book (ECB) mode, then it is a weakness
WC4	Identification of PseudoRandom Number Generator (PRNG)	If PRNG used for keys production, it is a weakness
WC5	Identification of CBC combined with PKCS5Padding	If PKCS5Padding used, then it is a weakness
WC6	Identification of any cryptographic operation exists	If application not used any cryptographic operation, then it is a weakness and impact will be high

3.2 Weak Implementations (WI)

If cryptographic algorithms are implemented in a non-secured manner, it will result weakness. For example, if developer modified the existing standard algorithm, then it will lead to WI1 weakness, i.e., re-implementation of the standard algorithm. Inappropriate usages of Weak Implementation are explained briefly in Table 2.

Table 2 Brief explanation of weak implementation (WI) misuses

Code	Name of the weakness	Explanation
WI1	Identification of any standard algorithm was re-implemented	If any standard algorithm was re-implemented, then WI1 weakness will arise
WI2	Identification of PBE with no salt	If Password Based Encryption (PBE) was implemented without salt, then it may vulnerable to brute-force attack
WI3	Identification of PBE with fewer than 1000 iterations	If PBE is implemented with fewer iterations, it will also lead to brute-force attack
WI4	Identification of static or reuse of PRNG seed	PRNG seed will not be reused
WI5	Identification of not processing the internal buffer's after encryption or decryption	Internal buffers which were not processed after encryption or decryption. If this happens, then it is W15 weakness
WI6	Identification of RSA with a padding other than OAEP	Combination of RSA with OAEP was recommended. While, other combinations were not recommended

Table 3 Brief explanation of weak keys (WK) misuses

Code	Name of the weakness	Explanation
WK1	Identification of any short keys used	If short keys are used in the implemented algorithm then it is a weakness
WK2	Identification of any hard-coded encryption keys	For encryption, hard-coded keys are not allowed
WK3	Identification of static/constant encryption keys	Don't use static/constant keys for encryption
WK4	Identification of hard-coded passwords for PBE	If hard-coded passwords are used in PBE, then it is a weakness

3.3 Weak Keys (WK)

If cryptographic keys are used improperly, then arisen weaknesses will come into this category. Suppose, if keys used for encryption generated by hard-coding, then it is WK2 weakness, i.e., identification of hard-coded encryption keys. Inappropriate usages of Weak Keys are explained briefly in Table 3.

3.4 Weak Cryptographic Parameters (WP)

While implementing cryptographic primitives in the applications by the developers, if they use weak parameters like cryptographic modes, constants, etc., then the raised weakness will belong to this category. Inappropriate usages of Weak Cryptographic Parameters were explained briefly in Table 4.

4 Methodology and Results

The authors have analyzed three categories of mobile applications with a combination of static and dynamic analysis. By using manual inspection and automated tools, the authors performed the static and dynamic analysis and identified the cryptographic primitives improperly. In the static analysis, java source files in apk file can be read by using dex2jar, Jadx tools, and manifest file, resource files are converted into a readable format by using apk tool. For better results, the authors performed dynamic analysis by using Burpsuite and MobSF tools, flaws are identified by executing the android applications. we presented the results after performing the static and dynamic analysis. Total 36 android applications were analyzed (Category1:14, Category2:12, Category3:10). Weaknesses found in each category are listed in Table 5.

After completion of this experiment, the authors observed that most of the applications have weaknesses in the first type. Most of the weaknesses belong to weak

Table 4 Brief explanation of weak cryptographic parameters (WP) misuses

Code	Name of the weakness	Explanation
WP1	Identification of block ciphers with Java's default cryptographic mode	Using Java's default cryptographic mode was not recommended
WP2	Identification of CBC encryption mode together with a non-random IV	Always use CBC encryption with non-random IV. Other than this combination, it can be treated as weakness
WP3	Identification of CTR encryption mode together with a static counter value	CTR with static counter value is not a safe approach
WP4	Identification of any hard-coded IVs	If any hard-coded IVs are there, then it is a weakness
WP5	Identification of constant IV	If constant IV used in the implementation, it is insecure
WP6	Identification of deriving TVs from keys or messages	Deriving IVs from keys may be predictable. So, it will lead to vulnerability
WP7	Identification of generating IVs from cipher's block size	Generating IVs by using random class is a good practice
WP8	Identification of predictable PRNG seeds	PRNG seeds must be in good strength so that PRNG seeds cannot be predictable

Table 5 Results—weaknesses found in ethre categories of Applications

S. No.	Category of weakness	Category1 Apps	Category2 Apps	Category3 Apps
1	Weak cryptography	28	39	42
2	Weak implementations	18	21	23
3	Weak keys	7	11	15
4	Weak cryptographic parameters	12	17	15

cryptography (WC) (66.7%), 58.3% of applications are misusing the cryptographic primitives related to weak implementations (WI). Similarly, WK weaknesses were found in 33.3% of applications and WP weaknesses were found in 36.1% of applications. The authors represented the applications by using magic quadrant shown in Fig. 2. Each category of applications was represented by using the classification scheme and shown in Figs. 3, 4, and 5. Mobile banking applications are Category1 named as A_1 to A_{14}. Mobile wallet applications are Category2 applications and named from B_1 to B_{12}. Similarly, e-commerce applications are category3 and named from C_1 to C_{10}.

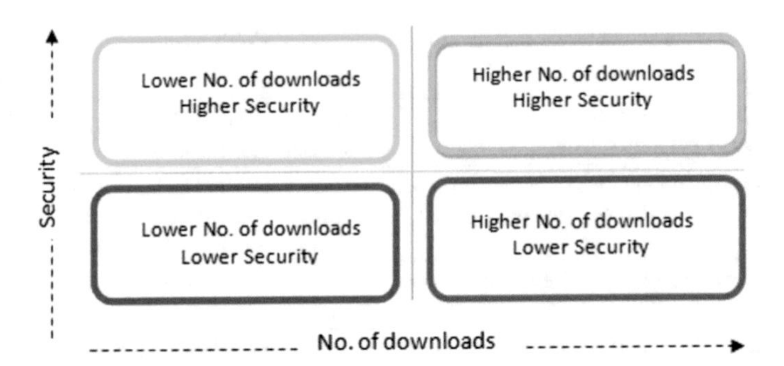

Fig. 2 Applications classification by using magic quadrant

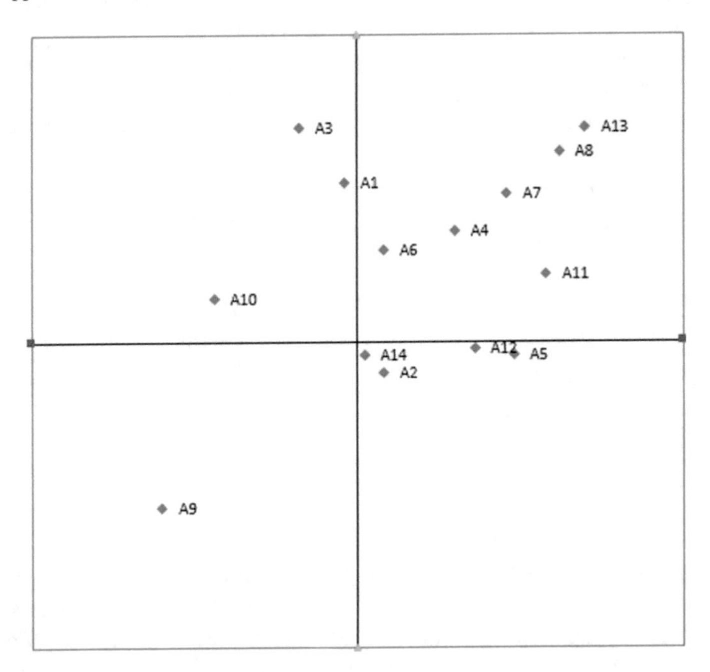

Fig. 3 Category1 (mobile banks) applications

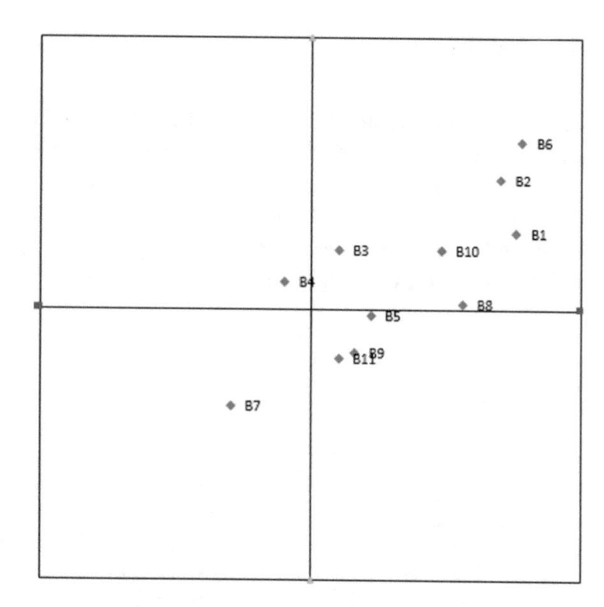

Fig. 4 Category2 (mobile wallets) applications

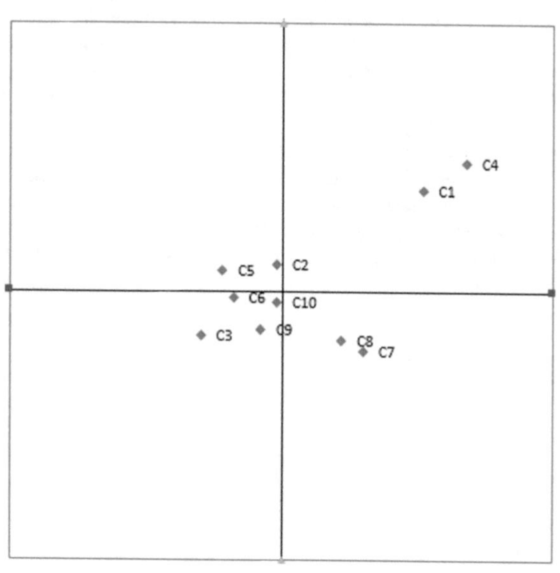

Fig. 5 Category3 (E-commerce) applications

5 Conclusion

The authors found that 24 inappropriate usages were there in cryptographic primitives and categorized then into 4 groups, namely weak cryptography, weak implementation, weak keys, and weak cryptographic parameters. The authors selected 36 android applications which were 3 different categories of android applications like mobile banks, mobile wallet, and e-commerce applications. The authors performed static and dynamic analyses, and identified inappropriate usages in these 36 applications.

References

1. Nadi S, Krüger S, Mezini M, Bodden E (2016) Jumping through hoops: why do Java developers struggle with cryptography APIs? In: Proceedings of the 38th international conference on software engineering. ACM, pp 935–946
2. Egele M, Brumley D, Fratantonio Y, Kruegel C (2013) An empirical study of cryptographic misuse in android applications. In: Proceedings of the 2013 ACM SIGSAC conference on computer & communications security, pp 73–84
3. Mitchell JC, Mitchell M, Stern U (1997) Automated analysis of cryptographic protocols using Murphi. In: Proceedings of the IEEE symposium on security and privacy, pp 141–151
4. Krüger S et al (2017) Crysl: validating correct usage of cryptographic apis. arXiv:1710.00564
5. Das S et al (2014) IV = 0 security: cryptographic misuse of libraries. Massachusetts Institute of Technology
6. Shuai S, Guowei D, Tao G, Tianchang Y, Chenjie S (2014) Modelling analysis and auto-detection of cryptographic misuse in android applications. In: IEEE 12th international conference on dependable, autonomic and secure computing (DASC), Aug. 2014, pp 75–80
7. Lazar D et al (2014) Why does cryptographic software fail? A case study and open problems. In: Proceedings of 5th Asia-Pacific workshop on systems. ACM

A Nature-Inspired Solution to Managing Activities in the Cloud with Equal Time by Using Machine Learning Approach

S. Naga Lakshmi, Madhuri Paul, and S. Nagaprasad

Abstract Cloud computing is one of the strongest paradigms in computing. The advancement of cloud storage infrastructure renders things complicated due to the huge growth within the field. The load balancing from a virtual computer plays a significant role in effective planning. The natural solution named "healthy bee feed" (LB-HBF), a system for the automated load balancing of machinery, is suggested here. The job and machine assignment method is the same time assignment in the context of this LB-HBF. This strategy assigns virtual machines activities primarily and reduces the relocation of work through virtual equipment. Load balancing is considered one of the key mechanisms for the efficient distribution of cloud services. The issue of load balancing would in future involve entirely autonomous distributed networks. In this case, an Osmosis Load Balancing (OLB) technique has been implemented. In order to program functions on Virtual Computers, OLB works on an osmosis principle. The method is based on the Digital Hash Table (DHT) chord overlay scheme. The overlay of chord is used to track the state of clouds and organic workers. In a variety of heterogeneous and homogeneous clouds, the proposed algorithms demonstrate better efficiency by simulation experiments.

Keywords Cloud computing · Machine learning · Load balancing · OLB · DHT

S. N. Lakshmi (✉)
Department of M.C.A., St. Ann's College, Mehdipatnam, Hyderabad, Telangana, India
e-mail: lakshmi.msccs@gmail.com

M. Paul
Department of Computer Science, St. Joseph's Degree and P.G. College, Hyderabad, Telangana, India
e-mail: madhuripaul.2020@gmail.com

S. Nagaprasad
Faculty of CS and CA, Tara Govt. College (A), Sangareddy, Telangana, India
e-mail: nagkanna80@gmail.com

N. Chaki et al. (eds.), *Proceedings of International Conference on Computational Intelligence and Data Engineering*, Lecture Notes on Data Engineering and Communications Technologies 56, https://doi.org/10.1007/978-981-15-8767-2_20

1 Introduction

Cloud computing is a network infrastructure, which provides rapid growth in communications technology, providing customers with different requirements with applications with the help of remote computer resources. Technology and facilities, technology production systems and testing experiments as tool [1, 2] are in storage. These resource allocations are rendered by energy providers. The second was put under the heading "Server Application Systems" (SaaS) and server Network as System (SaaS), respectively, [3] although the first is classified under Cloud Data Infrastructure (Iaas). Cloud storage is an on-demand cloud network that combines services with the pay-as-go (PAYG) [4]. Any of the major players for this growth are Amazon, Microsoft, Twitter, SAP, Oracle, VMware, IBM, and other big players [1, 2]. The vendors are mainly professional IT firms. For the cloud storage platform, two different headings are shown. The first is the provision of information defined by the form of operation of a conventional cloud provider. That explains why there is common usage of three primary SaaS, PaaS, and IaaS [5, 6] types. The other is the scale, relationship, management, and complexity of the cloud model and its visibility. The NIST overview offers four cloud systems, proprietary, public, group, and hybrid clouds [7]. Cloud Networking approved NIST definition.

One of the key issues of virtualization is configuration balancing. Configuration control. There are big load management experiments being carried out, but cloud infrastructure continues to be an important topic and various work initiatives are in progress [2]. That is because the cloud architecture is generic and the issue is normal. Traditional charge equilibrium algorithms can only be used with homogeneous and committed capital, so cloud computing cannot work successfully [3]. The sophistication, nuances, and versatility are also common aspects of cloud infrastructure, but cannot be extended specifically to cloud infrastructure with traditional load balancing algorithms.

Load balancing relates to how operations are fairly spread around the storage infrastructure of data centers of order to improve the performance of cloud computing. The primary function of load balancing that focus on the client or service provider, and can be identified by the user, independent of some other network activity, attempts to minimize the effect of its operation. The goal of the service company is to increase the turnaround period and effectively allocate available funds. The question is categorized into four steps that reflect a practical solution to handling loads. (1) Load calculation: an estimate of the load is required to evaluate the load disequilibrium first. Calculation of workloads involves different activities to assess the operation for the number balance. (2) Start-up load balance: if there is a discrepancy, when loads for all the VMs are defined. And the costs for load imbalance over load harmony, along with load equilibrium, are higher. (3) Task selection: tasks will be selected based on the details provided to switch from one VM to another VM in these measures. (4) Job Migration: job transfer will be begun after role collection from one VM to another. The algorithm requires to be preserved in the above measures. In this article, we propose to use the algorithm of honey bee load balance [4] to classify automatic machinery operations for federated clouds.

The problem of load imbalance is an unexpected occurrence on the CSP side, undermining the capacity and reliability of machine services along with a promise of service quality (QoS) under the negotiated Service Level Agreement (SLA). In these conditions, load balancing (LB) is important, and researchers are particularly interested in this subject. The load balancing is achievable at the physical system or VM stage [2] in cloud computing.

A job requires the resources of a VM and as a number of tasks arrive at a VM, the resources are drained, meaning that no new work requests are available. The VM is said to have reached an overwhelmed state when such a situation arises. At this stage, activities are either hungry or end in stalemate with no chance of accomplishing them. As a consequence, tasks on other VM will move to another tool.

The method of moving workloads requires three simple steps: load balancing, which tests the existing system load, resource selection, and another sufficient resources and workload migration. Three systems generally known as the load handling systems, the acquisition of services and the transfer of activities are used for these operations.

2 Literature Survey

Cloud computing is recognized as one of the latest technologies in the area of cloud computing and is developed for enterprise rather than university use. This cloud infrastructure offers virtualized, centralized, and scalable tools as end-user applications. For fact, it has a big advantage to fully support computation as a business. The cloud helps consumers to access the computational resources of providers by leasing policies. These operations are performed in a virtualized environment with the potential to deliver resources to hundreds of thousands of computers. Throughout the cloud world, manually assigning resources is not feasible, so we focus on the virtualization principle. Cloud infrastructure provides an innovative option for equipment maintenance, certified applications, and employee preparation. Cloud computing is entirely Internet based, millions of machines are Internet linked to the web. Online computing delivers servers, bandwidth, applications, network, and even other resources.

The definition of virtualization flexibly applies the cloud services to the clients. Virtualization is the principle behind cloud storage. In order to optimize the power, virtualization combines massive computational power. Foster et al. (2008) proposed that cloud computing consists of four layers.

- System layer
- Centralized infrastructure layer
- Base layer
- Structure layer.

Framework layer comprises computing resources, device resources, and network resources. The single property layer comprises the virtualization technique's representation of hardware. The application layer controls the midware framework for end users. The server layer contains the user interface on the cloud floor.

2.1 Survey on Load Balancing of Tasks in Cloud Computing

Load management is one of the key issues in conjunction with virtualization. Key studies in the field of load balancing, but cloud infrastructure is also an important topic and various work initiatives are ongoing. That is because cloud infrastructure is common and that the problem itself is distinctive.

The classic algorithms of load balancing can only be used with standard, committed services, so the cloud infrastructure cannot work correctly. There are other different aspects of cloud infrastructure, such as complexity and flexibility that cannot explicitly be utilized with cloud computing through traditional load balances. M. Randles et al. investigated a decentralized load balancing strategy focused on honeybees, which is a naturally inspired solution to self-relation.

It controls loads through neighboring application operations. The program's execution is enhanced with an increased range of features, though device scale improves the performance or not. This is ideally tailored to the circumstances in which specific service user population is needed. Z. Emilia et al's. load balancing solution was introduced in a transparent distributed computing system that cantered on the ant colony and on dynamic network theories. This method overcomes heterogeneousness, is scalable to diverse conditions, offers brilliancy in faulty tolerance and has high adaptability which helps boost the efficiency of the system. The system uses small worlds without the quality of a complex load balance.

A. Sarkozy et al.'s standard load balancing strategy has been introduced for VMs in cloud computing. It uses global state knowledge to make choices on load balancing. Load balancing increases average efficiency equally, and fault mitigation is not taken into account. R. Hamilton et al. proposed a carton solution that incorporates the load balance and the distributed rate management method which functions as a cloud management integration mechanism and utilizes load balance to minimize costs and a distributed resource allocation limit A. Columbia et al. VectorDot methodology has been applied for data centers with optimized cloud virtualization and storage. The dot product is used to distinguish nodes according to the commodity criteria.

The algorithm in the graph helps to address the question of load balancing for the distribution of capital. The method does not, however, tackle costs elimination, i.e., the expense of the load allocation, which can require more time than the real measurement period spent. Few analysis [H. Many test Shan and so forth. 2004] proposed latency algorithms to minimize transfer costs and gain from decreased data transmission in order to process internal data. Nevertheless, this type of algorithms requires concurrent applications for data processing and migration in order to maximize the data distribution and migration with the linear algorithm, and it implement master Slave Load balance (master slave load balancing). Yet only static load balancing is discussed in this algorithm. This suggests an efficient functional load balancing conversion algorithm based on estimation of the Lagrange multiplier for the transmitted weight in Euclidean form. The purpose of load balancing is to coordinate computing functions over virtual computers. It just operates in a homogenous setting and does not operate on heterogeneous grids. Calculation functions are named "building block" to reduce implementation time. Moreover, in distributed systems, makespan minimization issue is common; we also name it NP complete. In this way, reducing make-up is not just the duty of balancing loads, but also the need to cope with contact costs.

3 Load Balancing Model

The two-tier design model of load balancing is provided in imbalanced clouds for better load draining as shown in Fig. 1, in this section, that is, Gupta et al.'s updated design [16]. This concept is an example of the virtual machine manager and virtual machine control. At the physical machine (PM) stage, the first phase load balancing is carried out and the second level at the VM point. On this basis, two migration tasks are open.

The code generator generates user operations which require computer resources to be complied with by software requirements. Data Center Manager Process Management. For a single system task assignment, the Load Balancer controls the VM. The

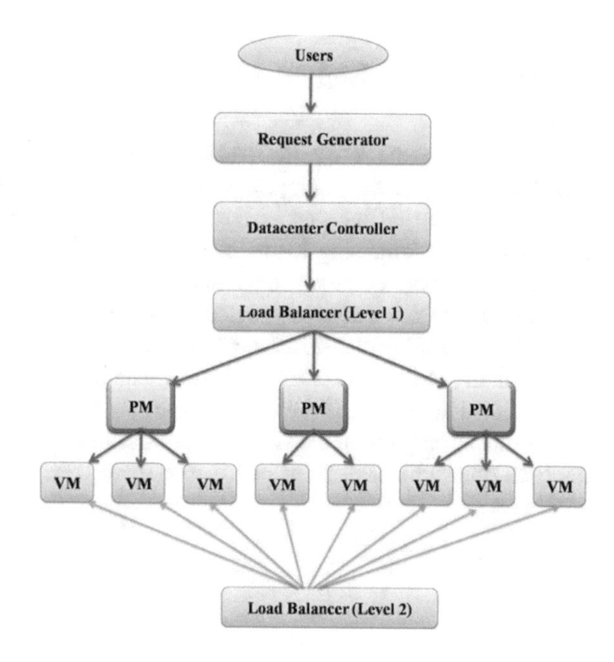

Fig. 1 Level load balancing architecture

first stage load balancing compensates by dividing the workload across the respective virtual machines. The second load balancer spreads the load across several virtual machines and various physical machines.

3.1 Identifying a VM's Resource Knowledge

The resource data in a file is checked. The existing usage of VM resources and unallocated resources is thus made. Since of this step, the status of VM may be calculated similarly with respect to a threshold.

3.2 Task Scheduling

After the resource specificities of a VM are specified, operations in proper VMs are scheduled by a programming algorithm.

Resources allocation For scheduled projects and facilities is reserved. It is achieved by way of distribution of capital scheme. A considerable variety of government formulation and distribution are discussed in the literature. Calendars are critical for improving performance, but allocation approach to handle resources properly and boost resource quality is introduced. The reliability and allocation technique of this scheduling system describe the strength of the load balance algorithm.

4 Research Methodology

To specifically hit the core of the load balance process, a suitable testing methodology was introduced. In the background of the general research program, the literary analysis was carried out descending the methods, theories, formulas, tactics, and paradigms of imbalancing loads. In accordance with CGF methodology [33], we studied the question of load imbalance to establish the triggers, factors, and load balance parameters. The study recommendations for Systematic Literature Review (SLR) for load balance inside the cloud have further strengthened literature work by Kitchenham [26, 27, 34]. An SLR is a repeated analysis that can be reproduced by other studies to obtain additional experience.

In order to emphasize the significance of load balancing in the cloud computing, multiple questions have been posed to resolve core concerns and load mismatch problems.

4.1 Proposed Classification of Load Balancing Algorithms

Algorithms for load balancing have been listed in this segment according to different parameters. A top-down approach is introduced and implemented in the classification process. The drawback in existing analysis papers is the absence of adequate and comprehensive hierarchy for load balancing algorithms. The role of a certain algorithm in taxonomy is very difficult to decide. A full-scale analysis of LB algorithms was first carried out in literature, which was scarce in the previous research, in the analysis. The different parameters used to classify include algorithm architecture, 'algorithm status, "load balancing table," load balancing type' and "load balancing technique." Depending on the design of the method, the load balancing algorithms are positive or reactive. This is our first clear taxonomic categorization where no prior literature studies have demonstrated. Depending on the device configuration, the LB algorithms are rigid, dynamic, or hybrid. LB algorithms are referred to as algorithms for planning and deployment, based on load balancing capabilities. The LB algorithms include LB task algorithms, LB network weight algorithms, and cloud loading algorithms. Based on functionality, load balancing types are categorized as hardware load balancing, which is also known as network load balance. The network describes load balancing algorithms, machine learning, inspired biological architecture, science algorithms, and swarm processing techniques.

4.2 Algorithm Design

The first categorization was carried out in the present research with the creation of the load balance algorithm. The LB algorithms are categorized as constructive approaches and reactive approaches. However, in other areas of technology, particularly communication and mobile ad hoc network networking (MANETS) [36], the essence of communications routing protocol was extensively studied.

A constructional strategy for LB, which requires behavior to alter and not only response as it happens, is an approach toward algorithmic architecture. The intention is to deliver a positive outcome and prevent a dilemma at an early stage rather than waiting for a crisis.

Proactive intervention aims to recognize and optimize opportunities to prevent future threats to issues. In reality only a few constructive solutions remain and that historically lack innovative ideas restrict current strategies. Dragonfly optimization and limit measurement-based LB was suggested in cloud storage by VMs with limited power usage through predictable load distribution. Xiao et al. proposed a game theory-conscious LB algorithm to reduce the estimated time to respond while preserving order and fairness. The game's Nash Control Point is the best price.

4.3 Random Server Load Balancing

The network must select the IP randomly from each link list [10, 11] It effectively relies on all customers having equal loads and the Law of Large Number [11] to maintain a fair low load spread through servers. That enables a system to use the random load balancing strategy. It was reported that caching issues with round-robin DNS are due to a random load balancing on the client side, that in broad DNS caching servers the distribution appears to be skewered for round-robin DNS, whereas random sorting is unexpected on the client side, regardless of DNS caching [11]. If an "intelligent device" is used, it often offers fault tolerance if the randomly chosen host is identified and linked arbitrarily.

The state of the LB algorithms is generally graded as static, complex, and hybrid based on the machine status of the algorithm. Recent literature work has shown that this method is more commonly utilized in the classification of LB algorithms. The bulk of the comparative charge balance work starts by putting the category above taxonomy in the algorithmic taxonomy. Traffic loads are distributed uniformly around servers for effective load balancing. This emerges from a well-informed model about computers' capabilities and activities. The static LB algorithm application is to VM for compilation periods. The downside of the static solution though is that the job does not pass through a load balancing system but would remain heavily obstructed. Static algorithms do not accept the existing device constraints, which include details regarding tasks such as task resource settings, time of contact, loading of nodes, energy, electricity, and bandwidth. The biggest downside of the LB static solution is that it is impossible to change and that a global network such as the cloud is not necessary because device state changes dynamically.

Collaborative algorithms are often labeled as "Offline," usually defined as Internet, ton and live mode as the picture indicates. The role is only assigned to a VM until the scheduler is reached in such pre-defined circumstances in a lot mode, electronically constructed. The hierarchical algorithms of the load balance are relatively complicated in contrast with their counterparts, and can regulate the movement of input traffic at any moment. Knowledge of the real network setting and capacity to accommodate diverse loads of the system are part of complicated load control. The benefit of dynamic load balancing is that operations can move from a complete system to a loaded system dynamically, but are extremely complicated in nature and very challenging to create compared with static LB algorithms (Fig. 2).

However, LB's dynamic algorithms are strong in efficiency, precision, and versatility. Dynamic load balancing algorithms operate well when node load variations are small but cannot be done in various load environments. Figure 3 demonstrates the taxonomic load balancing by design and state of the algorithm.

The algorithms are known as programming and delivery algorithms in this group. Role of load balancing. Cloud algorithms and allotments can be static or fluid, based on the current VM environment. Activities and implementation processes play an important role throughout the delivery of services and cloud control. Three related activities shall be divided into training policies, job management, work planning,

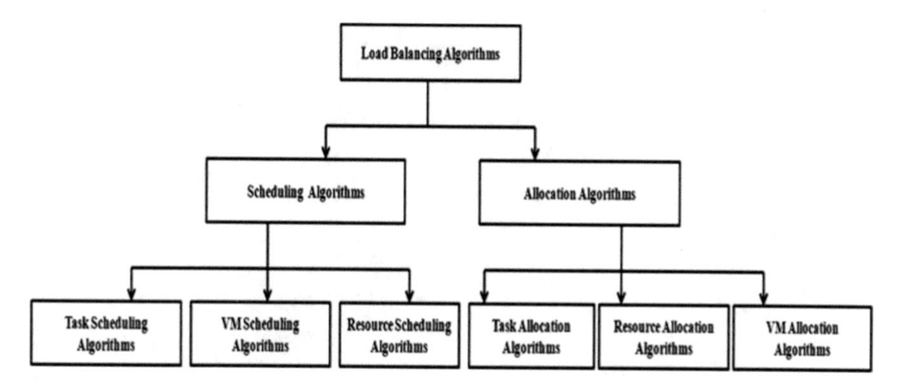

Fig. 2 Load balance algorithms focused on the system's existence and environment

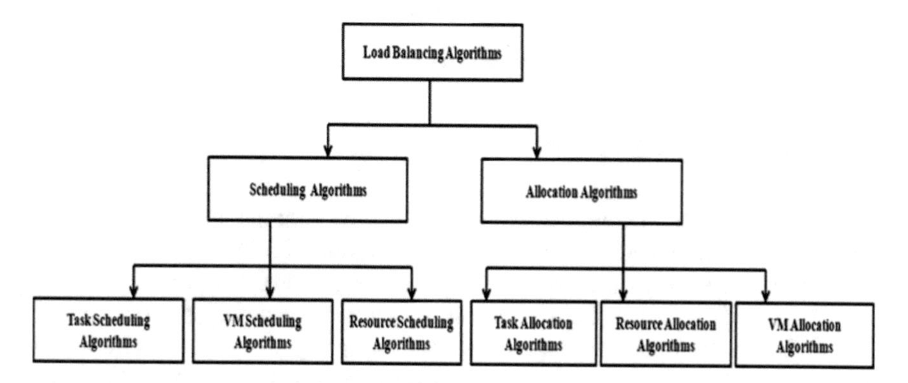

Fig. 3 Load balance depending on the attribute used

and preparation of VMs; the expense policies also shall be divided into duty sharing, operation transitions, and VM delivery.

Projects are organized to assign the computing resources required for the user job while the project is scheduled, programmed, or monitored. VM preparation consists of creating, deteriorating, and maintaining the VMs on physical hosts and handling the domain migration of VMs. Job is the main function of the unit. The usage of a work device is wealth control. The reciprocal replacement is the transition of responsibilities and the distribution of profits. For changes to other machinery applications or software groups, VM assignment occurs. Figure 4 demonstrates the equations for load control for the specified technologies.

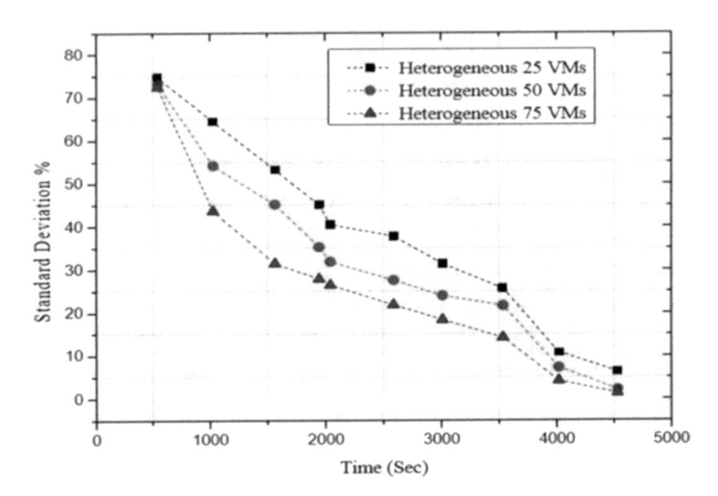

Fig. 4 The distribution of load in a heterogeneous world

5 Load Balancing Setup

Within the table, the load balancers are classified by configuration as elastic load balancers and mechanical load balancers. The equilibrium in hardware load was related to the hardware load distribution, e.g., memory, control, and processor. Elastic Load Balancing spread user traffic dynamically across various sites, such as Amazon EC2, bins, and IPs. In one or more mobility areas, appropriate consumer traffic can be accommodated. Elastic load balancing offers three forms of load balancing, each delivering high consistency, efficient scaling, and reliable control of tolerance. File Space Load Balancer user-level routing goals, containers, and IP addresses on the basis of the complexity of the request (Layer 7).

The load balancer network provides a high-technology, container-based network networking that is perfect for new applications and is appropriate for HTTP and HTTPS balanced load-balance. Load Balancer Server allows the usage of the most modern SSL/TLS ciphers and protocols simpler and safer for device protection. In the transport layer of the OSI architecture, network load balances are introduced. It can connect every second with millions of users.

The whole deployment cycle includes network load balancing in Microsoft's azure and AWS programs. The mechanism for network load balancing includes the transfer of data from servers utilizing the TCP/IP protocol. The standard load balancer offers a quick load balancing both at request level and at link level, with specific Amazon EC2 specifications and capabilities. New load balancers are required for applications developed in the EC2-Classic network.

6 Experimental Simulation Setup

Throughout this phase, we built 75 VM on a chord overlay. The roles are applied until all VMs are connected to the overlay. The load concentration in VMs is passed to the remaining machine every 30 s and the osmosis is done inside the remaining machine in total 60 s. A systematic review of standards is yet to be carried out and leaves for future analysis.

All VMs have the same features and the capacity to do a certain kind of job at maximum in a single task in a homogenous setting.

VMs and tasks provide various forms of configuration in a heterogeneous environment. The execution outputs of each VM are unique to the others. Task execution may differ from homogenous conditions. The standard deviation for load balancing calculations of the algorithms specifies the OLB parameters. The real standard deviation of VMs is taken into account in heterogeneous conditions and the overall standard deviation value is taken into consideration. The efficiencies of the algorithm is determined by the amount of tasks that move from VMs that we normally call.

7 Results

The following analysis of the proposed algorithm will cover the section. In the field of simulation, these values are preserved. In that relation, 1 Norm variations are preferably below 80% for all cases. For example, in heterogeneous conditions the same method is higher.

In the sense of the ring and data center, the recognition agent's main function is to provide reliable information on VM output and capacity [18]. The parents, heirs, and randomly picked VMs are sent daily by means of the Data Center. By using the finger list, the VM foremen and descendants may be listed (Fig. 5).

The model 25, 50, and 75 VMs are tested in both heterogeneous and homogenous environments and its default variance is measured. The findings are shown in (Fig. 6).

It is known that the VM objective is assigned to this unit on the basis of Algorithm 1, to move the VM objective (formerly α, following β and random μ). The agent acquires the surrounding VMs and output information and uses the VM goal to adjust awareness. The functionality of the identification agent is concentrated in two factors: one, VM load recognition and output data; and the other, VM file transmission by detection of file-based VMs.

Load balancing of cloud infrastructure is considered a major challenge. The even distribution of workload through VMs benefits from cloud load control. This improves the network reliability and the successful performance of the system such that consumers are more happy and helpful. This also guarantees a fast and decent delivery of the computing resource (Fig. 7).

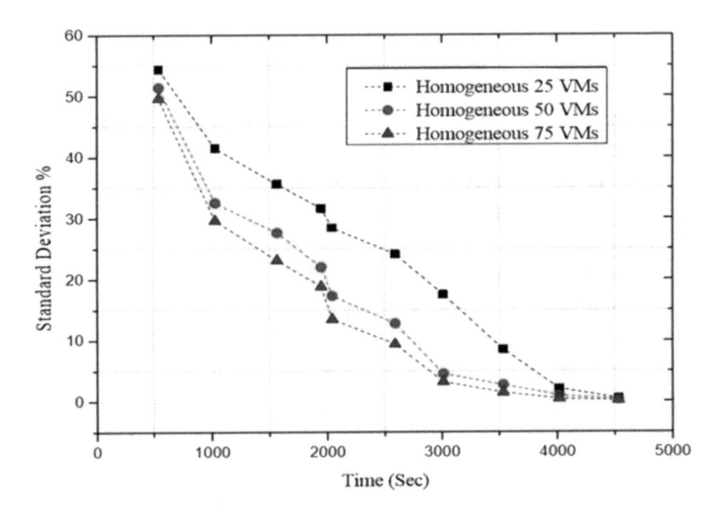

Fig. 5 Load equilibrium in standardized condition

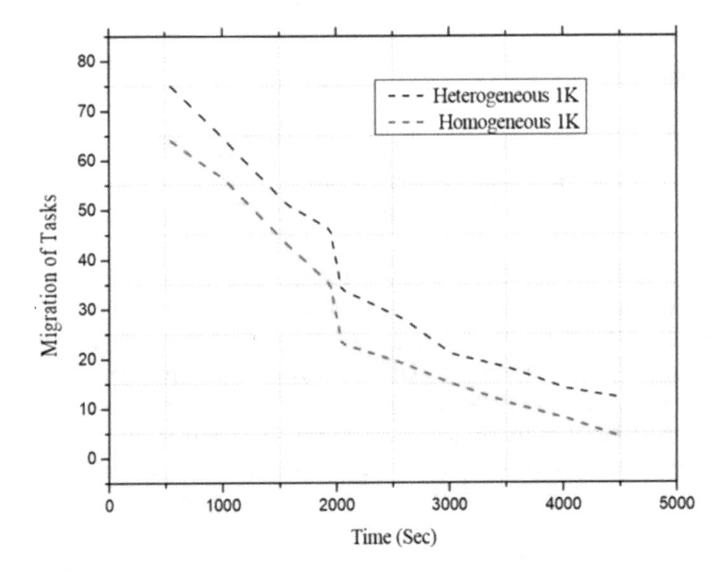

Fig. 6 Migration task versus time

The program is reliable and uniform in its scalability. Increased personal preferences suddenly that there is the accumulation of social disparities. The movement of activities is very sluggish in the homogenous environment as opposed to heterogeneous environments (Fig. 8).

The next thing is the issue of system mistakes due to load faults whether specific modules or several modules are not enough, resources specifications are installed or disabled without failure. The second part is the issue. It also guarantees that all

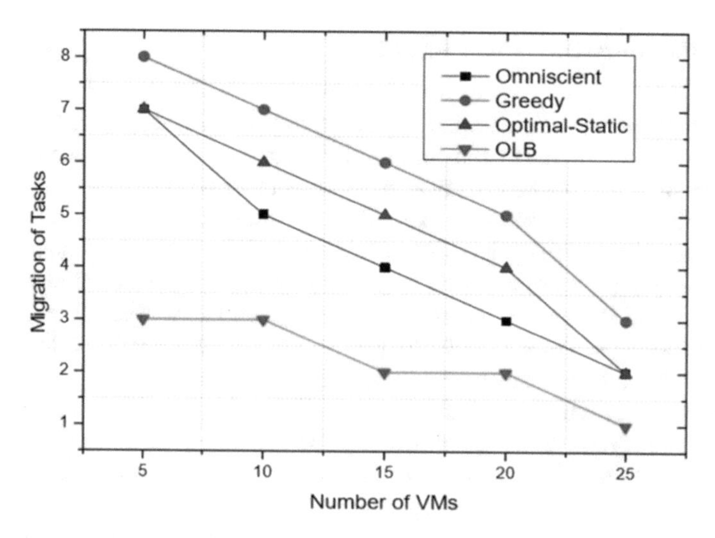

Fig. 7 Amount of work motions via VMs while the tasks are 20

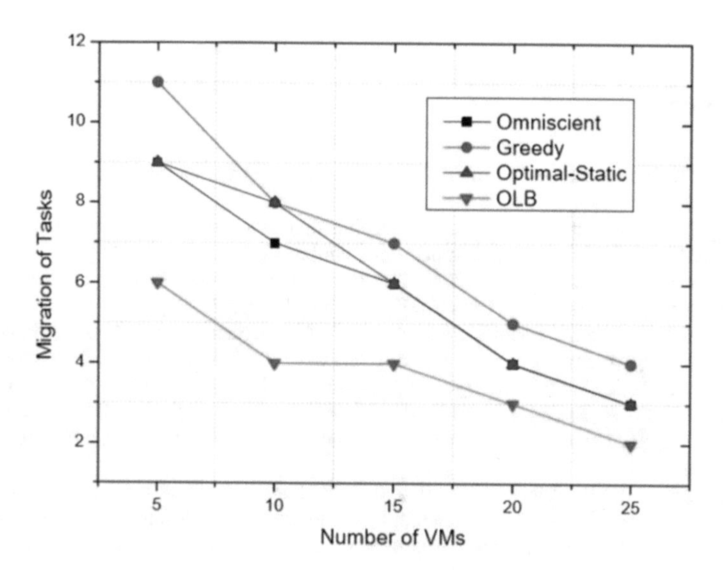

Fig. 8 Amount of work motions via VMs while the tasks are 40

estimation tools have been properly and fairly distributed. The cloud infrastructure ends with broad heterogeneous networks that combine smaller and big providers in order to facilitate the growth of on-demand services. Modern structures for the full execution of activities are also required. To cope with this, we have an integrated and controlled bio-inspired solution [20] that is used for a completely distributed VM load equalization [19, 20]. Our approach relies on non-inspired agents and load balancing through VMs focused on the general osmosis regulations.

8 Conclusion and Future Scope

In this article, we have discussed the Robin Round modified algorithm, a short move toward a proposed model that is optimized. Algorithms are not designed to control the behavior of the method (equivalent to scheduling results). How long will the round robin approach take to be appropriate? In consideration of the accuracy and usefulness of the Robin cycle, this article addresses the problem by quantum time rather than a quantum of a certain length. It requires time to accomplish a target. Load balancing is known as one of the key frameworks for efficient utilization of cloud computing services. In order to deal with problems of load equilibrium, total autonomous distributed networks will be needed. We introduced an Osmosis Load Balancing Device (OLB). With this in mind, OLB uses the principle of osmosis for virtual machine reprogramming functions. The solution is based on the hash table distributed with the chord overlay method (DHT). The Chord Overlay is used to monitor the cloud status and organic personnel. The proposed algorithm demonstrates stronger results both in heterogeneity and in uniform clouds in various situations by simulation analysis.

References

1. Naha RK, Othman M (2014) Brokering and load-balancing mechanism in the cloud—revisited. IETE Tech Rev 31(4):271–276
2. Fang Y, Wang F, Ge J (2010) A task scheduling algorithm based on load balancing in cloud computing. In: Web information systems and mining. Springer, pp 271–277
3. Maguluri ST, Srikant R, Ying L (2012) Stochastic models of load balancing and scheduling in cloud computing clusters. In: INFOCOM 2012 proceedings IEEE, pp. 702–710
4. Randles M, Lamb D, Taleb-Bendiab A (2010) A comparative study into distributed load balancing algorithms for cloud computing. In: 2010 IEEE 24th international conference on advanced information networking and applications workshops (WAINA), pp 551–556
5. Lucas-Simarro JL, Moreno-Vozmediano R, Montero RS, Llorente IM (2013) Scheduling strategies for optimal service deployment across multiple clouds. Future Generat Comput Syst 29:1431–1441
6. Wickremasinghe B, Calheiros RN, Buyya R (2010) Cloudanalyst: a cloudsim-based visual modeller for analysing cloud computing environments and applications. In 2010 24th IEEE international conference advanced information networking and applications (AINA), pp 446–452
7. [online] Available https://newsroom.fb.com/Key-Facts
8. Hu J, Gu J, Sun G, Zhao T (2010) A scheduling strategy on load balancing of virtual machine resources in cloud computing environment. In: Parallel architectures algorithms and programming (PAAP) 2010 third international symposium, pp 89–96
9. Naha RK, Othman M (2014) Evaluation of cloud brokering algorithms in cloud based data center. In: International computer science and engineering conference (ICSEC), pp 78–82
10. Mahajan K, Makroo A, Dahiya D (2013) Round Robin with server affinity: A VM load balancing algorithm for cloud based infrastructure. J Info Process Syst 9:379–394
11. Calheiros RN, Ranjan R, Beloglazov A, De Rose CA, Buyya R (2011) CloudSim: a toolkit for modeling and simulation of cloud computing environments and evaluation of resource provisioning algorithms. Software Pract Experi 41:23–50

12. Domanal SG, Reddy GRM (2014) Optimal load balancing in cloud computing by efficient utilization of virtual machines. In: 2014 sixth international conference on communication systems and networks (COMSNETS), pp 1–4
13. Wang SC, Yan K-Q, Liao W-P, Wang S-S (2010) Towards a load balancing in a three-level cloud computing network. In: 2010 3rd IEEE international conference on computer science and information technology (ICCSIT), pp 108–113
14. Venkata Krishna P (2013) Honey bee behavior inspired load balancing of tasks in cloud computing environments. Appl Soft Comput 13:2292–2303
15. Florence AP, Shanthi V (2014) A load balancing model using firefly algorithm in cloud computing. J Comput Sci 10(7):1156–1165

A Machine Learning Approach to Analyze COVID 2019

BKSP Kumar Raju, D. Sumathi, and Bhargav Chandra

Abstract COVID 2019 is a family of Human Corona Virus and it is disrupting human lives across the world. It even started affecting countries other than China at higher rate of transmission. The origin of COVID 2019 is not yet clear and no scientific medication is available for cure. We analyzed the Corona dataset which is of more than 3000 X 5 dimensions by applying Time Series Analysis and Regression Models. We could predict the futuristic trend and further propose a design of the dataset for getting more insights of the pandemic unanswered questions.

Keywords COVID 19 · Multi-linear regression · Time series

1 Introduction

In the December month of 2019, an epidemic which currently prevails worldwide was initially called as severe acute respiratory syndrome coronavirus-2(SARS-CoV-2). The quick and rapid identification of the new virus has been done by the public health organizations, scientific and clinical societies by sharing the gene sequence to the world in order to create awareness to the people. The World Health Organization named the disease activated by the coronavirus as Coronavirus Disease(COVID-19).

It is categorized under zoonotic virus since the reservoir seems to be bats; however, the transitional host is not yet determined. There is a possibility of transmission through the fomites and droplets in the interim close contact amid the infective person and the infector. From various reports, it has been identified that there is an

B. Kumar Raju (✉) · D. Sumathi · B. Chandra
VIT AP University, Amaravathi, Andhra Pradesh, India
e-mail: bksp.kumar@vitap.ac.in

D. Sumathi
e-mail: sumathi.d@vitap.ac.in

B. Chandra
e-mail: bhargavachandra.mutyala@vitap.ac.in

N. Chaki et al. (eds.), *Proceedings of International Conference on Computational Intelligence and Data Engineering*, Lecture Notes on Data Engineering and Communications Technologies 56, https://doi.org/10.1007/978-981-15-8767-2_21

efficient existence of human-to-human transmission and the probability of spreading of 2019-nCoV has been increased [1]. Yet, further investigations are required to discover the source and the way of spreading among the humans.

The assessment factor is reproduction number (R0) which is related to the viral transmissibility. It is defined as the average number of people who will encounter with a disease from the contagious person. The likelihood of spreading or declining of a disease could be expressed in three ways [2]

- If the reproduction number (R0) is less than 1, then the prevailing infection will lead to less than 1 new infection.
- If the reproduction number (R0) is 1, there will be no chance of epidemic but the disease is expected to stay there.
- If the reproduction number (R0) is greater than 1, then an exponential growth might arise and it might lead to epidemic or pandemic.

Due to the viruses' international exposure, panic in people created lot of myths like using surgical masks is mandatory [3], eating non-veg should be avoided, should not order goods from other countries, etc. According to AIMS and FSSAI, properly cooked non-veg will not have any chance of virus transmission [4]. The virus will just sustain only for few hours in goods or non-living objects.

Since there is no vaccine available as of now, it is highly important to take precautionary measures and significant among them are (1) Avoid touching your face and Wash hands frequently, (2) Proper Hydration, (3) Exercise, (4) Homely Diet, (5) Sun Exposure, (6) Taking foods with ingredients like Amla, Turmeric, and Giloy sticks.

2 Related Work

When analyzing data for COVID-19, it is observed that the affected are adults more than the children. From various reports, it has been identified that mostly 80% of the infected cases are found to be mild, 15% are considered as severely infected who are in need of oxygen and 5% are critically infected which requires ventilation [5].

In [6], a nowcast has been done which provides the information about the nature of the epidemic in size, distinguishing the challenges of earlier unidentified pathogen through the uncertain clinical range, even after the detection of the uncertain case. Forecasting the probability of the diseases that spreads both domestically and internationally has started in January 2020.

The epidemic could be reduced through the appropriate analysis for isolation and combined intrusions. As of now, the present period from the inception of the disease to the isolation is 6–14 days. As per the study done by Chen 2020 [7], the examined fatality rate for 2019-nCoV is 3 % when compared with SARS and MERS-CoV. The likelihood of survival or death is based on the situation or the person. The fatality rate might be higher when the outburst occurs in public which possess a mediocre health system. Several models have been developed right from the initiation of the epidemics [8].

Table 1 Comparison of various methods for epidemic analysis

Method	Period and R0 (Average)	Credible interval (95%)	Fatality rate	Epidemic doubling rate
By applying the Markov Chain Monte Carlo methods, the basic reproduction number is computed [6]	2.68 (from Dec 31, 2019, to Jan 28, 2020)	2.47–2.86	–	6.4 d
Mathematical model has been implemented, which is dynamic in nature and comprises five partitions such as individuals who are recovered, individuals those who receive treatment, susceptible cases, and during the incubation period, certain individuals might be symptomatic and certain individuals with indications [9]	6.49 (from 12–22 January 2020)	6.31–6.66	–	–
Through the implementation of Exponential Growth (EG) and maximum likelihood estimation (ML) , the value is computed [10]	As of January 23, 2020, 2.90	2.32–3.63 (2.90) 2.28–3.67 (2.92)	Out of 1965 suspected individuals and 1287 confirmed cases, 41 fatality had been stated	Before Jan 9, it was 6.7 d and after Jan 19, it has been reduced to 0.7 d.
A deterministic SEIR metapopulation transmission model is applied [11]	(Air travel alone is considered) From Jan 1–22, 2020, 3.1	0.4–4.1	–	–

(continued)

Table 1 (continued)

Method	Period and R0 (Average)	Credible interval (95%)	Fatality rate	Epidemic doubling rate
The Incidence Decay and Exponential Adjustment (IDEA) mode is applied [12]	From December 8, 2019 through January 26, 2020. 2.0–3.1	–	–	–
A simple disease-transmission model is applied [13]	starting in mid-November 2019, with a serial interval of 7 d (time between cases) 2.3	–	–	–
Using the Maximum-Likelihood (ML) the value is computed. It has been stated that 3711 people on board in the Diamond Princess cruise ship was identified with the epidemic of the novel coronavirus [14]	From 20 Jan to Feb 16, 2020-2.28	(2.06–2.52)	–	Feb 17–26, 2020
Simulations with one index as the first step is used [15]	On 18 January 2020, 0.8–5.0	–	–	–

A lot of health sector journals published more than 500 research papers in a short span of time. But less emphasis is made on data analytics point of view. We summarized the prevalent technical existing work on COVID-19 and the same is shown in Table 1.

3 Proposed Method

3.1 *Motivation*

As per the World Health Organization(WHO), a statistical analysis has been carried out and the number of confirmed and suspected cases due to the COVID-19 based on the provinces, regions and cities. From the furnished details declared by WHO, it has been understood that the cumulative confirmed cases have been increased and the number of deaths also has been increased. As many digital technologies come into existence, their intrusion in public health applications is widespread nowadays. Adoption of digital tools, data analytics also plays a vital role in the outburst of COVID-19 by which people get more awareness. Application of technologies in the health domain explores the data and its usage in decision-making. With the advancement of technologies like Artificial Intelligence and Machine learning, an exemplar swing occurs in data analytics, development of vaccines, drug molecules, and study of genomics; researchers, industries and academicians work with the available concepts in their domain. Hence, this motivated us to apply the Machine learning algorithms on the available dataset for supporting the factors.

3.2 *Proposed Machine Learning Approach to Analyze COVID-19*

With the current developments in analytics, data that has been collected from various sources are progressively used to forecast the upcoming trends in an extensive series of distributions. When considering the time series prediction, the metric which is used represents the record of similarly spaced disease count over the time. Even though, various time series predictions have been extensively used in the health care, many methods are available through which the best periodicity could be detected through the implementation of autocorrelation [16].

We initially assessed the existing impact of COVID-19 by taking reliable sources as base like, WHO Situation reports. It is clear from Fig. 1 that the number of confirmed cases were drastically getting increased. However, the death rate is almost constant due to the quarantine facilities initiated and followed by the countries' public health organizations and governments. Mortality rate was calculated and the same is shown in Fig. 2. For a unit of time, the death rate or the mortality rate is defined as the count of deaths in a specific population which is in extent to the size of the population.

We applied time series analysis on the COVID-19 data [17]. The structure of the data is

(ObservationDate, Province, Country, LastUpdate, Confirmed, Deaths, Recovered)

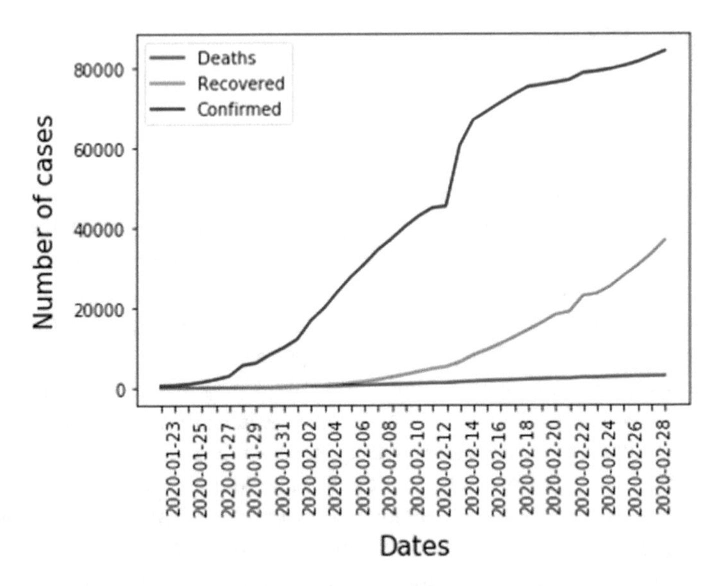

Fig. 1 Confirmed versus recovered versus deaths of COVID-19

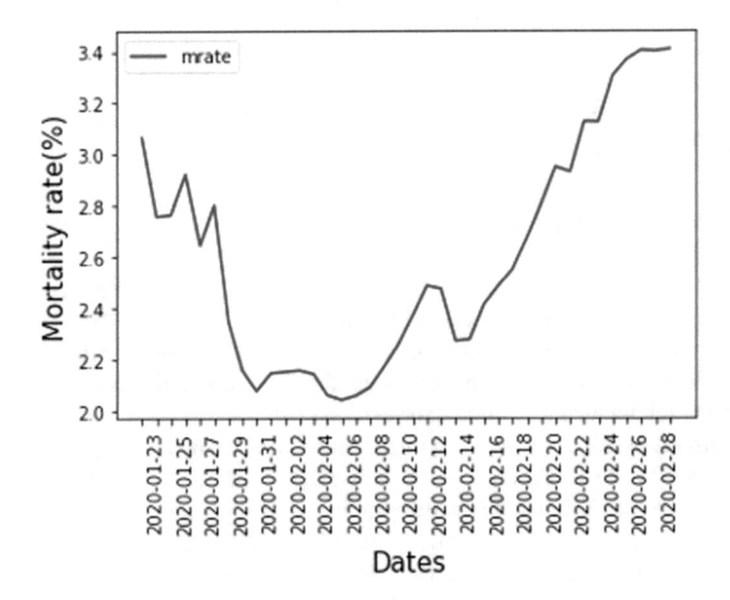

Fig. 2 Mortality rate of COVID 19

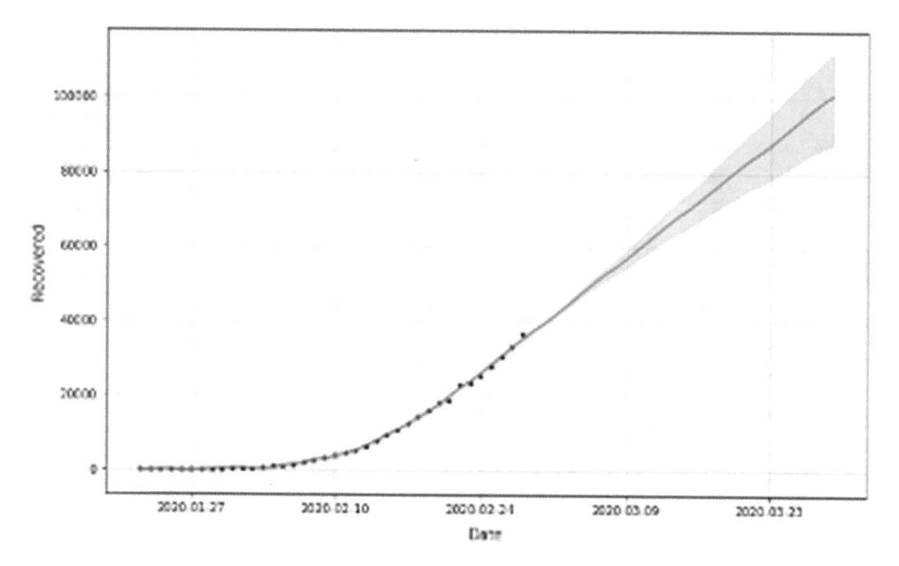

Fig. 3 Recovered trend predictions

We used data-cleaning techniques to handle inconsistencies in the above dataset. Also, we have identified that the attribute-"LastUpdate" does not create an impact on the prediction accuracy. Considering all the other attributes, we applied two different machine learning algorithms, namely, (1) Time series Analysis (2) Regression Models.

Applying Time Series Analysis on COVID 19 Data Time Series analysis would identify the trend in the data with respect to the date and time and extract meaningful insights out of it. We applied additive model based non-linear time series approach as it can handle abnormal trend deviations for analyzing the pandemic COVID 19.

The time series algorithm is applied on 'Recovered' attribute by considering other attributes. We observed that the trend estimation line is following the actual data line (Fig. 3). The same approach is used on 'Death' and 'Confirmed' features (Figs. 4 and 5). The trend analysis for confirmed cases is not effective when compared with other predictive attributes (Recovered and Deaths).

Applying Regression Model for COVID-19 To quantify the predictions, we use multi-linear regression. Linear regression is used to identify the dependent values based on the weighted sum of all independent variables.

$$y = b1*x1 + b2*x2 + b3*x3 + \cdots + c$$

The different values of y, x1, x2 and x3 would be known during the training stage of model. The model is trained, and therefore the best optimal value of b1, b2, and b3 are identified.

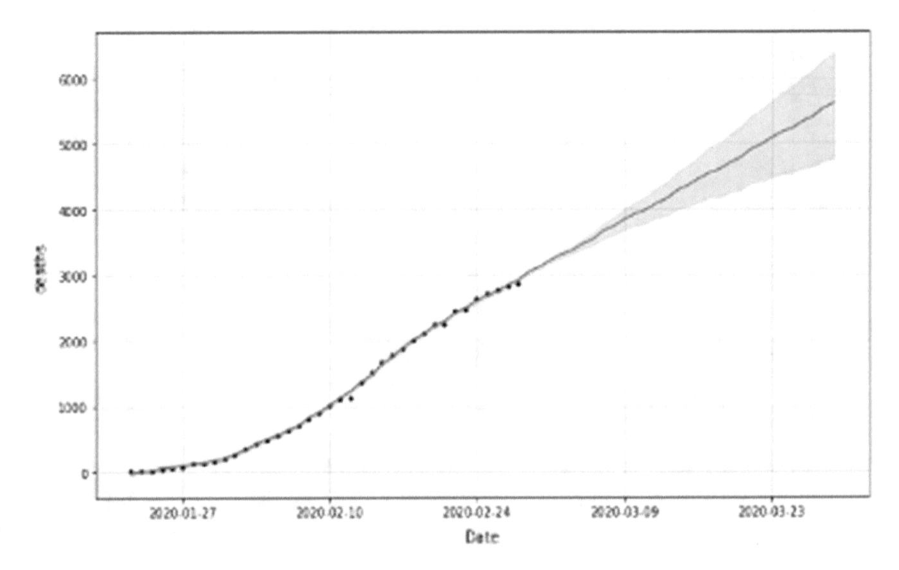

Fig. 4 Trend based predictions on number of deaths

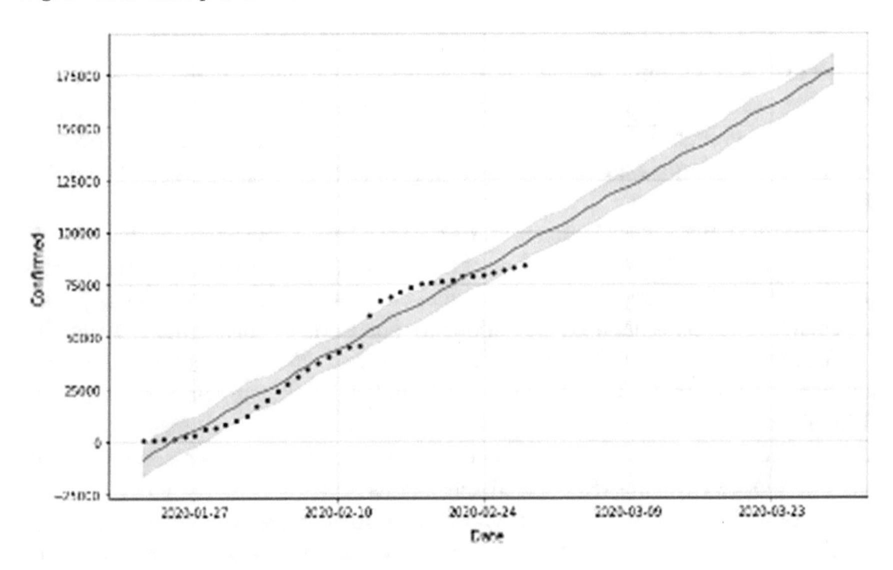

Fig. 5 Confirmed cases trend predictions of COVID 19

We have taken recovered as dependent variable and other useful features as independent attributes (Fig. 6). The same approach is iterated by considering confirmed and deaths as dependent attributes (Figs. 7 and 8).

We identified that the predictions for recovered and deaths were descent and average R^2 value of 60.03% was observed.

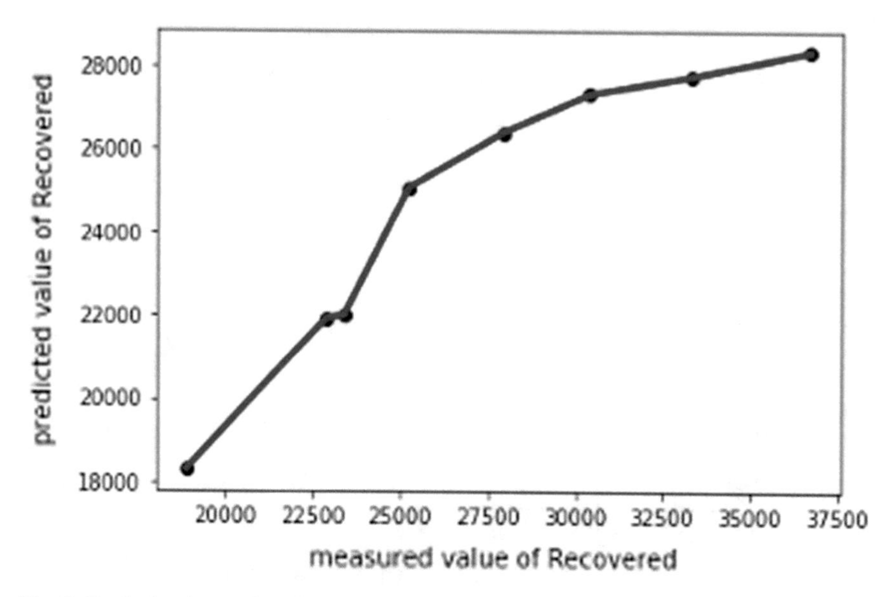

Fig. 6 Predicting the number of recovered cases

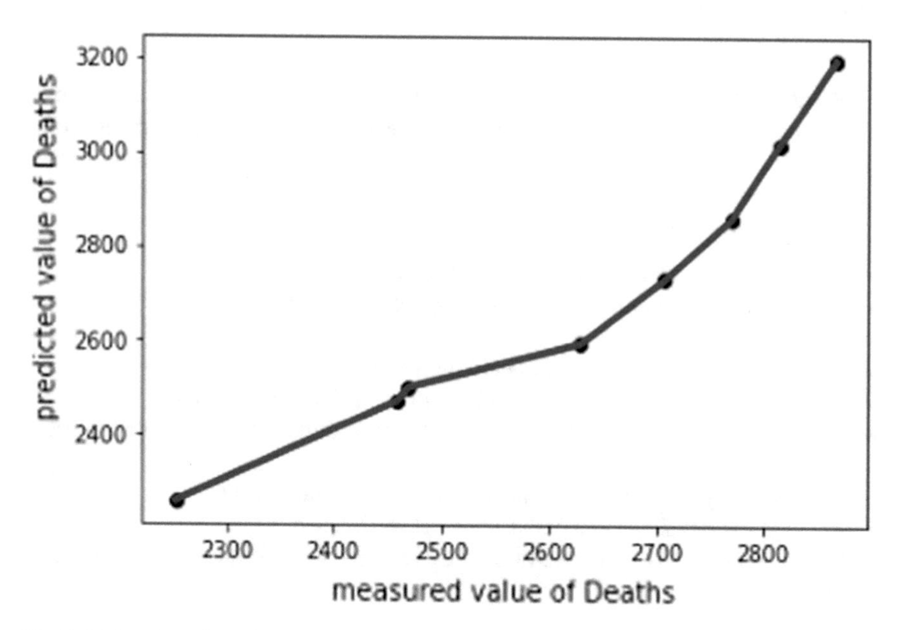

Fig. 7 Estimation of number of death cases

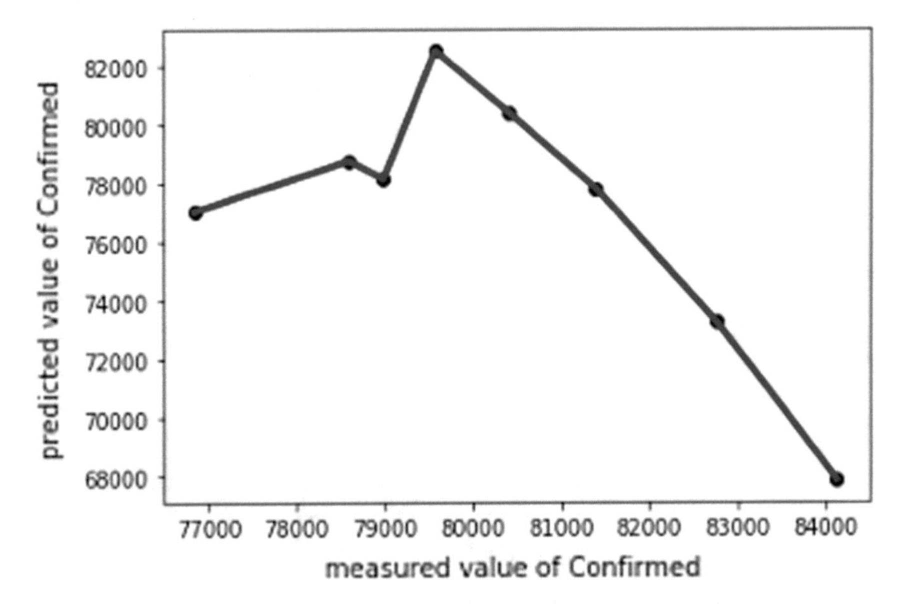

Fig. 8 Confirmed cases prediction

We applied other machine learning algorithms like Naïve Bayes, Decision Trees , SVM, and Random Forest but none of them gave good accuracy. This happened only because the dataset available for COVID 19 is not sufficient and does not include predictive attributes.

With the current dataset, lot of questions remains unanswered:

- How travel is impacting the COVID 19 spread?
- Are the precautionary measures reducing the transmission?
- Origin of the Virus?
- Which symptoms are critical in deciding the virus?
- Can the disease attack healthy person (Without BP, Diabetes, etc.)
- How genome sequence is changing from the date of detection to death confirmation?
- Will immunity boosters reduce the chances of getting affected?
- What are the precise medical reasons for death?
- What is the effective sequence of medical treatment in the quarantine health centres?

To answer most of these questions, we suggest the structure of dataset with various categorical and continuous attributes (Table 2).

As future work, we will collect the proposed dataset structure and test all the above hypothesis individually. We also plan to apply different time series algorithms and have detailed comparison of the dynamic change in the COVID 19 trends.

Table 2 Comparison of various methods for analysis of epidemics

Feature	Categorical/Continuous	Hypothesis
Age	Continuous	Only higher aged patients are likely to have chances of death
Gender	Categorical	To find out category wise
Travel History	Categorical	Travel history increases the risk of getting affected with virus
Sea Food intake	Categorical	Sea food is not a high deciding factor for virus transmission
Attended Public gathering	Categorical	Public gathering would increase the chance of spread
Weather	Continuous	Hot weather will reduce the virus impact
Psychological characteristics	Categorical	Psychologically weak patients will increase the criticality of the virus in the patient
Other health issues (Diabetes, Hypertension, liver disease etc.,)	Categorical	Chances of Death is high in virus infected patients with other health issues
Health details related to women	Categorical	Details of pregnancy, other issues also could be helpful in order to check

4 Conclusion

COVID 19 is a severe epidemic of the twenty-first century and WHO declared it as International Emergency. Since cure is not available as of now, a proper data analysis would help in speeding up the process of drug inventory. However, a comprehensive dataset is not yet available to identify the prevailing reasons behind the pandemic. With the current dataset, initial efforts were made to identify the trend analysis using time series models. We also applied regression models to predict the number of recovered, deaths, and confirmed cases and observed R^2 average score of 60.03%. We further proposed structure of the dataset which should be made available for answering the hidden patterns of the COVID 2019.

References

1. Liu Y, Gayle AA, Wilder-Smith A, Rocklöv J (2020) The reproductive number of COVID-19 is higher compared to SARS coronavirus. J Travel Med https://doi.org/10.1093/jtm/taaa021
2. Ramirez VB (2016) What is R0?: gauging contagious infections, 10 March 2020. https://www.healthline.com/health/r-nought-reproduction-number

3. Taylor DB (2020) New York times surgeon general urges the public to stop buying face masks, 15 March 2020. https://www.nytimes.com/2020/02/29/health/coronavirus-n95-face-masks.html
4. Ayyangar GSG (2020) The economic times coronavirus does not spread through chicken, mutton, seafood: FSSAI chief, 12 March 2020. https://economictimes.indiatimes.com/news/politics-and-nation/coronavirus-does-not-spread-through-chicken-mutton-seafood-fssai-chief/articleshow/74492046.cms
5. World Health Organization, 'Coronavirus Situation Reports', 2020. [Online]. Available:https://www.who.int/docs/default-source/coronaviruse/situation-reorts/20200306-sitrep-46-covid-19.pdf?sfvrsn=96b04adf_2. [Accessed: 7- Mar- 2020]
6. Wu JT, Leung K, Leung GM (2020) Nowcasting and forecasting the potential domestic and international spread of the 2019-nCoV outbreak originating in Wuhan, China: a modelling study. The Lancet
7. Chen J (2020) Pathogenicity and Transmissibility of 2019-nCoV-A quick overview and comparison with other emerging viruses. Microbes Infect. https://doi.org/10.1016/j.micinf.2020.01.004
8. Rabajante J (2020) Insights from early mathematical models of 2019-nCoV acute respiratory disease (COVID-19) dynamics
9. Shen M, Peng Z, Xiao Y, Zhang L (2020) Modelling the epidemic trend of the 2019 novel coronavirus outbreak in China. bioRxiv 2020. https://doi.org/10.1101/2020.01.23.916726
10. Liu T, Hu J, Kang M, et al (2020) Transmission dynamics of 2019 novel coronavirus (2019-nCoV). bioRxiv 2020. https://doi.org/10.1101/2020.01.25.919787
11. Read JM, Bridgen JRE, Cummings DAT, Ho A, Jewell CP (2020) Novel coronavirus 2019-nCoV: early estimation of epidemiological parameters and epidemic predictions. medRxiv 2020. https://doi.org/10.1101/2020.01.23.20018549
12. Majumder, M, Mandl, KD (2020) Early transmissibility assessment of a novel coronavirus in Wuhan, China, 27 Jan 2020. https://papers.ssrn.com/abstract=3524675
13. Tuite AR, Fisman DN (2020) Reporting, epidemic growth, and reproduction numbers for the 2019 novel coronavirus (2019-nCoV) epidemic. Ann Int Med [Epub ahead of print 5 February 2020]. https://doi.org/10.7326/M20-0358
14. Zhang S, Diao M, Yu W, Pei L, Lin Z, Chen D (2020) Estimation of the reproductive number of novel Coronavirus (COVID-19) and the probable outbreak size on the Diamond Princess cruise ship: a data-driven analysis. Int J Infect Dis https://doi.org/10.1016/j.ijid.2020.02.033
15. Riou J, Althaus CL (2020) Pattern of early human-to-human transmission of Wuhan 2019 novel coronavirus (2019-nCoV), December 2019 to January 2020. Euro Surveill 25(4):2000058. https://doi.org/10.2807/1560-7917.ES.2020.25.4.2000058
16. Otunba R, Lin J (2014) KSI research, APT: approximate period detection in time series, 21 Dec 2018. https://ksiresearchorg.ipage.com/seke/seke14paper/seke14paper_9.pdf
17. https://www.kaggle.com/sudalairajkumar/novel-corona-virus-2019-dataset#covid_19_data.csv

An Intelligent and Smart IoT-Based Food Contamination Monitoring System

M. Harshitha, Ch. Rupa, B. Bindu Priya, Kusuma Sowmya, and N. Sandeep

Abstract It is no wonder that one in six Indians fall ill due to food poisoning. Food Poisoning can lead to various diseases like diarrhoea, fever and many other health issues. Also, there are no current mechanisms and devices to effectively monitor the quality and freshness of food. This calls for a device to accurately monitor and report the freshness of food. In this work, we proposed a mechanism to effectively supervise the status of food. The main issues identified by the literature survey are that they have used mechanisms that are purely based on one parameter. We have resolved this issue by proposing a mechanism that monitors by taking into account all the parameters that can actually predict the status of food like moisture, pH and odour. The main strength of this work is that it is cost-effective and at the same time produces accurate and reliable results when compared to the existing mechanisms in this area.

Keywords Food poisoning · Moisture · pH · Odour · Food freshness

1 Introduction

In today's digital world, we are excessively dependent on electronic circuits and other scientific equipment that make our day-to-day life more productive and simple. Nowadays, food poisoning is the biggest problem faced by everyone. Almost everyone is affected in one way or the other due to this problem, and it may be due to eating contaminated food in restaurants, road side food stores where there is lack of supervision on quality of food and its nutrient values and freshness.

Recent data revealed by the Union Health Ministry's Integrated Disease Surveillance Program (IDSP) has stated that food poisoning is one of the most common

M. Harshitha (✉) · Ch. Rupa · B. B. Priya · K. Sowmya · N. Sandeep
Department of CSE, VR Siddhartha Engineering College (A), Vijayawada, India
e-mail: harshithamedara@gmail.com

Ch. Rupa
e-mail: rupamtech@gmail.com

N. Chaki et al. (eds.), *Proceedings of International Conference on Computational Intelligence and Data Engineering*, Lecture Notes on Data Engineering and Communications Technologies 56, https://doi.org/10.1007/978-981-15-8767-2_22

outbreaks reported until 2017. According to sources, Acute Diarrheal Disease (ADD) was the cause of 312 cases out of the total 1649 outbreak cases registered until the third week of December 2017.

The ISDP stated that most of the ADD cases and food poisoning cases were reported in places where large amount of food is being cooked very often like mid-day meals, hostels, wedding ceremonies as lack of proper supervision on freshness of food and its cooking methods exists there [1, 2]. The main problem here is the lack of proper techniques or mechanisms or devices to effectively forecast the threat of food poisoning and alert the concerned officials or organizers about the issue.

Food-borne diseases in India cost the country Rs. 1,78,100 crores which is approximately around 0.5% of the country's Gross Domestic Product (GDP) every year and at the same time leaves thousands of people hospitalized or even dead, as revealed by a study named 'Food for All' which is carried out with the partnership of the World Bank Group and the Netherlands Government. So experts suggest that India should start investing in food safety measures and mechanisms to reduce this economic burden of food-borne diseases and ensure food safety to the people [3, 4].

The proposed system mainly consists of sensors which measure the physical parameters of food and analyze its status. The sensors are connected to an Adruino Uno which is a micro-controller board equipped with sets of digital and analog input or output pins which are used to read input and display output on the LCD screen connected to it [5]. The sensors are programmed using the Adruino software customized for the Adruino board; the programming language used is C++ along with some other special functions and methods.

The remaining part of the paper is arranged in the following order: Literature survey is presented in Sect. 2. The detailed description of the proposed prototype and its functions is presented in Sect. 3. In Sect. 4, the results of our proposed system are presented.

2 Literature Survey

Mustafa and Andreescu [1] proposed a method called as 'Chemical and Biological Sensors for Food-Quality Monitoring and Smart Packaging'. In this work, biological or biosensors were used to investigate the freshness of food and crops. The biosensors interpret the status of food based on properties of bio-molecules present in the food such as enzymes, aptameters and antibodies. These properties are useful in enhancing the selectivity of target to analyze when integrated with the physical transducer. The main limitations of this work are (i) Moisture content of food that is not considered while testing the food freshness and (ii) Odour and Gas sensors that are not considered which are essential to differentiate the spoiled food from fresh food item.

According to Shahzad, et al. [2], selection of food is complex as it is influenced by a variety of factors and understanding these factors keenly is also required as the population's dietary change depends on this analysis. A variety of mechanisms have been proposed in this paper, which analyzes such factors of food which ultimately determine the status of it. The mechanisms which categorize the factors affecting food selection like physical/chemical properties, nutrition content, and the person involved in the selection process, and the psychological and sensory attributes of the person making the selection also plays a significant role. The economic and social environment around the person making the choice is also important because it affects various attributes of the product like its price, availability, brand and amount of the product.

Kuswandi [3] proposed a work named 'Freshness Sensors for Food Packaging'. In his work, he classified the freshness sensors into two categories, i.e. direct sensors and indirect sensors. The direct freshness sensor considers the detection of a specific analysis as a direct marker for determining the freshness of food. The indirect freshness sensor on the other hand mimics the change in a parameter of food when it is exposed to temperature or time.

Popa et al. [5] stated that in the fast-growing pace of modernization and standardization of food, the crave for ready-to-eat food is growing all over the globe. But the ready-to-eat foods pose a great threat of food poisoning due to various microbes contained in them, and the stringent processing conditions simulate the deterioration of the food quality. The author suggests that ultra sound technology can prove to be a solution to this problem as it is highly environment friendly and can be used in various processes of ready-to-eat food manufacturing like disinfection, sterilization, tenderization, dehydration and cooking of these foods. The main drawbacks of this work are (i) The free radicals formed during cavitations may cause harmful effect on the consumer (ii) Budding technology (iii) Ultrasound may cause physicochemical effect which may be responsible for off-flavour, discoloration and degradation of the components.

3 Proposed Methodology

We propose a system to accurately measure the parameters related to food poisoning and store them for further use also. For this purpose, we use sensors like gas sensor, moisture sensor and pH sensor [6]. These sensors are integrated on the Adurino Uno. This device is fitted onto the container in which the food items are being stored, and then it detects whether or not the food is fit for consumption or not. Based on the observations it makes the food assessed for freshness. This can help hotels to assure quality service to their customers and also the customers can be sure of what they are eating, and food safety officers can also access this information and can take legal action against food poisoning [7]. Table 1 depicts the various hardware/sensors used in checking the freshness of food and also their usage and their range.

The circuit diagram depicts the way in which the sensors and the LCD screen are connected to the Adruino as shown in Fig. 1. The moisture sensor is connected to the ground and voltage 3.3v pins, and the input analog pin connected is A0. The pH sensor is connected to the ground pin and for voltage an external 9 V battery is connected to take input from the sensor or to the sensor the analog pin A1 is used.

Table 1 Hardware/sensors used

Hardware/sensor used	Usage	Range
Adruino Uno	Micro-controller board to interface different sensors	20 digital input/output pins
LCD display	Display the result	0 C to +50 C (C = Celsius)
Moisture sensor	Measure the moisture content in food	0–1023 (in ADC value)
pH sensor	Measure the pH content in food	0–399.98 (in mV) (pH 7 = 0 mv) (pH 14 = −399.98 mv) (pH 0 = +399.98 mv)
Odour/gas sensor	Measure the gas content in food	0.05−10 mg/L Concentrations of alcohol

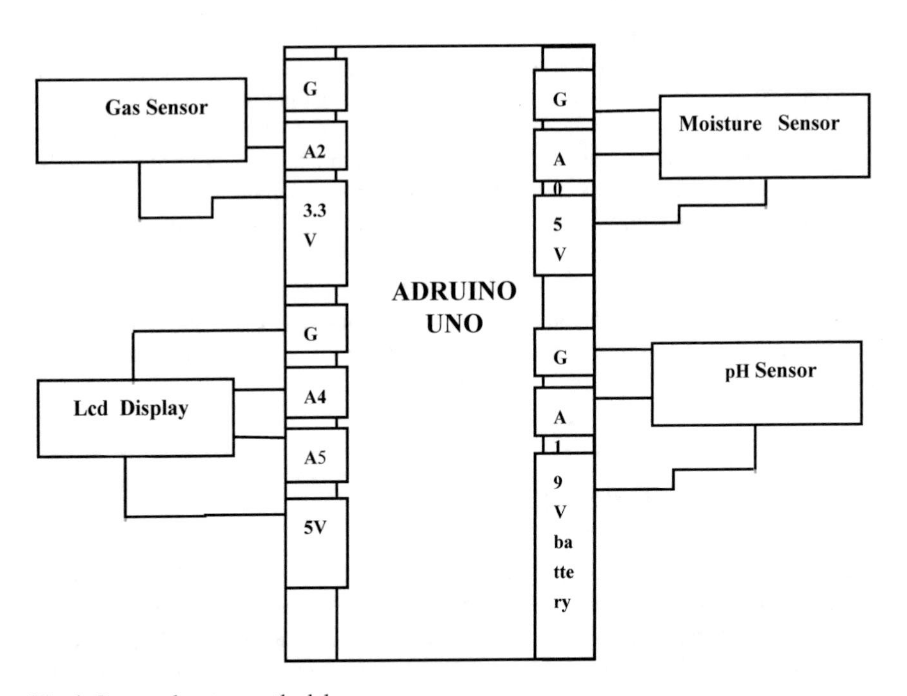

Fig. 1 Proposed system methodology

The gas sensor is connected to the ground and voltage 5v pins and the analog pin A2 is used for input and output. The LCD display screen connected to the A4 and A5 pins displays the status of food being tested. The ground pin of Adruino is connected to ground of LCD display. The LCD display is soldered to the Adruino cable to provide external power required to run it.

Freshness of food is predicted based on many parameters, but a predominant characteristic of food can be considered to effectively determine its freshness [8–10]. For example, we can determine the freshness of milk based on its pH value. The sensors are independent of each other because we consider the most predominant characteristic of the given food rather than considering all the parameters and causing ambiguity [11]. The above diagram explains the mechanism of testing the freshness of food.

The food to be tested is brought in contact with the corresponding sensor, which measures its predominant characteristic like moisture or pH to predict its status. The sensor transmits the readings to the serial monitor in the Adruino software, which then displays it on the screen. The LCD display connected to the Adruino displays the final result whether it is fit to consume or not.

Figure 2 depicts the food freshness checking mechanisms. When food is supplied as input to the sensors, and the sensors detect the corresponding parameters like moisture sensor detects moisture content and by comparing with the standard values of food the result is produced whether the food is suitable to consume or not. This approach has five modules which have been described as the following subsections.

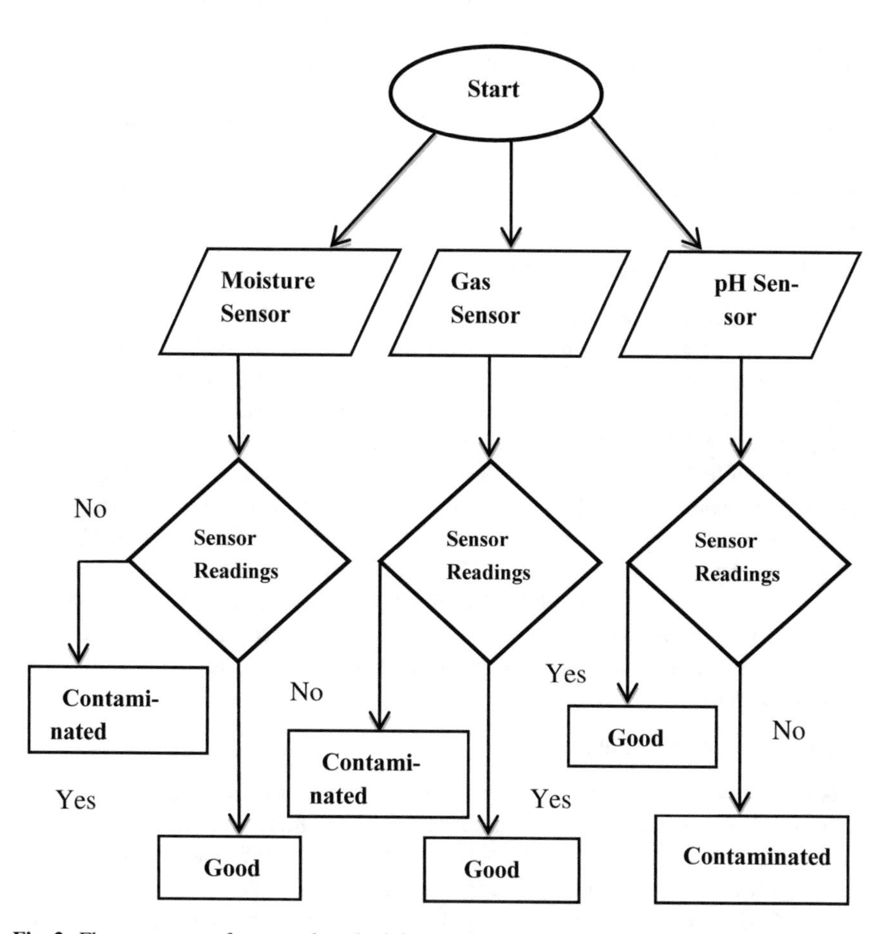

Fig. 2 Flow sequence of proposed methodology

Module 1: Moisture Measurement in Rice

The moisture content of rice is the measurement of water content in it. In freshly cooked rice, the standard value of moisture content is approximately around 65–68%, but this value can vary based on different regions and various cooking methods [12]. For measuring moisture content in freshly boiled rice, we used a moisture sensor which measures the moisture content of rice by calculating the difference in resistance between its two conducting probes. The measuring range or output range of this sensor is 0–1023 in terms of ADC. The analog output of the sensor is processed using ADC. The moisture content is displayed in terms of percentage on the serial monitor.

Module 2: Moisture Measurement in Bread

The moisture content of bread is the measurement of water content in it. In freshly baked bread, the standard value of moisture content is approximately around 35%, but this value can vary based on different regions and various cooking methods. For measuring moisture content in freshly baked, we used a moisture sensor which is equipped with two conducting probes and the difference in resistance between these two probes expressed in a range of 0–1023 (in ADC value) is the measurement of moisture content of bread being tested. The analog output of the sensor is processed using ADC. The moisture content is displayed in terms of percentage on the serial monitor.

Module 3: pH Measurement in Milk

Milk consists of lactose sugar which plays a vital role in lowering the pH of milk over passage of time. This reaction takes place due to the lactic bacteria which converts lactose sugar into lactic acid as time passes by. The lactic acid content in milk increases when placed at room temperature and it reaches a threshold value. After that the pH level decreases to such an extent that the lactic acid simulates the growth of bad bacteria which at last results in contamination of milk [13]. The accepted or standardized pH range of milk is 6.5–6.7. When a sample of fresh milk is stored at room temperature its pH value decreases over time which indicates that milk is getting sour or contaminated. The trend in decrease of pH value depends upon the type of impurity contained in the sample. The pH sensor measures the voltage of the solution in millivolts (mV) and the actual pH reading between 0 and 14 is obtained by converting this voltage value into pH using standard formulae and various operations.

Module 4: pH Measurement in Curd

Milk is the key ingredient of curd or yogurt so curd also exhibits the same pH trend as that of milk. Milk is converted into curd by the fermentation process. This process is a result of the series of reactions between lactic acid bacteria and casein. As lactic acid plays a vital role in forming curd it also plays a significant role in spoiling it too. When a fairly good sample of curd of pH value 4.4 (standard value) is left at room temperature, the lactic acid in the curd starts to lower the pH level in curd resulting in sour or spoilt curd.

Module 5: Ethanol Measurement in Banana

Banana, the most loved fruit across the world gets spoilt in no time due to the emission of ethanol or otherwise called alcohol. This gas emission also results in dark black spots on the fruit's peel and foul smell, and the fruit also becomes soft during this phase. The MQ3 gas sensor detects ethanol from the spoiled fruit and based on the concentration of ethanol it detects that the banana is spoiled.

4 Results and Analysis

The main aim of this device is to detect the freshness of food in order to curb food poisoning and as well as to avoid food wastage. The food is tested for parameters like moisture, odour, pH values. The testing mechanism can be divided into several modules based on the food materials tested. Figure 3 shows the quality of rice on consideration of various samples by following the approach mentioned in module 1.

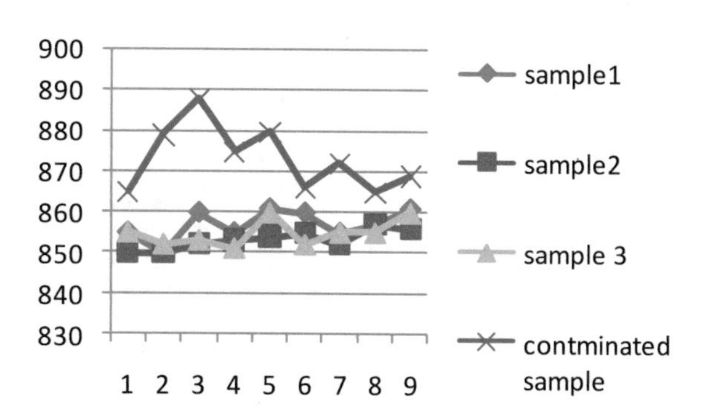

Fig. 3 Moisture measurement in rice

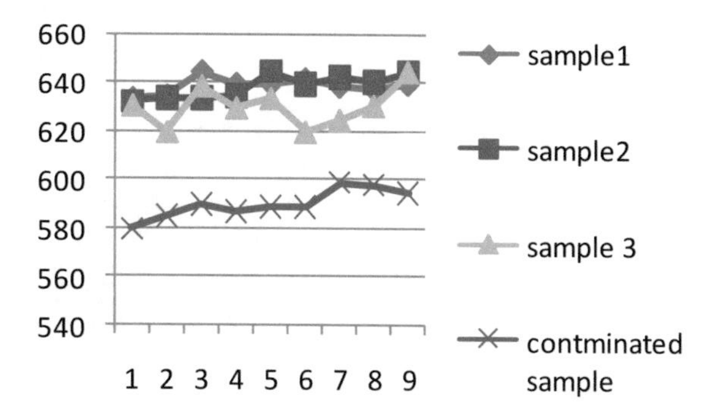

Fig. 4 Moisture measurement in bread

Figure 4 shows the quality of bread measurement by considering various bread samples as input. Module 2 approach followed here to test and analyze the quality of bread.

Figure 5 shows the milk quality by considering the approach mentioned in module 3.

Figure 6 shows the curd quality measurement which considered module 4 approach to test the quality.

Figure 7 shows the quality of banana in terms of contaminated or not. It follows module 5 approach to test whether it is contaminated or not.

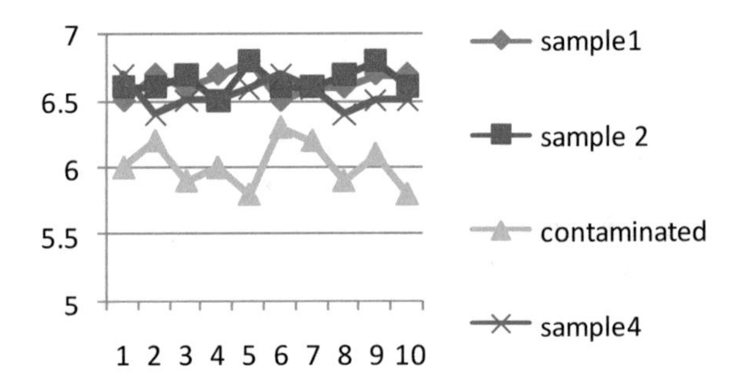

Fig. 5 pH measurement in milk

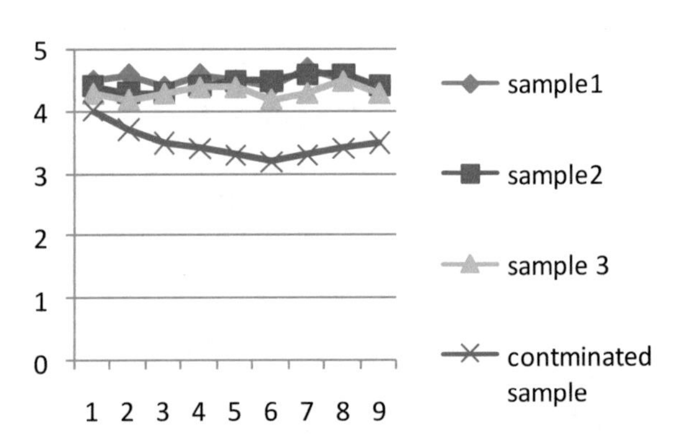

Fig. 6 pH measurement in curd

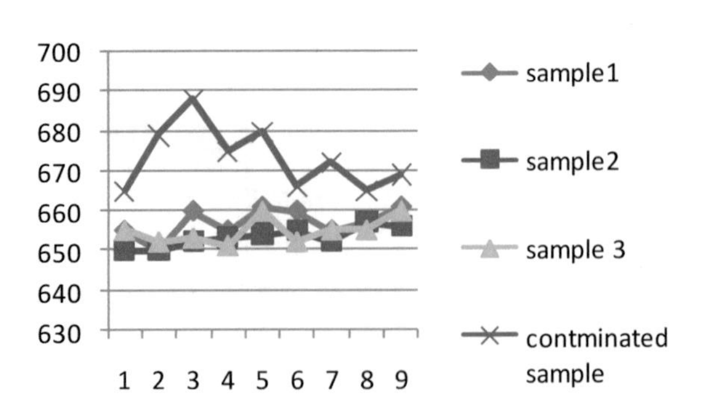

Fig. 7 Ethanol measurement in banana

Table 2 Comparison analysis of food items

Input	Sensor used	Value	Output
Rice	Moisture sensor	>850	Contaminated
Curd	pH sensor	>4.4	Contaminated
Milk	pH sensor	>7	Contaminated
Bread	Moisture sensor	>650	Contaminated
Banana	Gas sensor	>660	Contaminated

Table 2 shows the comparison results for various food items with their considered threshold value [14] to check whether the food has contaminated or optimized item.

5 Conclusion and Future Scope

Food poisoning is a major threat to the society because it may be fatal when it is caused by dangerous bacteria and fungus. There is an alarming need to effectively detect and curb food poisoning, and the proposed system in this work can lead us in the path of preventing food poisoning. The various parameters of food based on which we determine its status are predominant characteristics of the food items and a small change in these parameters can lead to its decomposition and hence when such food is consumed by humans can become fatal for them. This mechanism predicts the freshness of food by measuring the parameters like moisture, odour and pH with the help of their corresponding sensors and display the result on the LCD screen connected to circuit. In future, we would like to make it as a mobile app system.

References

1. Mustafa F, Andreescu S (2018) Chemical and biological sensors for food-quality monitoring and smart packaging, pp 1–25
2. Shahzad N, Khalid U, AtifIqbal, Ur-Rahman M (2018) eFresh—a device to detect food freshness. Int J Soft Comput Eng (IJSCE) 8(3):1–20
3. Kuswandi B (2017) "Freshness sensors for food packaging", postharvest handling (3rd edn), 2014, Reference module in food science, pp 1–25
4. Ali S, Wang G, Bhuiyan MZA, Jiang H (2018) Secure data provenance in cloud-centric internet of things via blockchain smart contracts. In: IEEE smart world, ubiquitous intelligence & computing, advanced & trusted computing, scalable computing & communications, cloud & big data computing, internet of people and smart city innovation, pp 991–998
5. Popa A, Hnatiuc M, Paun M, Geman O, Jude Hemanth D, Dorcea D (2019) An intelligent IoT-based food quality monitoring approach using low-cost sensors. J Symm MDPI
6. Yousefi H, Ali MM, Su HM, Filipe CD, Didar TF (2018) Sentinel wraps, "Real-time monitoring of food contamination by printing dnazyme probes on food packaging." ACS Nano 12:3287–3294

7. Giuffrida A, Giarratana F, Valenti D, Muscolino D, Parisi R, Parco A, Marotta S, Ziino G, Panebianco A (2017) A new approach to predict the fish fillet shelf-life in presence of natural preservative agents. Italian J Food Safe 6(2):88–92
8. Sheth A, Jaimini U, Yip HY (2018) How will the internet of things enable augmented personalized health? IEEE Intell Syst 33:89–97
9. Dutta J (2019) A next generation sensing and monitoring platform for quality assessment of perishable foods. IEEE Beyond Standard
10. Shroff R (2019) Smart food monitoring with smart shelves, IoT and AI. Software solutions
11. Paul A, Kant K (2020) Smart sensing, communication, and control in perishable food supply chain. ACM Trans Sensor Netw 16(1)
12. Witjaksono G, Rabih AAS, Yahya NB, Alva S (2018) IOT for agriculture: food quality and safety. IOP Conf Series Mater Sci Eng 342
13. Hu R, Ya Z, Ding W, Yang LT (2020) A survey on data provenance in IoT, Springer—WWW 23:1441–1463
14. Pal A, Kant K (2018) IoT-based sensing and communications infrastructure for the fresh food supply chain. IEEE Comput 51(2):76–80

Optimized Candidate Generation for Frequent Subgraph Mining in a Single Graph

D. Kavitha, D. Haritha, and Y. Padma

Abstract Mining frequent subgraphs from graph databases is a basic task with broad applications. Frequent subgraph mining is defined as finding all subgraphs that appear more than specified threshold value. It consists of mainly two steps, candidate generation and frequency calculation. In candidate generation step, most of the existing work starts with a frequent edge or vertex to generate frequent candidate patterns. This process is not scalable due to exponential number of candidate patterns generation. In this paper, an optimized algorithm is presented to generate candidate patterns for mining frequent subgraphs from a large single graph. The proposed algorithm starts and extends candidates with frequent subgraphs. The proposed algorithm uses graph invariant properties and symmetries present in a graph to generate candidate subgraphs thus reducing generation of enormous amount of candidate subgraphs. Subgraphs are extended by adding another frequent subgraph determined by the symmetry mapping of subgraph there by reduces the complexities involved in candidate generation and frequency counting. An evaluation study on datasets explores the strengths and limitations of the proposed work. The results make sure that, this is an optimized approach to generate candidate subgraphs directly using invariant properties.

Keywords Graph mining · Frequent pattern · Candidate generation · Partitioning · Graph invariants · Symmetry

D. Kavitha (✉) · Y. Padma
PVPSIT, Vijayawada, India
e-mail: kavitha_donepudi@yahoo.com

Y. Padma
e-mail: padmayenuga@gmail.com

D. Haritha
Department of CSE, SRKIT, Vijayawada, India
e-mail: harithadasari@rediffmail.com

N. Chaki et al. (eds.), *Proceedings of International Conference on Computational Intelligence and Data Engineering*, Lecture Notes on Data Engineering and Communications Technologies 56, https://doi.org/10.1007/978-981-15-8767-2_23

1 Introduction

Mining frequent subgraph patterns is a well-studied problem in graph mining to mine and analyze data in applications such as protein interactions, cheminformatics, drug discovery, social net, and web interactions [1–4]. Frequent subgraphs are also extensively useful for other mining tasks such as indexing, clustering, and classification. Frequent subgraph mining [5] problem can be defined as it is a process of identification of a set of frequent subgraphs from graph database that may be a set of small or a large single graph with an occurrence frequency is not less than a given threshold. In this paper, a single graph scenario is considered and a subgraph can be considered as frequent if it has at least T occurrences in graph.

The difficulty in frequent subgraph mining algorithms lies in candidate generation and calculating the support of subgraphs. Designing algorithms is a computationally challenging and data-intensive task to mine frequent subgraphs from a large graph. It is due to the (a) size of the graph and (b) the enumerated subgraph space increases in exponential with the size of graph when finding frequent patterns. In recent years, several approaches are developed to detect patterns in a large graph [6–10]. In general, these algorithms start with a frequent edge or vertex and extend the graph by adding a new edge or by adding a new vertex to the existing graph recursively until all frequent subgraphs are generated. However, the number of subgraphs generated in mining process is exponential in proportion to the graph size.

Along with the above issues, one fundamental problem that needs to consider in single large graph is overlapping of subgraphs shown in Fig. 1. In general, one can consider two subgraphs as different if they differ just by a single edge. As a result, we may find number of overlapped subgraphs as candidates. On the other hand, two subgraphs are said to be dissimilar if they don't share any or just by sharing single edge or node. A critical step in obtaining such no overlapping or minimum overlapping frequent subgraphs is to find maximum common subgraph or maximum independent set which itself is NP-complete.

In this paper, an optimized algorithm is presented to generate candidate subgraphs based on the symmetries present in a graph. This algorithm starts candidate generation with a frequent subgraph contrasting with other algorithms which start with a frequent

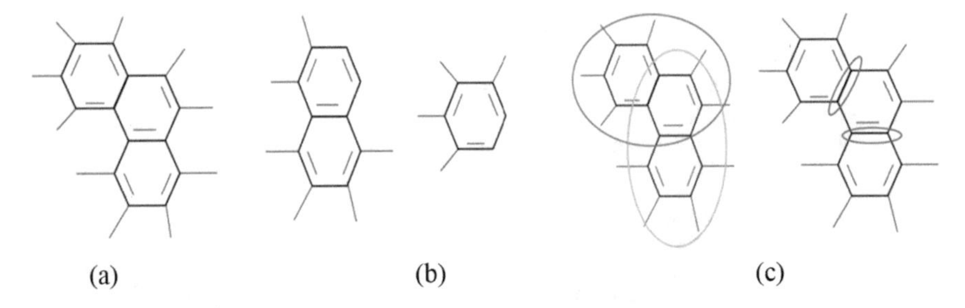

Fig. 1 Frequent subgraphs in a single graph. **a** Graph **b** subgraphs in a graph **c** subgraphs with maximum and minimum overlapping

vertex or edge. It generates candidate subgraphs with minimal overlapping subgraphs and avoids enumeration of all subgraphs in candidate generation. The algorithm uses symmetries present in a graph to generate candidate subgraphs. Unlike most algorithms, this algorithm extends candidates by adding another frequent candidate subgraph determined by the symmetry mapping of subgraph.

1.1 Our Approach

Figure 2 depicts the outline of our approach. At first, partition the vertices of graph by identifying symmetries present (vertex triples) in the graph using graph invariants. If the size of a part is more than specified threshold, generate first-level candidates by enumerating the associated graph of each vertex in the part. Then apply subgraph extension algorithm to find frequent candidates. If the part size is less than threshold support, then identifies symmetries using vertex dual approach, label-based candidate enumeration to generate frequent candidates. Adjacent list is used to represent graphs which eases the traversal of graph. Subsequent sections explain the proposed algorithm in detail.

With this approach, firstly, we address the issue of scalability and complexity during candidate generation. Secondly, isomorphism is almost avoided which is required in frequency counting and duplicate candidate identification.

The rest of this paper is organized as follows. Section 2 presents the terminology used in the problem formulation and preliminary concepts. The proposed technique to

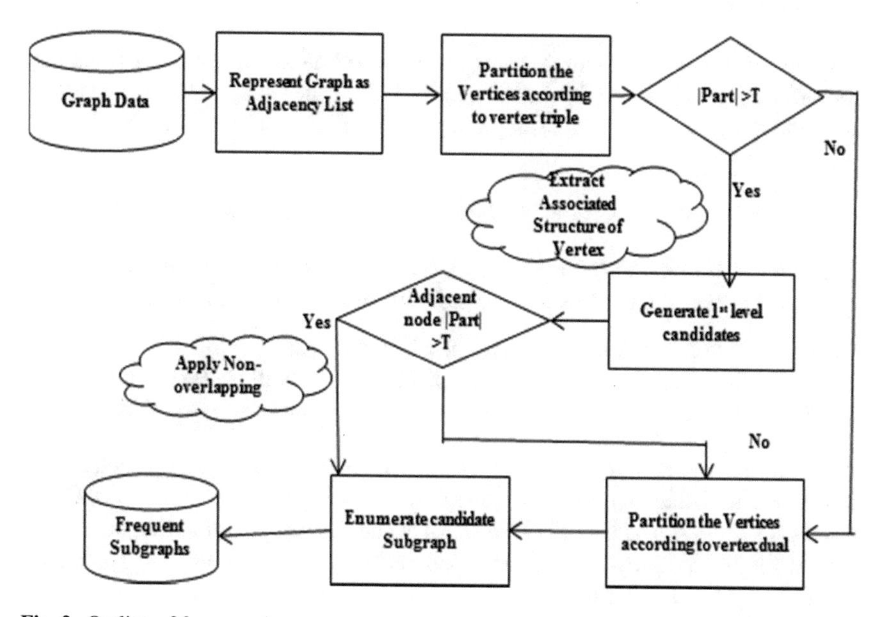

Fig. 2 Outline of framework

mine frequent subgraphs in a single large graph using optimized candidate generation process is discussed in Sect. 3. We study an optimized approach of candidate generation using graph invariants is discussed in this section which are in turn use them to find frequent patterns. Section 4 provides experimental evaluation. The related work is reviewed in Sect. 5 and work conclusions are presented in Sect. 6.

2 Preliminary Concepts

Definition 1 Graph: A labeled graph G is denoted as (V, E, L, l), where V is a set of vertices and $E \subseteq V \times V$ is a set of edges connecting vertices. Vertices and edges have labels and represented with L, and a function l assigns a unique label to each vertex of G.

Definition 2 Subgraph: Two graphs $G = (V, E, L, l)$ and $g = (V', E', L', l')$. Graph g is a subgraph of G iff $V' \subseteq V$, and $E' \subseteq E \wedge ((v_1, v_2) \in E' \rightarrow v_1, v_2 \in V')$ and labels of edges are identical along with corresponding vertices labels. $g \subseteq G$ denotes that g is a subgraph of G.

Definition 3 Isomorphism of Subgraphs: Let two subgraphs g and h, these g and h are isomorphic iff there is an injective function f: $V(g) \rightarrow V'(h)$ such that $\forall\, v \in V(g)$, $L(v) = L'(v')$ and $\forall(u,v) \in E(g) \Leftrightarrow (f(u), f(v)) \in E(h)$ and $L(u, v) = L'(f(u), f(v))$.

In a large single graph setting, the problem formulation for a frequent subgraph is defined as follows:

Definition 4 Frequent Subgraph: Let input graph is G, a subgraph of G is g, frequency threshold T, and observed frequency of g in G is f, then subgraph g is a frequent subgraph if frequency of subgraph g is greater than or equal to given threshold T, i.e., if $f \geq T$.

The storage order of vertices in adjacency list is defined as follows:

Definition 5 Vertex *order*: For a vertex v, let A be the set of m adjacent vertices u_1, $u_2, \ldots, u_m$, the order of vertices u are, which satisfies the following criteria:

(1) For each $u_k \in A$, $lbl(u_k) \geq lbl(u_{(k+1)})$, $\forall$ k, $1 \leq k \leq m$
(2) For each $u_k \in A$, $deg(u_k) \leq deg(u_{(k+1)})$, $\forall k$, $1 \leq$ k $\leq$ n iff $lbl(u_k) = lbl(u_{(k+1)})$
(3) For each $u_k \in A$, $lbl(u_1, v) \leq l(u_2, v)$, $\forall k$, $1 \leq$ k $\leq$ n iff $lbl(u_k) = lbl(u_{(k+1)})$

Definition 6 An *ordered partition* V is a sequence $\{V_1, V_2, \ldots, V_p\}$ of non-empty subsets of V such that $\{V_1, V_2, \ldots, V_p\}$ is a *partition* of V. The subsets $V_1, V_2, \ldots, V_p$ are called *cells* of V. A *trivial cell* is with size one and *discrete partition* has only *trivial cells*, while the *unitpartition* has only one *cell* and refers to complete symmetric graph. An *ordered partition* is an *equitable ordered partition* iff the vertices are partitioned based on the *Definition* 5. In a regular graph, the *unit partition* is equitable.

Definition 7 Vertex *triple*: For a vertex v with m adjacent vertices $u_1, u_2, \ldots, u_m$, vertex triple $T(v)$ is a string defined as: Td_v+ Tl_v+ Te_v, where "+" is string concatenation, and which satisfies the following criteria:

(1) For each $v \in A$, $Tl_v = (lbl(u_1) + lbl(u_2) + \ldots + lbl(u_m))$, where $lbl(u_k) \leq lbl(u_{(k+1)})$, for all k, $1 \leq k \leq m$
(2) For each $v \in A, Td_v = (deg(u_1), deg(u_2), \ldots deg(u_m))$, where $deg(u_k) \geq deg(u_{(k+1)})$, for all k, $1 \leq k \leq m - 1$ *and* $lbl(u_k) \leq lbl(u_{(k+1)})$, for all k, $1 \leq k \leq m$
(3) For each $v \in A$, a sequence $Te_v = (lbl(u_1, v) + lbl(u_2, v) + \ldots + lbl(u_m, v))$, where u_i is the in the order of l_v.

Vertex *dual* $D(v)$ is a string defined as Tl_v+ Te_v which is composed with 1 and 3 of above.

Definition 8 Symmetry Group: Let $Sym(g)$ denotes a group that contains all symmetries present in a graph g under functional composition. Symmetry of a graph g is identified by permutation of vertices of g's that also conserves edge relationships of g's, i.e., $g' = g$. The $Aut(g)$ of a graph g is the set of all automorphisms of g with permutation composition as group operation.

Lemma 1 *Given two vertices v_1 and v_2 with their triples $T(v_1)$and $T(v_2)$ of a labeled graph Gg, v_1 is symmetric to v_2 if and only if $T(v_1) = T(v_2)$.*

Lemma 2 *Given two vertices v_1 and v_2 with their duals $D(v_1)$ and $D(v_2)$ of a labeled graph G, v_1 is symmetric to v_2 with their partial associated graph if and only if $D(v_1) = D(v_2)$.*

As these vertices are symmetrical, the associated subgraph of a vertex with its adjacent vertices is also symmetrical. This symmetry property of vertices can be used to enumerate subgraphs just by connecting an edge between subgraphs of vertices. By this Lemma 1 and Lemma 2, the problem of mining frequent symmetrical nodes is equivalent to mining their corresponding subgraphs consequently candidate subgraphs to enumerate frequent subgraphs.

3 Frequent Pattern Mining

The implementation of finding frequent subgraphs algorithm is as follows: (i) The algorithm starts by executing *partition*() algorithm that partition the set of vertices V of graph G into *parts* and are further refined into *orbits* based on the symmetry of vertices which have vertex-transitive property. This algorithm is further explained. (ii) Find vertex orbits whose $|S_k| > f$ and store all associated subgraphs of vertices as primary candidates. Here, the subgraph associated with each vertex of an *orbit* becomes the candidate subgraph. (iii) These candidate subgraphs are further extended by *candgextn*() algorithm to enumerate frequent subgraphs. Finally, subgraph extension algorithm *candgextn*() is recursively executed to find out all the frequent subgraphs.

A. Frequent Mining Framework

Algorithm: *Frequent Subgraph Mining*
Input: A graph G and frequency threshold f
Output: *result* ←The frequent subgraphs of G
Begin

1. Obtain equitable ordered partition of the vertices V of a graph G. $S \leftarrow partition(V(G))$
   $S \leftarrow S_1, S_2, ...S_n$
2. For each vertex orbit $S_k \in S$ do
   if$|S_k| > f$ then
   $result \leftarrow g_i$ of $v \in S_k$
3. *result* ←*result* Uc *and g extn*(S_k)

End

B. Partitioning

Algorithm:*partition*()
Input: $V \leftarrow$A set of vertices of graph G
Output: $S \leftarrow$Refined ordered partition of graph G
Begin

1. Compute $V \leftarrow V_1, V_2 ... V_n$ the initial partition according to definition 5
2. For each cell $V_i \in V$ do s.t $|V_i| >$f
   For each $v \in V_i$ do
   Compute Tri_v *and* $Dual_v$(definition 7)
3. Let V_{ik} be the sub partition of V_i
   Refine V_i into $V_i \leftarrow V_{i1}, V_{i2} ... V_{ij}$ s.t $\forall$u, v $\in V_i T(u) = T(v)$
4. For each V_{ij}do
   if $|V_{ij}| > f$
   S ←S $\cup V_{ij}$
   Else If $|V_{ij}| < f$
   Refine V_i into $V_i \leftarrow V_{i1}, V_{i2} ... V_{ij}$ s.t $\forall$u, v $\in V_i$
   $D(u) = D(v)$
   S ←S $\cup V_{ij}$

End

Initially, the algorithm starts by forming an equitable ordered partition of vertices according to Definition 5, thereby extracting all the label and degree information. In order to find symmetrical vertices, other graph theoretical information such as degree of adjacent vertices and labels of adjacent vertices and edges are exploited. For each vertex, v in the cell V_i triple is calculated. These vertices are then divided into groups of equal triples or duals according to the symmetry present, forming vertex orbits V_{ij}. The process is iterated for each cell V_i of the initial partition V. *partition*() algorithm eliminates the cells that do not support the frequency threshold. As we all know that according to the anti-monotone property, if a candidate is infrequent, its extensions are also infrequent. Finally, the partition S, returned by this refinement procedure contains the vertices of resultant vertex orbits that will become basic subgraphs and can be further extendible. At this position, a subgraph of a vertex, i.e., a vertex along with its adjacent vertices in each orbit will return a kind of subgraph.

C. Candidate Generation and Extension

The algorithm generates candidate subgraphs by extending a frequent subgraph with its frequent children subgraphs. In the candidate generation process, the subgraph associated with a node has been discovered if the node has been identified as frequent, and all subgraphs that its descendent frequent nodes represent must have been discovered too. This process of identifying frequent nodes with their associated substructure is algorithmized in the next section.

Algorithm: *candgextn(C)*
Input: A candidate set C
Output: The frequent subgraphs of G
Begin
1. for each vertex $v \in C$ do
$ext \leftarrow sg(v)$ //a subgraph of vertex v in the candidate set C
mark v visited
2. for each vertex $v \in C$ do
Find $Adj(v) \leftarrow$set of adjacent vertices of v
Repeat
if ($Adj_i \in S_i \backslash C$) and (vertex *i unvisited*)
$ext \leftarrow ext <> sg(i)$
mark vertex *i visited*
else
$ext \leftarrow ext <> e(v,i)$
mark vertex *i visited*
until no more vertices to visit
3. return *ext*
End

In the proposed approach, the subgraph enumeration is carried out by extending the previously enumerated subgraph with newly identified extendible frequent subgraph. The input to this algorithm set S_i contains only the vertices which are symmetrical according to the Definition 6 and frequent. So each vertex v of the set S_i (i.e., the element of frequent set) can be associated to a small subgraph with its adjacent vertices. To extend the subgraph, now check the signature of its first adjacent vertex with other vertices first adjacent vertex. If they are same then connect the vertex v to the subgraph of first adjacent vertex v_1. Repeat this procedure until no more matching's found or all the vertices are visited. Execute this procedure recursively for the next level adjacent vertices also if the extensions are possible to further extend the subgraphs. This algorithm will return the frequent subgraphs that can be enumerated from each vertex orbit.

D. Frequency Count

There are two things in the candidate generation process explained in the above section. One is the selection of candidate seeds and the other is frequency evaluation of candidate subgraphs. The associated subgraphs of frequent vertices in an orbit are the first level of frequent subgraphs and candidates that are to be enumerated. The vertices in an orbit could be frequent if the size of the orbit is greater than threshold frequency. Based on symmetry property, the frequency computation using isomorphism is pruned. Coming to the second step, the extension of candidates is also carried out if their siblings are also frequent. But we have to check the occurrence of each extended subgraph type. There we need isomorphism testing. When compared with other algorithms, generated candidate subgraphs are promising frequent subgraphs. The only requirement for frequency checking is type checking the candidate subgraph is constructed by extending a parent subgraph with one frequent sibling vertex associated subgraph. It may be a single edge or it may be another subgraph. Since the children are also frequent, the associated subgraph can be added to generate new candidate. But extending one of a child of all candidate subgraphs may not result same new subgraph. At this point, we may require isomorphism testing.

4 Experimental Evaluation

The performance of optimized candidate generation for Frequent SubGraphs in a Single Graph (FSSG) is evaluated in this section and it is extended version of [11]. Three standard data samples are employed to perform experimentation. From these, aviation graph is very large and sparse which have number of distinct node labels. Other two datasets are from protein interactions which exhibit high symmetry in their nature. Datasets are described briefly.

Aviation data [12]: The aviation dataset is obtained from SUBDUE downloads. It is actually from the aviation safety reporting system database. It contains a set of records reported for each event of the flights. A record represents an event. Events are plotted in terms of graphs. Nodes represent the events and labeled with event ids. Edges represent information concern to those events. Graph consists of 100 K vertices and 133 K edges. It is a sparse graph as on average it contains one edge per one node.

Protein-protein interaction data [13]: It is a dataset of proteins and the interactions between them. Database of Interacting Protein (DIP) is the source for the Saccharomyces cerevisiae protein interaction data for experimental analysis. Experimentally determined protein interactions data is provided by them. It contains 1274 protein nodes and 3222 interactions between proteins randomly selected from dataset. Each node corresponds to a type of protein or one of its main property and an edge represents the interaction between them. For some of interactions, protein structure is also implemented as we are retrieving based on the symmetries.

Another protein dataset is from DIP [14]. Various molecules collection is maintained. Nodes may be proteins and these proteins are extended with their original structure. This is about 1178 proteins which are classified into 691 enzymes and 487 non-enzymes. The graph is dense when compared with aviation dataset graph containing 285 nodes and 715 edges on average. Eighty-two various vertex labels are present.

The performance of the proposed algorithm (fssg) with its variation (fssg-l) based on only labels (dual) on above-mentioned datasets in comparison with the implementation of existing approach SIGRAM [6] named gns is presented below. Other existing algorithms employ approximate matching, sampling approach, etc., to retrieve patterns. The performance is compared in terms of (a) execution time to compute frequent graphs with respect to threshold and (b) number of frequent subgraphs found.

The performance of algorithms with respect to frequency threshold and time for different datasets is shown in Fig. 3. In general, the execution time required to generate frequent subgraphs at low threshold is exponential when compared with the time required at high threshold. It happens as the amount of generated frequent subgraphs increases exponentially when the threshold is decreasing. So that the required time to execute also increases. The algorithm shows a linear increase in execution time against threshold as it identifies the frequency at the time of partitioning. For other algorithms, the rate of increase in runtime is exponential when the frequency threshold decreases. Unlike other algorithms, the proposed algorithm does not need to generate all intermediate subgraphs. Consequently, the computation required to perform isomorphism testing for all intermediates is pruned, thus, it is efficient.

Frequency threshold: By observing the results, the support threshold plays key role to determine frequent subgraphs. As the threshold value decreases, there is an exponential number of candidate subgraphs that leads to exponential runtime. An efficient algorithm should be able to generate frequent subgraphs for lower threshold values in an effective execution time.

Results indicate that the proposed algorithm execution time increases linearly with the decrease in the frequency threshold in all three datasets, aviation, protein interaction and chemical data. But at higher threshold, the algorithm takes more time when compared with others. The reason behind it is the algorithm performs partitioning. Partitioning computation requires certain time irrespective of threshold that leads to good performance in lower threshold and general performance at higher thresholds.

Symmetry: The proposed algorithm accomplishes mining frequent subgraphs by exploiting the symmetry properties present in the given graph. Consequently, the proposed algorithm performs well for graphs that have symmetry. For protein interaction and chemical set, the performance is good when compared to aviation data with respect to other algorithms as those two datasets contain high symmetry in a graph. Actually, traditional algorithms suffer to generate candidates for symmetrical graphs, whereas fssg performs well shown in Fig. 4.

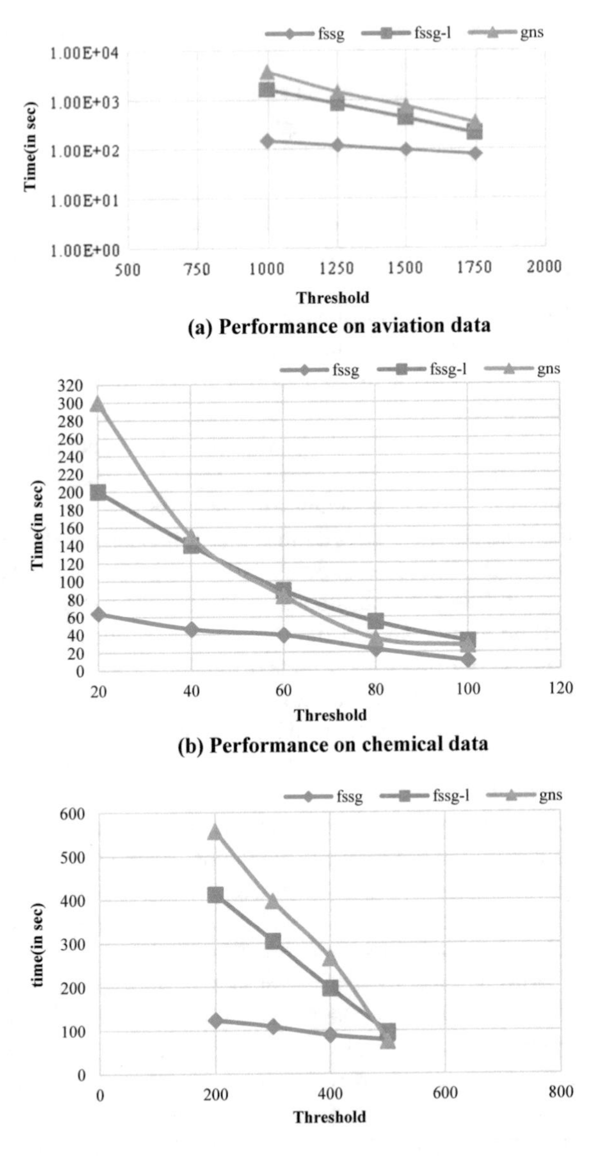

Fig. 3 Performance evaluation of algorithm

Coming to the number of frequent graphs generated during the process, all algorithms show nearly similar performance except small variations because of the process adopted to generate frequent graphs shown in Fig. 4. The proposed fssg algorithm allows subgraphs with a single vertex or single edge overlapping in candidate subgraphs generation, whereas other algorithms are not. And it may show a slight difference in the size of frequent subgraph also.

Fig. 4 Frequency of subgraphs

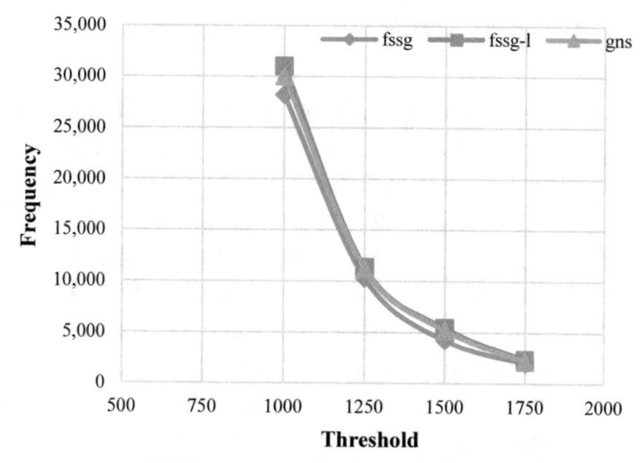

(a) Frequency of subgraphs in aviation data

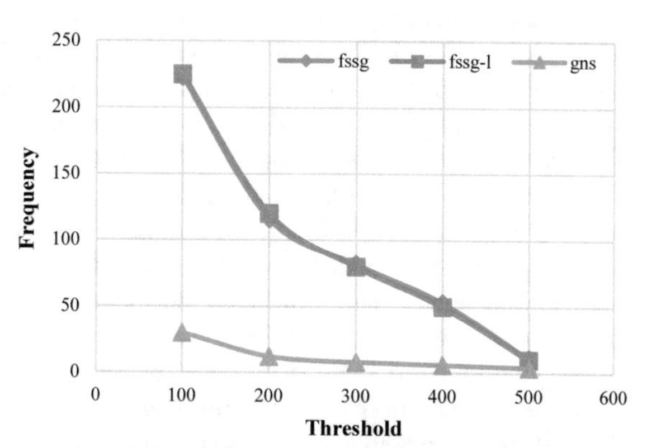

(b) Frequency of subgraphs in chemical data

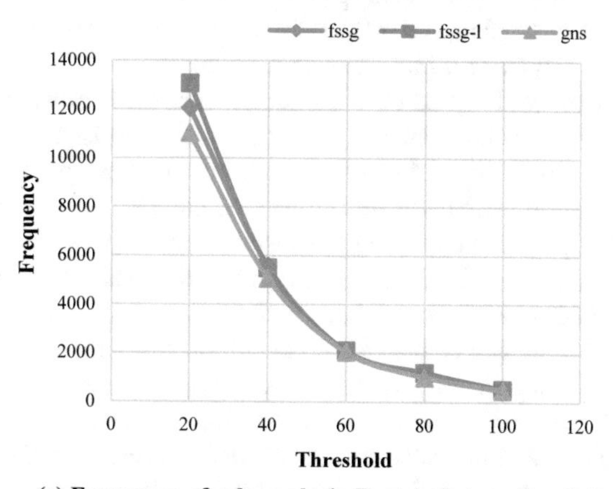

(c) Frequency of subgraphs in Protein Interaction data

5 Related Work

One of the first frequent substructure detection algorithms that fall under the single input graph category, was SUBDUE, an approximate and greedy search algorithm proposed by Cook and Holder [15]. SUBDUE finds repetitive patterns from the graph data by using background knowledge and identifies patterns of the compressed input graph by employing the minimum description length concept. It identifies subgraphs that compress the original data by maximum. As a side effect, it can build a concept hierarchy in the input data based on the substructures. The method has been successfully applied to chemical compound data, CAD circuits, etc. This algorithm is incomplete in terms of retrieving all frequent subgraphs. Later, Gb-Subdue and Db-Subdue algorithms address the limitations of Subdue algorithm.

SIGRAM [6] is another algorithm proposed by Kuramochi and Karypis that mines frequent subgraphs in a labeled, sparse, and single large graph. They proposed HSIGRAM and VSIGRAM algorithms to find frequent subgraphs in breadth-first and depth-first manner. Apriori principle is adopted in those algorithms to generate the candidate subgraphs and MIS metric to determine edge-disjoint embeddings of a graph. SIGRAM needs to enumerate all embedded subgraphs and it is expensive as the computation of maximum independent set is NP-hard. The same authors developed GREW [7], a heuristic approach that find long vertex-disjoint embeddings in a large graph. It is able to identify multiple patterns concurrently by employing heuristics and maintains location of identified frequent subgraph by rewriting the given graph. But it is unable to identify all frequent patterns. Another frequent subgraph mining algorithm GRAMI [8], proposed by Mohammed et al., identifies minimal set of subgraphs that satisfy threshold and avoids enumeration of all frequent subgraphs. GRAMI generalizes the concept of frequent pattern mining by allowing distance-constrained paths in the patterns. He mapped CSP problem with the subgraph isomorphism problem and solved CSP problem to find frequent subgraphs. Zhou and Holder [9] proposed a random sampling approach for very large graphs that don't fit in memory. They identified frequent subgraphs efficiently in a single large graph by sampling at random areas in graph. Other works have been proposed approaches like NODAR [16] and G-Miner [17] based on the pattern growth approach. Although these approaches mine completely, the discovered subgraphs are semantically very complex and require high consumption in terms of time and computing resources.

As we all know, candidate generation, candidate extension, and isomorphism to avoid duplicate candidates is a complex task in mining graph data. And in all the previous works, candidate generation starts with a node or an edge. The candidate extension also takes place by extending a single node or edge. Isomorphism of graphs is required to find frequency and also duplicate graphs. The proposed algorithm avoids all these costs by using invariant properties of graphs. For example, candidate may not start with a single edge or node. But starts with a frequent node with its associated graph which is also frequent(which satisfies the given conditions). So the candidate seed graph may be as large as it satisfies the first-level adjacent nodes satisfy the conditions. And here, only frequent nodes and edges are used for extension.

Isomorphism is mostly avoided as we are using invariant properties for generation and extension.

6 Conclusion

An algorithm to provide an efficient and fast computational approach to generate candidate subgraphs in a large graph using invariants of a graph is presented in this paper. The proposed algorithm achieves the specified outcome by exploiting the symmetries present in a graph and makes use of them to generate candidate subgraphs and to reduce isomorphism testing. This algorithm performs partitioning based on signature identifies the symmetries present in a graph and make available to enumerate frequent subgraphs faster than existing techniques. The obtained experimental results confirmed the effectiveness of the proposed algorithm. This algorithm is applicable to search structures in graph datasets for general applications. The algorithm can also be extended to find frequent subgraphs for unlabeled graphs.

References

1. Kazius J, Nijssen S, Kok J, Back T, Ijzerman AP (2006) Substructure mining using elaborate chemical representation. J Chem Inf Model 46:597–605
2. Taewookkim ML, Ryu KH, Shin J (2012) Prediction of protein function from protein-protein interaction network by weighted graph mining. In: 4th international conference on bioinformatics and biomedical technology, IPCBEE, vol 29
3. Jiang C, Coenen F, Zito M (2010) Finding frequent subgraphs in longitudinal social network data using a weighted graph mining approach. In: Cao L, Feng Y, Zhong J (eds) Advanced data mining and applications ADMA, pp 405–416
4. Donato D, Gionis A (2010) A survey of graph mining for web applications. Managing and mining graph data, advances in database systems, vol 40. Springer
5. Inokuchi A, Washio T, Motoda H (1998) An apriori-based algorithm for mining frequent substructures from graph data. In: Proceedings of the 2000 European symposium on principle of data mining and knowledge discovery (PKDD'00), pp 13–23
6. Kuramochi M, Karypis G (2005) Finding frequent patterns in a large sparse graph. Data Mining Knowl Discov 11(3):243–271
7. Kuramochi M, Karypis G (2004) GREW—a scalable frequent subgraph discovery algorithm. In: Proceedings of the fourth IEEE international conference on data mining (ICDM'04)
8. Elseidy M, Abdelhamid E, Skiadopoulos S, Kalnis P (2014) GraMi: frequent subgraph and pattern mining in a single large graph. In: Proceedings of VLDB endowment, vol 7, pp 517–528
9. Zou R, Holder LB (2010) Frequent subgraph mining on a single large graph using sampling techniques. In: Proceedings of workshop on mining and learning with graphs, pp 171–178
10. Bordino I, Donato D, Gionis A, Leonardi S (2008) Mining large networks with subgraph counting. In: IEEE ICDM conference
11. Kavitha D, Prasad KV, Murthy JVR (2016) Finding frequent subgraphs in a single graph based on symmetry. Int J Comput Appl 146(11):5–8
12. Subdue Databases. http://ailab.wsu.edu/subdue/
13. Salwinski L, Miller CS, Smith AJ, Pettit FK, Bowie JU, Eisenberg D (2004) The database of interacting proteins: 2004 update. Nucleic Acids Res 32:D449–D451

14. DIP. http://dip.doe-mbi.ucla.edu
15. Cook J, Holder L (1994) Substructure discovery using minimum description length and background knowledge. J Artif Intell Res 1:231–255
16. Hellal A, Romdhane LB (2013) NODAR: mining globally distributed substructures from a single labeled graph. J Intell Inf Syst 40(1):1–15
17. Jianga, Xiong H, Wang C, Tan A-H (2009) Mining globally distributed frequent subgraphs in a single labeled graph. Data Knowl Eng 68(10):1034–1058

A Utility of Ridge Contour Points in Minutiae-Based Fingerprint Matching

Diwakar Agarwal, Garima, and Atul Bansal

Abstract The performance of any biometric matching system depends on the features extraction of the used biometric identifier (fingerprint, face, iris, voice, retina, etc.) during enrolment and its matching while identification. The minutiae-based fingerprint matching systems are primarily dealing with the problem of the resulting number of false minutiae correspondences between query fingerprint and the template stored in the database during the identification process. This affects the system performance and eventually decreases the fingerprint matching accuracy. This paper presents a proposed algorithm as the solution incorporating the use of ridge contour points (level 3) with minutiae at feature level in order to reduce the false minutiae correspondences. The set of ridge contour points are detected in the vicinity of the minutiae involved in the matched minutiae pairs. The false correspondences are reduced by minimizing the Euclidean distance between the sets of ridge contour points associated with the minutiae pair with the help of Iterative Closest Point (ICP) algorithm. The fingerprint matching by utilizing the proposed algorithm achieves high recognition accuracy and improved error rates in comparison to the minutiae-based only fingerprint matching when applied over FVC 2006 database.

Keywords False correspondence · Fingerprint · Ridge contour · Matching · Minutiae

D. Agarwal (✉) · Garima · A. Bansal
GLA University, Mathura, UP 281406, India
e-mail: diwakar.agarwal@gla.ac.in

Garima
e-mail: garima.gla_mt18@gla.ac.in

A. Bansal
e-mail: atul.bansal@gla.ac.in

N. Chaki et al. (eds.), *Proceedings of International Conference on Computational Intelligence and Data Engineering*, Lecture Notes on Data Engineering and Communications Technologies 56, https://doi.org/10.1007/978-981-15-8767-2_24

1 Introduction

The fingerprint of an individual is described by three levels of features, namely, level 1 which includes ridge flow and pattern type, level 2 such as ridge bifurcation and ending called minutiae points, and level 3 which includes ridge dimensional attributes such as ridge width, ridge deviation, pores, ridge contour, etc. [1]. In view of the advantages of the level 3 features, latent forensic examiners and FBI's recent Next Generation Identification (NGI) system work upon the extended feature set defined by the ANIS/NIST committee CDEFFS [2]. Although level 3 features provide robust fingerprint identification, large number of commercially available fingerprint-based biometric machines still utilized level 2 features. During an enrollment process, fingerprints of all legitimate users are stored as the reference (template) to form the large database where each print is described by the set of minutiae points extracted by the fingerprint matching system. The representation of minutiae must be rotational and translational invariant since the spatial and angular alignment of the query fingerprint may be different from the reference fingerprints of the database. During an identification, the system extracts the minutiae set of the query fingerprint (either genuine or imposter) and determines the correspondences with the stored minutiae points of the reference fingerprints. The decision of accepting or rejecting the query fingerprint depends on the matching score determined mathematically from the number of total minutiae correspondences, i.e., matched minutiae pairs. These correspondences may contain some false correspondences which could occur due to the incorrect representation of the minutiae points. The false correspondences make an imposter fingerprint wrongly accepted as the genuine one and a genuine fingerprint wrongly rejected as an imposter fingerprint. This leads to an erroneous decision which may cause unfavorable conditions.

In order to reduce false correspondences, the main concept is to utilize the higher level features in conjunction with the minutiae points. The proposed algorithm utilizes the sets of ridge contour points as level 3 features around minutiae that form the initial matched pairs. The Euclidean distance between two sets of each pair is determined by applying an iterative ICP algorithm [3]. Then, the initial minutiae matched pairs satisfying the threshold constraint are selected as the final pairs to form the refined minutiae correspondences. The main contributions of the proposed method are as follows: (i) Reduces the false minutiae correspondences which are the main reason behind the disappointed fingerprint matching performance in most of the minutiae-based matchers. (ii) Improves the fingerprint matching accuracy through the fusion of level 3 feature with level 2 at feature level. The proposed algorithm is applied on FVC 2006 database [4].

2 Related Work

Jiang and Yau [5] proposed the use of both local and global structure of the minutiae for fingerprint alignment and matching. The local structure was defined by relating each minutia by the neighboring minutiae and global structure was defined by the best correspondences resulted from local minutiae matching. Tico and Kuosmanen [6] have introduced the use of non-minutiae features as the new representation of the minutia. Minutiae correspondences were improved by utilizing the fingerprint pattern around the minutia in addition to the minutia triplet. Feng [7] have proposed texture-based and minutiae-based descriptors for the minutia for fingerprint matching. Later, alignment-based greedy matching algorithm was applied in order to establish the minutiae correspondences and similarity computation. Chen and Gao [8] have proposed the method to align two fingerprints since the main reason behind false correspondences is the misalignment of fingerprints. The Minutiae Direction Map (MDP) of a query and template fingerprint was computed and the transformation parameters were determined by using phase correlation between two MDPs. Cappelli et al. [9] have introduced another approach of fingerprint representation categorizing itself as the fixed radius based local minutiae matching. Minutiae were described by the Minutia Cylinder Code (MCC) and their local structure was encoded to form rotation and translation invariant fixed length feature vector code. Medina-Pérez et al. [10] have introduced the concept of m-triplet representation of the minutia which was sensitive to the reflection of the traditional minutia triplet. The m-triplet contains several attributes like clockwise arrangement of minutiae, Euclidean distance between minutiae, minimum, medium, maximum distances, and angles to rotate the direction of minutiae points.

3 Proposed Method

The main objective of the proposed method is to reduce the number of false minutiae correspondences. Figure 1 shows the block diagram representation of the proposed method. The query fingerprint and the template fingerprint from the system database undergoes ridge contour extraction process and minutiae-based matcher. The matching is performed by applying Abraham et al. [11] method. The matching result contains the set of matched minutiae pairs where each pair is said to be an initial minutiae correspondence. The false correspondences in the set are reduced by using ridge contour points and initial correspondences in the ICP algorithm [3] to generate refined minutiae correspondences.

The following steps summarize the proposed algorithm.

1. The set of minutiae points M^Q and M^T is extracted for the given query fingerprint Q and the template fingerprint T, respectively. The set M^Q is defined by (1)

$$M^Q = \{x_i, y_i, \theta_i\}, i = 1, 2, \ldots, N^Q \tag{1}$$

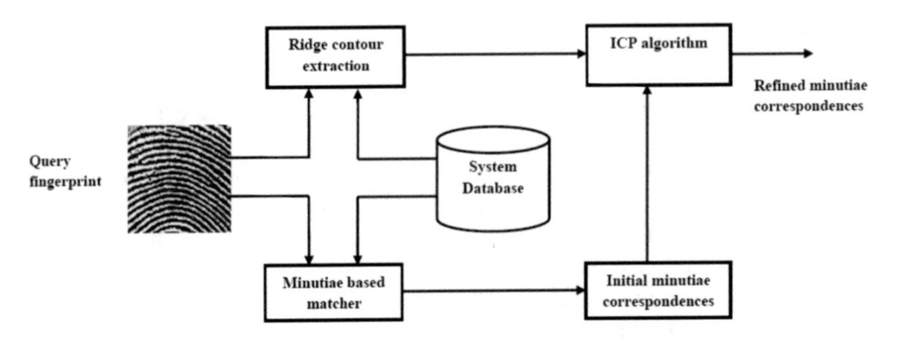

Fig. 1 Block diagram representation of the proposed method

where (x_i, y_i) is the spatial location, θ_i is the direction of the ith minutia, and N^Q is the total number of extracted minutiae in Q. The set M^T is defined by (2)

$$M^T = \{x_j, y_j, \theta_j\}, j = 1, 2, \ldots, N^T \tag{2}$$

where (x_j, y_j) is the spatial location, θ_j is the direction of the jth minutia, and N^T is the total number of extracted minutiae in T.

2. The matching of the minutiae set M^Q with the minutiae set M^T is performed. Based on the matching results, the initial correspondences are determined between the minutia $m_i \in M^Q, i = 1, 2, \ldots, N^Q$ and the minutia $m_j \in M^T, j = 1, 2, \ldots, N^T$. Let there be N^{QT} number of correspondences where each correspondence is defined by the minutiae pair $MP_t = (m_i, m_j), t = 1, 2, .., N^{QT}$. Thus, the set of initial minutiae correspondences is defined by (3)

$$C^{initial} = \{MP_1, MP_2, \ldots, MP_{N^{QT}}\} \tag{3}$$

3. The ridge contour of both Q and T is extracted by applying Jain et al. [1] method.
4. Then each minutia of the matched pair MP_t is represented by the set of its neighboring ridge contour points. For example, let the minutiae pair $(m_a, m_b) \in MP_t$ where $m_a \in M^Q$ and $m_b \in M^T$ be one of the correspondences in $C^{initial}$. The set of ridge contour points surrounding the minutia m_a is defined by $P^Q = (x_Q, y_Q)$ and the set of ridge contour points surrounding the minutia m_b is defined by $P^T = (x_T, y_T)$. The coordinate $(x, y) \forall P^Q, P^T$ is the location of the ridge contour point.
5. For N^{QT} matched minutiae pairs, N^{QT} Euclidean distances E_t where $t = 1, 2, \ldots, N^{QT}$ between P^Q and P^T are computed and minimized by applying the ICP algorithm [3].
6. The Euclidean distances E_t are then compared with the pre-specified threshold value th.

The minutiae pair from the set of initial correspondences $C^{initial}$ which has the Euclidean distance less than the threshold value is selected to be the suitable candidate pair for the set of refined minutiae correspondences $C^{refined}$. For example, if MP_1 and MP_2 from the set $C^{initial}$ have Euclidean distance E_1 and E_2, respectively, less than *th*, then $C^{refined}$ is defined by $C^{refined} = \{MP_1, MP_2\}$.

3.1 Minutiae Extraction, Matching, and Initial Correspondences

In this work, the task of minutiae extraction and minutiae matching to form an initial minutiae correspondences between query and template fingerprints is accomplished by applying the algorithm proposed by Abraham et al. [11]. It is a hybrid approach based on local minutiae matching in which minutiae are represented by the shape and orientation descriptors. The one advantage of using this approach is that the contextual information provided by the ridge flow and orientation robustly handles the spurious and missing minutiae. The following steps summarize the extraction of the minutiae.

1. Apply global thresholding on the fingerprint enhanced image $I_{enh}(x, y)$ [12] to generate a binary image I_{bin} as given in (4).

$$I_{bin}(x, y) = \left\{ \begin{matrix} 1, I_{enh}(x, y) > 0 \\ 0, otherwise \end{matrix} \right\} \tag{4}$$

2. Apply morphological thinning operation (skeletonization) on I_{bin} in order to make binary ridges 1 pixel wide. Then each pixel *p* of the thinned binary image is analyzed in its 8 neighborhood for being the minutia point. For this purpose, Rutovitz crossing number [13] at each pixel *p*, i.e., *cn(p)* is determined by following the Eq. (5)

$$cn(p) = \frac{1}{2} \sum_{i=1,2....8} \left| val\left(p_{(i mod 8)}\right) - val(p_{i-1}) \right| \tag{5}$$

where $val \in \{0, 1\}$. If $cn = 1$, then the pixel p is the ridge ending whereas if $cn = 3$, then p is the ridge bifurcation. Figure 2 shows the 8 neighboring pixels of p that are traversed in the counter clockwise direction. The fingerprint border regions

Fig. 2 Locations of 8 neighbors of the central pixel *p*

p_3	p_2	p_1
p_4	p	p_0
p_5	p_6	p_7

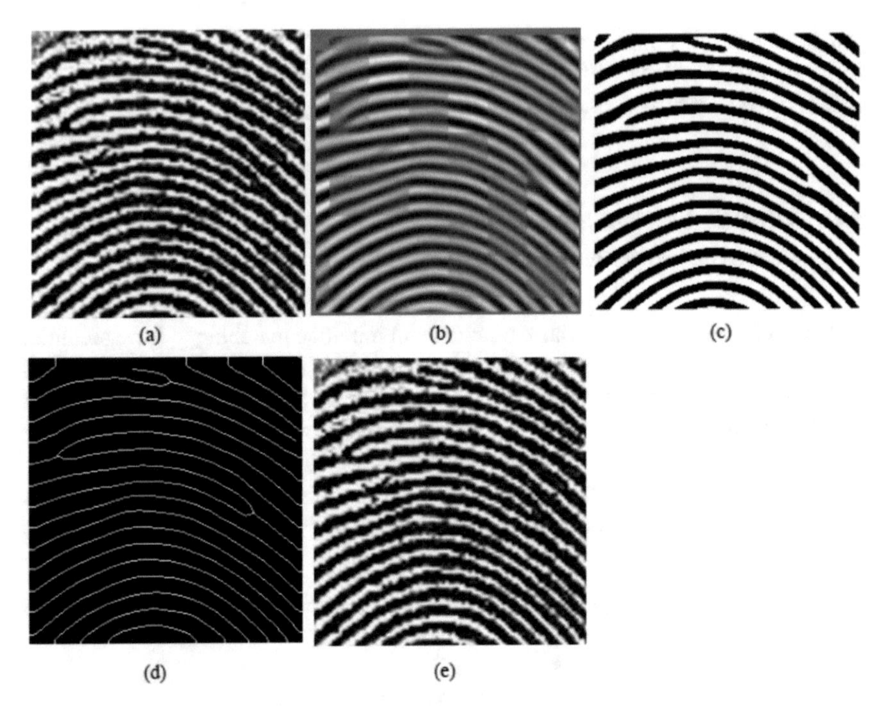

Fig. 3 Results of minutiae extraction. **a** Input fingerprint image 1_1 adopted from DB1-A of FVC 2006 database [4], **b** fingerprint enhanced image, **c** binarized ridge image, **d** thinned image, **e** minutiae points marked by red square indicators

undergoes crude filtering to remove short spurs and false minutiae. Figure 3 show the results of the minutiae extraction steps.

Now, each minutia of the query and template fingerprint is represented by the shape context descriptor [14] and orientation-based descriptor [6]. The shape context descriptor utilizes the contextual information about the type of the minutiae (bifurcation or ending) present in the surrounding of each minutia. This information is recorded in the polar histogram generated by representing the region around the minutia in log-polar coordinate system. The log-polar space is divided into concentric circles centred at the minutia position into equal distance bins and equal angular bins. The orientation-based descriptor includes the information about the relative orientation of the sampling points marked on the concentric circles with respect to the direction of the central minutia. The example of the neighboring minutiae and sampling points around the central minutia is shown in Fig. 4a, b.

After defining the descriptors, next task is to establish the set of initial correspondences $C^{initial}$ between M^Q and M^T. For each minutia $m_i \in M^Q$ and $m_j \in M^T$, the sets M^Q and M^T undergoes through the affine transform, respectively, by using rotation and distance offset where the minutiae m_i and m_j are involved in the matching process. Each minutia m_i of M^Q is compared with each minutia m_j of M^T in order to find the best matched minutia in M^T.

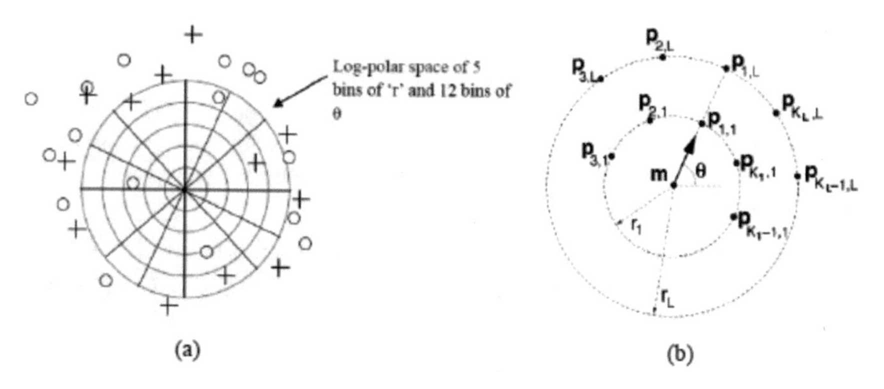

Fig. 4 **a** Log-polar space around central minutia with '+' indicates bifurcation and 'o' indicates ending. **b** Sampling points $p_{k,l}$ arranged in a circular pattern around the minutia point m. $p_{k,l}$ is the kth sampling point at the lth concentric circle

Figure 5 shows the initial correspondences including the matched minutiae between query and template fingerprints. Based on the number of initial matched minutiae pair, the score between two matching fingerprints is calculated by (6)

$$S^{initial} = \left(\frac{N^{QT} - 0.2(N^{T} - N^{QT})}{N^{T} + 1}\right) + \left(\frac{N^{QT} - 0.2(N^{Q} - N^{QT})}{N^{Q} + 1}\right) \quad (6)$$

where N^{QT} is the number of matched minutiae pairs, N^{Q} is the number of minutiae in the set M^{Q} and N^{T} is the number of minutiae in the set M^{T}.

Fig. 5 Initial minutiae correspondences between query fingerprint 1_1 and its genuine template fingerprint 1_5 (images are adopted from DB1-A of FVC 2006 database [3])

3.2 Ridge Contour

According to the Jain et al. [1], the fingerprint ridge contour is used as the valuable level 3 information to be utilized in the fingerprint matching. In this work, the algorithm implemented by Jain et al. [1] is applied to extract the ridge contour. The details of the algorithm are explained below.

1. Apply the Mexican hat wavelet transform on an input fingerprint image to enhance the fingerprint ridge details. The Mexican hat wavelet [15] is defined by (7). The value of σ is empirically chosen as 1.32.

$$\nabla^2 h = -\left[\frac{\left(x^2+y^2\right)^2-\sigma^2}{\sigma^4}\right]e^{-\frac{\left(x^2+y^2\right)^2}{2\sigma^2}} \tag{7}$$

2. Subtract the output wavelet response image from the fingerprint enhanced image I_{enh} of the minutiae extraction process. As a result, the ridge contour of the input fingerprint image gets further enhanced. Then, the binarization is performed on the resultant image by applying the global thresholding method.
3. Determine the 2D convolution of the binarized image and the filter H $=$ $[0, 1, 0; 1, 0, 1; 0, 1, 0]$ to locate the neighboring edge points of each pixel of the binarized image. The pixel of the convolved image is set to binary '1' if it holds the value of 1 and 2 to form the ridge contour image. Figure 6 show the results at various steps of the algorithm.

3.3 Refined Correspondences

The refined minutiae correspondences are derived from an initial correspondences by utilizing the fingerprint ridge contour. For the given matched pair (m_a, m_b) where $m_a \in M^Q$ and $m_b \in M^T$, let (x_a, y_a) and (x_b, y_b) be the spatial coordinate of the minutiae m_a and m_b, respectively. Consider the region of size 31×31 centred at the position $(\frac{x_a+\bar{x}}{2}, \frac{y_a+\bar{y}}{2})$ associated with m_a and the region of same size centred at the position $(\frac{x_b+\bar{x}}{2}, \frac{y_b+\bar{y}}{2})$ associated with m_b. For m_a, the $(\bar{x}, \bar{y})$ is the mean value of the spatial coordinates of the minutiae in the set M^Q and for m_b, the $(\bar{x}, \bar{y})$ is the mean value of the spatial coordinates of the minutiae in the set M^T. Generate the sets P^Q and P^T of the ridge contour points from the ridge contour image within the associated region of m_a and m_b. The ICP algorithm [3] is then applied to minimize the Euclidean distance between the sets P^Q and P^T which results to the final distance.

Therefore, for N^{QT} matched pairs, N^{QT} distances $E_t, t = 1, 2, \ldots, N^{QT}$ are obtained. These distances are then compared with the specific threshold value *th*. In this work, the value of *th* is considered as the mean value of E_t. The matched pair with the Euclidean distance between its associated set of ridge contour points less than *th* is ensured to be a part of the set of refined correspondences $C^{refined}$ as shown

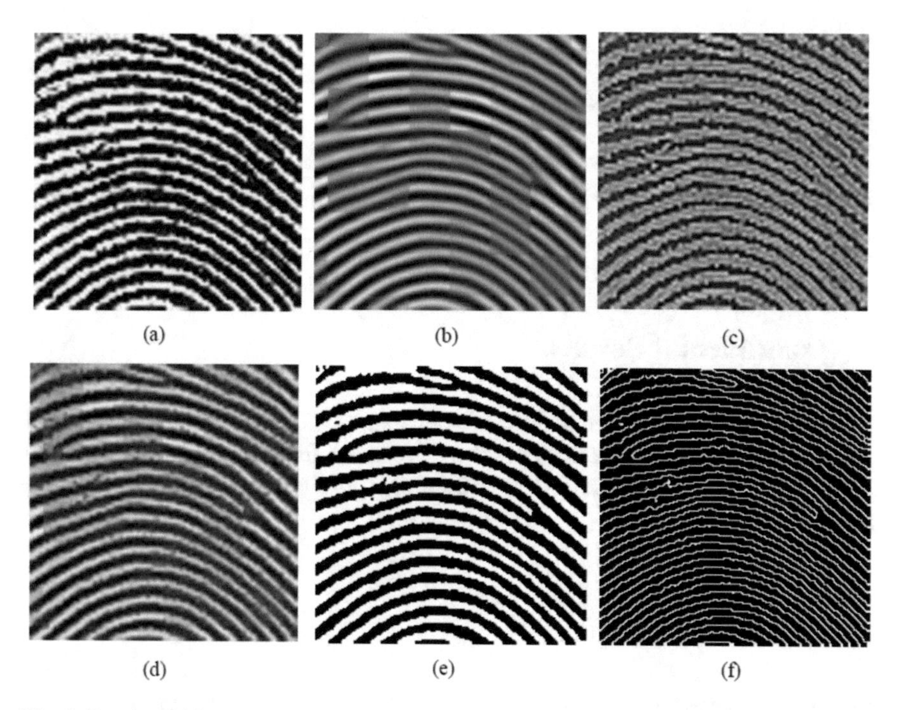

Fig. 6 Results of ridge contour extraction. **a** Input fingerprint image same as in Fig. 3a, **b** fingerprint enhanced image, **c** Mexican hat wavelet response image, **d** image after subtraction, **e** binarized image, **f** fingerprint ridge contour

in Fig. 7. Based on the refined minutiae correspondences, the new score is defined by (8).

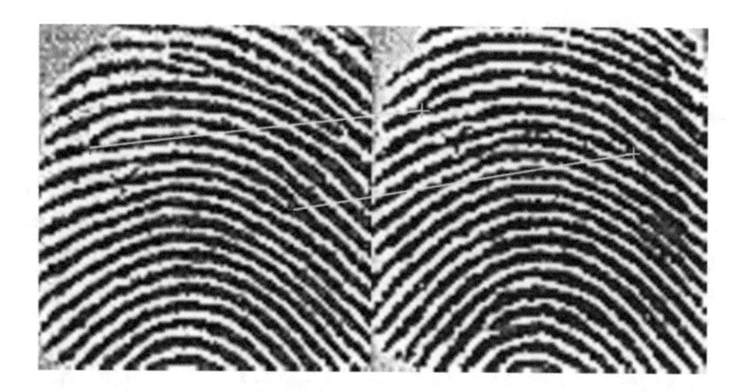

Fig. 7 Refined minutiae correspondences derived from initial correspondences of Fig. 5

$$S^{refined} = \left(\frac{N_R^{QT} - 0.2\left(N^T - N_R^{QT}\right)}{N^T + 1} \right) + \left(\frac{N_R^{QT} - 0.2\left(N^Q - N_R^{QT}\right)}{N^Q + 1} \right) \quad (8)$$

where N_R^{QT} is the number of matched minutiae pairs in the set of refined minutiae correspondences.

4 Experimental Results

The efficacy of the proposed method is evaluated by observing the performance of the minutiae-based matcher in two cases. Case 1: the fingerprint matching without using ridge contour points and Case 2: the fingerprint matching utilizing the ridge contour points.

4.1 Database Description

The FVC 2006 database [4] includes four fingerprint databases DB1, DB2, DB3, and DB4 collected from electric field sensor, optical sensor, thermal sweeping sensor, and SFinGe v3.0, respectively. Each database contains 12 samples of each 150 distinct fingers, i.e., 1800 fingerprint images. Each database is partitioned into two sets A and B, namely, DB1-A, DB1-B; DB2-A, DB2-B; DB3-A, DB3-B, and DB4-A, DB4-B. The set A of each database contain the fingerprints of first 140 fingers, i.e., 1680 fingerprint images. The set B contain the fingerprints of last 10 fingers, i.e., 120 fingerprint images.

4.2 Fingerprint Recognition

The performance of the proposed method is evaluated on the FVC 2006 DB1-A database. The following matches are carried out for the above mentioned two cases: Genuine match: Each sample is matched against the fingerprint samples of the same finger (excluding symmetric matches), thus total matches are equal to 9240 genuine tests. Imposter match: The first sample of each finger is matched against the first sample of other fingers; thus, total matches are equal to 9730 imposter tests.

The scores $S^{initial}$ and $S^{refined}$ lies in the closed interval [0, 100]. The score '0' means 'no match' and the score '100' means 'full match'. The decision of either accepting or rejecting the query fingerprint is made at an estimated threshold value. The query fingerprint is accepted if the fingerprint matching score is greater than the threshold value otherwise, it is rejected. Due to the appearance of the false minutiae correspondences, there may be the possibility that the genuine score falls below the threshold or the imposter score exceeds the threshold. This erroneous decision by the fingerprint matcher is quantified by False Rejection Rate (FRR) and False Acceptance Rate (FAR). The FRR is determined from the scores obtained from genuine tests and it is defined as a fraction of positive scores falling below the threshold. The FRR is given by (9)

$$FRR(\%) = \frac{FN}{TP + FN} = \frac{Genuinescorefallingbelowthethreshold}{Allgenuinescores} \quad (9)$$

where *FN* is False Negative and *TP* is True Positive. The FAR is determined from the imposter tests score and it is defined as a fraction of negative scores exceeding the threshold value. The FAR is given by (10)

$$FAR(\%) = \frac{FP}{FP + TN} = \frac{Imposterscoreexceedingthethreshold}{Allimposterscores} \quad (10)$$

where *FP* is False Positive and *TN* is True Negative.

Figure 8 shows the FAR, FRR curves plotted against the threshold values for the case 1 and case 2. The best matching result is obtained at the threshold value at which the FAR and FRR curves intersect. In case 1 (see Fig. 8a), the threshold value is observed as 1.2 and at this threshold the *FAR* is 64.6%. Figure 9a shows that after applying the proposed method, in imposter fingerprint matching the total number of matched minutiae pairs are reduced due to the reduction in false minutiae correspondences. This reduction shifts many imposter scores below the threshold value, therefore at the same threshold value the better *FAR,* i.e., 30.85% is observed in case 2 as shown in Fig. 8b.

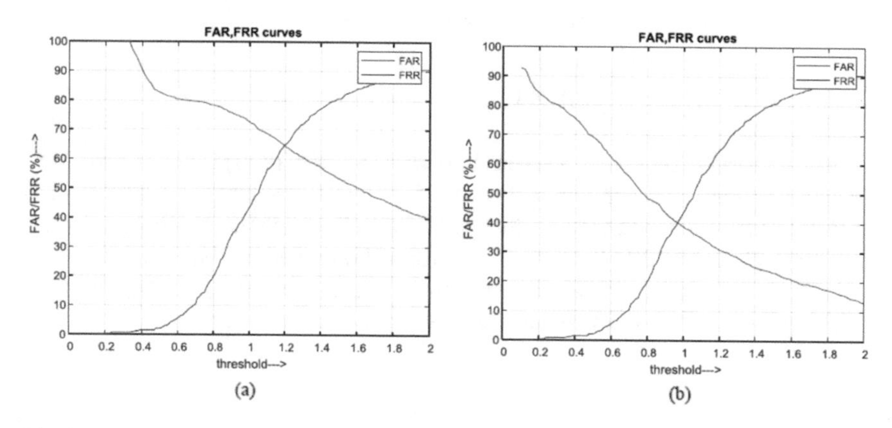

Fig. 8 FAR, FRR curves, **a** case 1 (without ridge contour points), **b** case 2 (with ridge contour points)

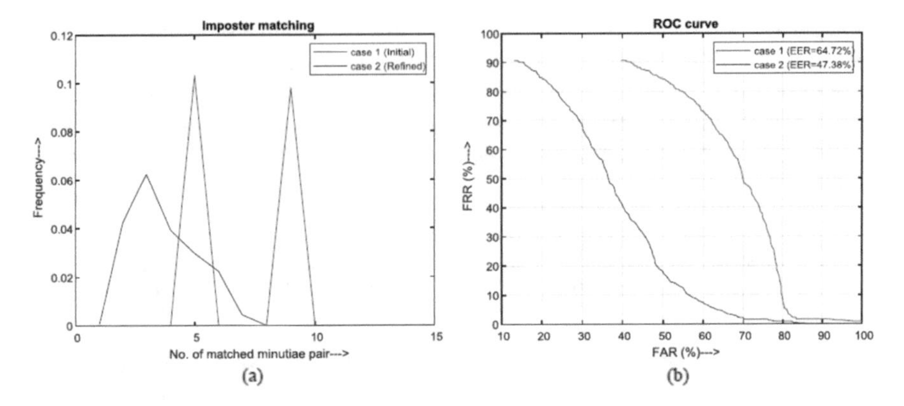

Fig. 9 ROC curves for case 1 and case 2

Table 1 FRR, *FAR, EER, ACC* at the threshold value of 1.2 for the case 1 and case 2

Cases	Threshold value	*FRR* (%)	*FAR* (%)	*EER* (%)	*ACC* (%)
Case 1 (without ridge contour)	1.2	64.84	64.60	64.72	35.28
Case 2 (with ridge contour)	1.2	63.91	30.85	47.38	52.62

At the observed threshold value, the equal error rate *EER* and the matching accuracy *ACC* are also calculated which are given by (11) and (12), respectively. Table 1 show the values of *FAR, FRR, EER, ACC* at the threshold of 1.2 for the two cases considered in this paper.

$$EER(\%) = \frac{FRR + FAR}{2} \tag{11}$$

$$ACC(\%) = 100 - EER \tag{12}$$

Figure 9b show the Receiver Operating Characteristic (ROC) curve of two cases. By utilizing the ridge contour points, the fingerprint matching accuracy reaches 52.62% (*EER* 47.38%) in case 2 which is 17.34% more than the matching accuracy of 35.28% (*EER* 64.72%) in case 1. This result shows that the ridge contour (level 3 features) information when utilized with level 2 features (case 2) improves the matching performance of the minutiae-based fingerprint matcher (case 1).

5 Conclusion

This paper presented the proposed method as the solution to the problem of false minutiae correspondences that arises during the fingerprint matching. The proposed method utilized the ICP algorithm to minimize the Euclidean distance between the set of fingerprint ridge contour points associated with the initial matched minutiae pair. The matched pair was subjected to the set of refined correspondences if it fulfilled certain criteria of thresholding otherwise, retained from the refined set. The experimental results so achieved are remarkable and proved the usefulness of fingerprint ridge contour as level 3 features in reducing false minutiae correspondences. The use of ridge contour points improved the matching accuracy of the minutiae-based fingerprint matcher in comparison to the matching without ridge contour. As advancement in future aspect, fingerprint ridge dimensional attributes such as ridge width and ridge deviation could be determined from the fingerprint ridge contour. Such extended level 3 features could aid the forensic investigators in developing the robust latent fingerprint matcher.

References

1. Jain AK, Chen Y, Demirkus M (2006) Pores and ridges: high-resolution fingerprint matching using level 3 features. IEEE Trans Pattern Anal Mach Intell 29(1):15–27
2. CDEFFS: The ANIS/NIST Committee to Define an Extended Fingerprint Feature Set. https://fingerprint.nist.gov/standard/cdeffs/index.html. Accessed 18 Mar 2020
3. Besl PJ, McKay ND (1992) A method for registration of 3D shapes. IEEE Trans Pattern Anal Mach Intell 14(2):239–256
4. Cappelli R, Ferrara M, Franco A, Maltoni D (2007) Fingerprint verification competition 2006. Biom Technol Today 15(7–8):7–9
5. Jiang X, Yau WY (2000) Fingerprint minutiae matching based on the local and global structures. In: Proceedings 15th international conference on pattern recognition (ICPR-2000), vol 2. IEEE, pp 1038–1041
6. Tico M, Kuosmanen P (2003) Fingerprint matching using an orientation-based minutia descriptor. IEEE Trans Pattern Anal Mach Intell 25(8):1009–1014
7. Feng J (2008) Combining minutiae descriptors for fingerprint matching. Pattern Recogn 41(1):342–352
8. Chen W, Gao Y (2007) A minutiae-based fingerprint matching algorithm using phase correlation. In: 9th biennial conference of the Australian pattern recognition society on digital image computing techniques and applications (DICTA 2007). IEEE, pp 233–238
9. Cappelli R, Ferrara M, Maltoni D (2010) Minutia cylinder-code: a new representation and matching technique for fingerprint recognition. IEEE Trans Pattern Anal Mach Intell 32(12):2128–2141
10. Medina-Pérez MA, García-Borroto M, Gutierrez-Rodríguez AE, Altamirano-Robles L (2012) Improving fingerprint verification using minutiae triplets. Sensors 12(3):3418–3437
11. Abraham J, Kwan P, Gao J (2011) Fingerprint matching using a hybrid shape and orientation descriptor. State Art Biom 25–56
12. Hong L, Wan Y, Jain AK (1998) Fingerprint image enhancement: algorithm and performance evaluation. IEEE Trans Pattern Anal Mach Intell 20(8):777–789
13. Rutovitz D (1966) Pattern recognition. J R Stat Soc 129:504–530

14. Kwan PW, Gao J, Guo Y (2006) Fingerprint matching using enhanced shape context. In: Proceedings of the 21st image and vision computing New Zealand (IVCNZ 2006), Great Barrier Island, New Zealand, pp 115–120
15. Rafael CG, Richard EW (2002) Digital image processing, 2nd edn. PHI

VeNNus: An Artificial Intelligence Accelerator Based on RISC-V Architecture

S. Harini, Aswathy Ravikumar, and Dhruv Garg

Abstract Developing prototypes for systems with custom chips is the most recent advancement in accelerator-centric architectures. The development of such prototype is a challenging task but with the emergence of new open-source software's and hardware's help in addressing this challenge to an extent by reducing implementation, design and effort. With increasing utility of Artificial Intelligence (AI) in various applications, there are great benefits to be reaped from faster training and computation of neural networks. The objective of this research is to develop an open-source architecture for acceleration of AI applications. We studied in detail the existing and available AI accelerators, and understood their key features. Their key functionalities were incorporated in the design of the VeNNus processor. This paper describes the architecture of the VeNNus processor, which uses the RISC-V Instruction Set Architecture (ISA). RISC-V ISA offers benefits such as flexibility, lower costs and high efficiency. However, to further improve the performance and energy efficiency of our Artificial Intelligence (AI) accelerator, 16 custom vector instructions were added in extension to the RISC-V ISA. The VeNNus processor includes redundant Arithmetic-Logic Units (ALUs), and uses quantization of training weights to 8-bit integers to deliver better performance, driving higher throughput. With the processor using vector instructions and quantization, we anticipate good acceleration of deep neural networks. The vital benefits and challenges faced with the RISC-V instruction set are described in this paper.

Keywords Artificial intelligence · Accelerator · Architecture · RISC-V · Vector instructions

S. Harini · A. Ravikumar (✉) · D. Garg
School of Computer Science and Engineering, VIT, Chennai 600127, India
e-mail: aswathy.ravikumar2019@vitstudent.ac.in

S. Harini
e-mail: harini.s@vit.ac.in

D. Garg
e-mail: dhruvshekar.bhargav2016@vitstudent.ac.in

N. Chaki et al. (eds.), *Proceedings of International Conference on Computational Intelligence and Data Engineering*, Lecture Notes on Data Engineering and Communications Technologies 56, https://doi.org/10.1007/978-981-15-8767-2_25

1 Introduction

The main research directions in the computer architecture field is the building of prototypes with custom-designed chips and systems [2, 8, 11, 17]. The principles can be validated by the development of prototypes and this helps to measure the system performance, efficiency, and to gain intuitions about the issues present in physical designs, and this in turn provides a direction for future research. The development of new prototypes is the base for new research directions. The increase in the need of accelerator-centric architectures has led to the development of new prototypes with dark silicon [4, 9, 10, 12, 15, 16, 18]. The mix of both programmable and specialized accelerators is implemented in these architectures. The design of prototypes with custom-designed chips is a tedious and challenging task. It takes almost months to get access to a reasonably modern technology node and that can require months of legal negotiation. The extension of the existing instruction set is legally prohibited due to licensing issues and this leads to the development of new instruction sets which is a really difficult task with a huge amount of effort. As mentioned above, modification is prohibited in the processor cores and this will lead to the need of developing, verifying, testing, and implementing general purpose processor cores. The prototypes developed can be either a board design or a FPGA, and to evaluate them there is a need of a system-level hardware infrastructure, and this task is a doable task. For the development of open-source hardware and software ecosystem, a new open-source instruction set RISC-V architecture (ISA) is serving as the base [1, 14]. This ecosystem includes the RISC-V ISA specification; on-chip networks (OCN) specifications; a complete software stack for both embedded and general purpose computing; various RISC-V processor and OCN implementations; and system-level hardware infrastructure for RISC-V processors. RISC-V ecosystem is not an effective solution for solving all the challenges in building research chips in academia, but it can to an extent help in reducing the design, implementation, and verification effort required for building accelerator-centric prototypes. A complete off-the-shelf RISC-V software stack (e.g., binutils, GCC, newlib/glibc, Linux kernel, Linux distributions) enables rapidly bringing up initial workloads on new prototypes before modifying and/or extending this software. The image applications like classification, identification, detection, and localization tasks are now widely implemented using Convolutional Neural Networks (CNNs).

In order to make the CNN effective, the hardware modifications are needed for accelerating the process. The pervious works are mainly focused on the requirement of large traditional on and off chip memories for storing the activation values and the weights and carefully hand-crafted digital VLSI architectures [3, 5, 19]. Binarized Neural Networks (BNNs) have demonstrated that initial weights and activations (i.e., $+1, -1$) can, in certain cases, achieve accuracy comparable to full-precision floating-point CNNs [6, 7, 13]. In this paper, we develop the design of VeNNus for hardware acceleration of CNN training and computations.

2 Related Work

To derive our own architecture to accelerate AI computations, we first studied and analyzed the existing AI accelerator architectures. These architectures gave us an insight into the working of the chips.

2.1 Radeon Instinct MI25

Radeon Instinct MI25 accelerator is a training accelerator, which uses AMD Vega 10 GPU architecture for large-scale machine intelligence and deep learning. The performance offered is 24.6 TFLOPS of FP16 and 12.3 TFLOPS of FP32 peak performance through its 64 Next-generation Compute Units (NCUs) with 4096 stream processors. The instruction throughput can be improved by combing with higher clock speeds. NCU array is capable of 13.7 teraflops of single-precision arithmetic throughput. The programmable NCUs at the core of “Vega” GPUs have been developed with a feature called Rapid Packed Math. With mixed-precision support, “Vega” can be used to improve the speed of the higher precision operations while maintaining full precision for the ones.

2.2 NVIDIA Tesla V100

Tesla V100 differentiates itself from other AI accelerators like Radeon Instinct MI25 by including dedicated Tensor cores. Tensors are data structures used in linear algebra, and like vectors and matrices, all arithmetic operations can be performed with tensors. On the chip dye, there are 672 tensors cores-8 per streaming multiprocessor, and there are 84 SMs in total. The 84 Volta SMs contain FP32, INT32, FP64, 8 Tensor cores, and 4 texture units. Along with the new SIMT threads that remove the limitation of current SIMT and SIMD, the mixed processing Tensor cores deliver 12X TFLOPS. Tensor core in NVIDIA Tesla operates on a 4×4 matrix performing matrix multiplications and neural network training. To enhance the performance, Tesla has a single memory block with combined data cache and shared, thereby giving low latency and higher bandwidth.

2.3 Google Tensor Processing Unit (TPU)

Google’s TPU delivered 15-30X higher performance than contemporary CPUs and GPUs. TPU is ideal for training deep neural networks since it effectively addresses the bottlenecks that increase the training time for neural networks. Major breakthrough in performance in TPUs was brought by the concept of quantization. TPU contains 65,536 8-bit integer multipliers, in contrast to few thousand 32-bit floating-point multipliers on other GPUs. By this quantization, we can speed up the processing by 25 times. There are places where Google TPU takes an approach contrary to other CPUs and GPUs. For example, the basic TPU instruction set is the Complex Instruction Set Computer (CISC). A CISC style design focusses on implementing high-level instructions that run more complex tasks. TPU also uses the concept of a systolic array where the result after computation is directly passed on to the next ALU, instead of being written back to the register each time. To compute hundreds of operations (matrix operations) in a single clock cycle, they designed MXU or the matrix processor. In order to reduce the power utilization and also to have a noticeable increase in throughput during all the massive matrix multiplications the intermediate results are passed directly between the ALUs without any memory access.

2.4 Shakti I Class Processor

Shakti processors are open-source RISC-V based high-performance cores. Their SoC configuration is mainly applied to parallel high-performance computing system and analytics of the workloads. The high-performance cores can be a combination of C or I class, single thread performance driving the core choice. I-Class processor has features which are mainly performance oriented like multithreading, non-sequential execution, pipelining, etc. The I-Class processors are targeted at the compute, storage, and networking the mobile and networking segments with target operating range between 1.5 and 2.5 Ghz. Much of our design is largely inspired by the Shakti I class processor.

2.5 LA Core

Linear Algebra is the foundation of High-Performance Computing (HPC). LACore design addresses the numerous shortcomings of modern accelerators like memory transfers, sequential mode and synchronization bottleneck. LACore instruction set is based on the RISC-V ISA with 68 new instructions to incorporate faster operations. Its execution units were thus designed to be mixed precision systolic data path, having memory units to read and write data streams to the data path FIFOs. To enhance computational performance, LACore has 8 parallel VecNodes and reduction units that form a binary tree with 7 ReduceNodes and one AccumulateNode. Owing to these optimizations, LACore significantly outperforms the x86. RISC-V and GPU for linear algebra applications.

2.6 Key Inferences

The key features and the working of the above accelerator chips were thoroughly analyzed and have inspired various elements of our VeNNus processor. The Radeon Instinct MI25 uses mixed precision which can accelerate operations which have no benefit from higher precision. This inspired us to think of performing operations such that lower precision is used, thereby reducing the memory overhead and accelerating computations. NVIDIA Tesla V100 contained separate FP32, INT32, FP64 and Tensor cores in their streaming multiprocessors, motivating us to think about how we would accelerate tensor operations in our processor (Fig. 1).

Google's TPU showed that in training deep neural network and performing neural network predictions, quantization can achieve 25X the previous performance without compromising the accuracy by more than 2. LACore showed the benefits of incorporating custom vector instructions, since they were able to outperform x86 and RISC-V

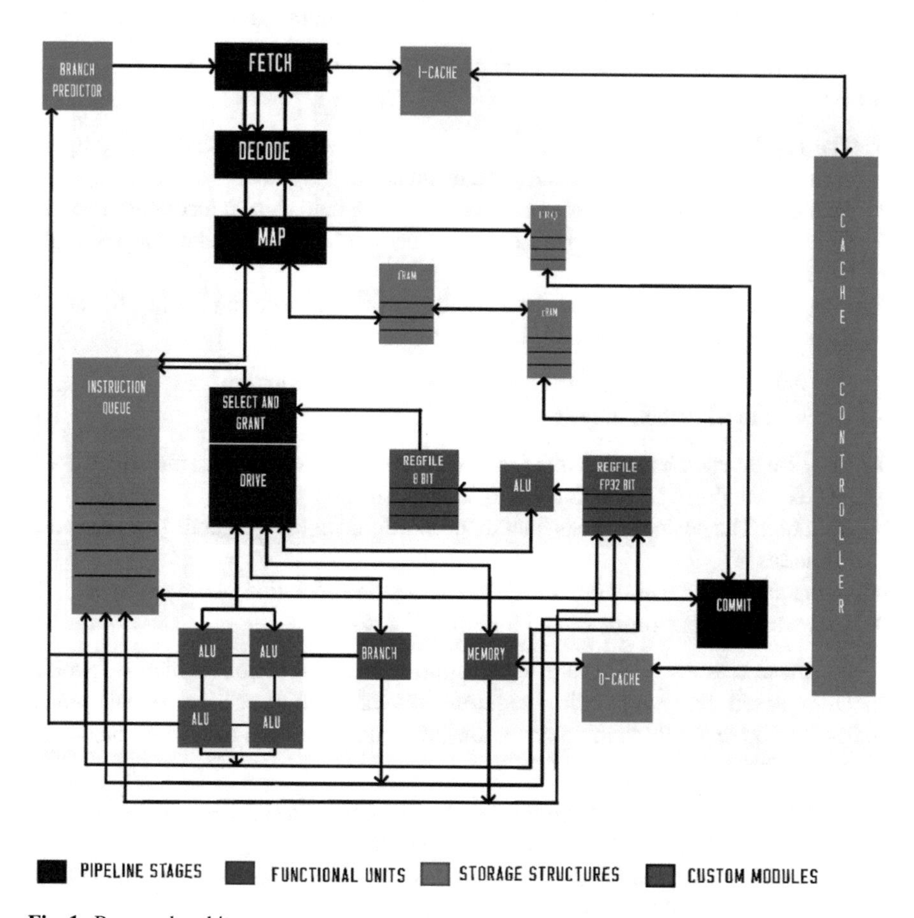

Fig. 1 Proposed architecture

on major benchmarks relating to linear algebra operations. This inspired us to think of incorporating vector instructions in extension to those provided in the RISC-V ISA. Lastly, Google TPU and LACore motivated us to employ a systolic array datapath in our processor, since systolic datapaths are able to reduce the memory access, thereby increasing throughput.

Since RISC-V Instruction Set Architecture (ISA) does not currently support vector instructions, VeNNus processor would be saddled by the same problems encountered with the conventional exploitation of Instruction Level Parallelism (ILP). The limits to the conventional Instruction Level Parallelism are

- Pipelined clock rate: during certain cases, an increase in clock rate will have an effect on the Clocks Per Instruction (CPI)
- Instruction fetch and decode: In certain situations, the fetching and decoding of more instructions in a single clock cycle is difficult.
- Cache hit rate: Both long-running (scientific) programs that have very large data sets accessed and continuous data streams (multimedia) have poor locality.

Using vector instructions solves many of the above problems. Some properties include

- The results are independent of each other. The longer pipeline and compiler ensuring no dependencies ensures high clock rate.
- Vector instructions make memory access with the help pattern recognition, amortization of memory latency over large number of elements. No data caches might be required.
- Single vector instruction has the advantage of few instruction fetches but at the same time the work involved is high.

Due to the above-mentioned properties, vector instructions also have many advantages. In a nutshell they are

- They are independent, use same functional unit, access contiguous memory words, and exploitation of the high memory bandwidth.
- Scalable: The performance is linearly proportional to the availability of hardware resources
- Compact: Single instruction for the effective representation of n operations.
- Expected high performance and multimedia ready.

For these reasons, we have a few custom vector instructions which we would include, based on computation requirements of training convolutional neural networks. There are three main operations to be done on each neuron of the neural network. They are

- Signal strength obtained from the product of input data (X) with weights (w)
- Aggerate of all the results to a single state is done.
- After aggregate operation the activation function (f) such as Rectified Linear Unit (ReLU) or Sigmoid or SoftMax or tanh to get the activity of neurons.

Supported instructions incorporated in the processor are as follows:

Table 1 Comparing existing AI accelerating architectures and VeNNuS

Radeon instinct MI25	NVIDIA Tesla V100	Google tensor processing unit (TPU)	LA core	Proposed VeNNus
Performance of mixed precision is very good. No benefit from higher precision	Dedicated tensor processing cores improves performance	Neural Network prediction and quantization improves performance greatly	Uses custom vector instructions to improve performance	Lower precision is used for operations. Dedicated FPGA elements for performance are used. Similar to TPU uses neural network prediction. Custom vector instructions are added to RISC-V ISA for improved performance like LA core

- 53 RISC-V instruction
- Custom vector memory instructions: (2) VLD—vector load
- VST—vector store
- Custom vector integer instructions: (5) ADDV, VADDI—addition
- SUBV—subtraction MULV—multiplication DIVV—division
- SLLV—shift
- Custom vector floating-point instructions: (5) ADDVF, VFADDI—addition
- SUBVF—subtraction MULVF—multiplication DIVVF—division SLLVF—shift
- Custom vector conversion instructions: (6) HICVTVF—integer to float (half precision) SICVTVF—integer to float (single precision) DICVTVF—integer to float (double precision) HCVTVF—float to float (half precision) SCVTVF—float to float (single precision) DCVTVF—float to float (double precision) (Table 1)

3 Working and Implementation

The VeNNus processor is aimed at performance seeking machine learning and artificial intelligence applications. The processor has been designed using Bluespec System Verilog (BSV) with modularization to enable us to modify the parts of the design with ease. It features a pipeline of 8 stages—Fetch, Decode, Map, Wakeup, Select, Drive, Execute, and Commit, where each of the stages take utmost a single cycle to execute. VeNNus supports standard RISC-V ISA instructions and we also have incorporated our own 16 custom vector instructions to speed up the matrix computations used in artificial intelligence applications. To speed-up the execution, instructions are issued to the execution units in out-of-order fashion, but are committed in-order. Register renaming or mapping is done through a merged register file approach. It stores both the architectural register values and the speculated values. With a total of 96 registers of which 32 are architectural registers and 64 are physical registers. A buffer (register alias table) maintains the map from architectural registers to physical registers. The type of Branch Predictor used for speculative branching is the Tournament Branch Predictor. Tournament Branch Predictor has Bimodal and Global predictors that contend between each other. Current design uses 5 Arithmetic and Logical Units, a single Branch Unit and Load Store Unit. The parametrized I-Cache and D-Cache use physical address for both index and tag. The cache is parametrized in terms of size of the cache, associativity and the number of blocks within a cache line.

3.1 Branch Predictor

The branch predictor used is a Tournament Branch Predictor that uses two branch predictor models: bimodal and global branch predictor. Both are 2-bit branch prediction models and have a branch predictor table of length 15 which stores a tag of 60 bits. The branch predictor table gets the actual jump data, whether branch is taken or not and PC value if taken, as training data from the branch unit. The training data obtained from branch unit is updated in the Branch Target Buffer (BTB), i.e., tb_branch_addr and tb_tag. For every Program Counter (PC) sent by the fetch unit, based on the slot indexed in rg_select, one of the two branch predictors is selected for prediction. Branch predictor module returns predicted PC and prediction status (branch taken or not) to fetch unit.

3.2 Instruction Fetch

The fetch module receives revert_back_end and revert_front_end from the main module. These are flush signals which when activated, fetch module that will be reverted to initial state and abandon its current operation. The fetch module is Dual Fetch. In this module, the program counter is read from rg_program_counter and the next PC is assumed to be rg_program_counter + 4 (Dual Fetch). Thus two instructions are received from instruction-cache as 64-bit packets. Each set of PC, instruction and prediction status are queued into a FIFO. If the prediction of PC in rg_program_counter recorded is taken, then data queued in ff_fetch_1 is tagged invalid. This module receives a pair of predicted PCs from Branch Predictor module. If any one of the first two PCs is predicted to be taken or PC in rg_program_counter is found to be not aligned with double word, rg_program_counter is updated with second of the two predicted PCs from Branch Predictor module. Otherwise rg_program_counter is updated with the other.

3.3 Decode Stage

It imports Prf_decoder which implements a function (a combinational block) that returns data of Decoded_info_type. This function identifies the types and subtypes of instructions for each functional unit. For Memory type instructions Prf_decoder identifies if it is load instruction or a store instruction, for Arithmetic instructions Prf_decoder identifies if it is a single cycle instruction, or a multiply instruction or a division instruction and for a Branch instruction it identifies if it is conditional or unconditional. The function also isolates the register addresses and immediate values encoded in the instruction. The data returned by module is queued into a FIFO of type Decoded_instruction. Prf_decode receives revert_back_end,evert_front_end from the top module which act as flush signals. JAL (an unconditional branch) instructions are executed in the decode phase. The module Prf_decode returns PC which is calculated from immediate for JAL instruction to Prf_Fetch. Then Prf_Decode queues information into the FIFO later dequeued by Prf_Map.

3.4 Map Stage

The module Prf_MAP does not consist of any storage structures or registers. It gets the decoded instruction dequeued from Prf_decode. The destination register address is assigned a new value by dequeuing from FRQ at head. Based on the register addresses from decode stage, this module maps source operand addresses from fRAM. When a pair of instructions are dequeued from decode stage, each of the instruction's destination register address is renamed. The registers address that are available at the head of the Free Register Queue (FRQ) are assigned as instruction destination register. This module marks an instruction operand ready if there is a broadcast from functional unit to same register or if the value in corresponding register is valid. After collecting the operand information this module fills the Instruction Queue entries at tail.

3.5 Grant and Select

This module consists of 8 FIFO units. There are two sets of four FIFOs each. Each FIFO in a set corresponds to a functional unit (4 ALUs, 1 load-store, 1 branch). Based on the priority encoder each entries and grants obtained from the functional units are selected from the Entry Reorder Buffer (EROB) and are enqueued to Data Read FIFOs (one of the two sets of FIFOs). In the next pipeline stage data is dequeued from the FIFOs. The operand values are read from Physical Register File and appended to the dequeued data. The operand values are later read from register file and queued in next set of 4 FIFOs called Payload FIFOs. The data read from Payload FIFOs are sent to functional units as inputs.

3.6 Commit Stage

The module Prf_commit does not have any storage structures or registers or FIFOs. In case of wrong branch prediction flush signals are activated and the instruction commit is aborted. When an instruction commits, the physical register address assigned to destination register (architectural register) of the instruction is copied to corresponding slot in rRAM and the entry in the EROB is tagged invalid and head pointer is incremented. The value that is being replaced is queued at the tail of FRQ. In case of flush, the data in rRAM is copied to fRAM.

3.7 Instruction Queue

The module Prf_instruction_queue has six storage structures, two large storage structures and four small ones (Bool type). Entry Reorder Buffer (EROB) allocates data as soon as it is dequeued from decode stage. It consists of 16 entries each of Entry_rob_type. imm_buf and allocates immediate value in the instructions if valid. It consists of eight entries and each of Imm_buf_type. It has six other storage structures each of 16 entries and of Bool type.

3.8 FRam

The module Prf_fRAM contains an array of registers of length 32 and stores a 6 bit value which corresponds to index of a physical register file. This module stores the mapping of destination registers of instructions which are just decoded. At the time of exception all the entries in the fRAM are flushed.

3.9 RRam and FRQ

This module Prf_rRAM has array of registers similar to that of fRAM but its functionality in the processor is different from that of fRAM. This module stores mapping of destination registers of instruction that are just committed. When an instruction commits, the value in rRAM which is being replaced is sent to FRQ. The module Prf_FRQ has an array of registers of length 64 each store 6-bit value. There are two more registers to indicate head and tail of the queue each of 6 bits. This module indicates the number of free registers which can be used for register renaming.

4 Performance and Analysis

As stated above, increase in performance of the VeNNus processor with existing RISC-V ISA will be limited with scalar operations. To improve performance and scalability, we have incorporated 16 custom vector extensions to RISC.

4.1 Benchmark

For the purpose of evaluating the proposed work, from the LinPack Bench MatMul is considered. We have customized this matrix multiplication for RISC-V processor and further analysis is done. Various matrix sizes are considered for the analysis purpose.

4.2 Analysis Explanation

Let us say we have the most basic, three-layer neural network of each a input layer, hidden layer, and output layer, respectively. The number of neurons in the three layers are given as (2, 3, 1). For training three neurons of the hidden layer, there are three steps of training involved, as shown below. There will be six total multiplications and six additions. The scalar code execution is

Step 1: Multiplication (Find product of inputs with their corresponding weights)

```
MOVI R0 = 6 1
MOVA R1 = A             1
MOVA R2 = B             1     //A, B: Input and weights
MOVA R3 = C             1
X: LD R4 = MEM[R1++] 2        //Auto-increment addressing
LD R5 = MEM[R2++]       2
MUL R6 = R4 * R5        2
ST MEM[R3++] = R6       2
DECBNZ –R0, X           2     //Decrement, branch if NZ Number of cycles
```

required:

4 + 6*10 = 64 cycles

```
Step 2: Addition (Sum the products)
  MOVI R0 = 61
 MOVA R1 = C              1
 MOVA R2 = C              1   //C: Products from Step 1
 MOVA R3 = D              1   //D: Store the sum of products
 X: LD R4 = MEM[R1++]     2 //Auto-increment addressing
 LD R5 = MEM[R2+3]        2
 ADD R6 = R4 + R5         2
 ST MEM[R3++] = R6        2
 DECBNZ –R0, X            2   //Decrement, branch if NZ Number of cycles
 required: 4 + 6*10 = 64 cycles
```

Step 3: Activation function based on user choice (from sigmoid, ReLU and tanh) is applied on the sum obtained in Step 2. Thus, excluding step 3 we can see that the normal scalar code will take 128 cycles to train the hidden layer.

We will compare this with vector execution time. The vector code execution is

```
Step 1: Multiplication (Find product of inputs with their corresponding weights)
  MOVI VLEN = 6       1
  MOVI VSTR = 1       1
  VLD V0 = A  1       //A, B: Input and weights
  VLD V1 = B  1
  MULV V2 = V0 * V1  2        //Considering 4 ALUs
  VST C = V2  2               //C:
  Store the products
  Number of cycles required: 4 + 4 = 8 cycles
```

```
Step 2: Addition (Sum the products) MOVI VLEN = 6      1
MOVI VSTR = 1       1
VLD V0 = C  1
VLD V1 = D  1
ADDV V2 = V0 + V1  2          //Considering 4 ALUs
VST E = V2   2 //E: Store the sum of products
Number of cycles required: 4 + 4 = 8 cycles
```

Step 3: Activation function based on user choice (from sigmoid, ReLU, and tanh) is applied on the sum obtained in Step 2. Thus, excluding step 3 we can see that the vector code will take only 16 cycles to train the hidden layer. Therefore, we can see that there is a tremendous reduction in training cycles (16 cycles compared to previous 128 cycles) of the neural network for the hidden layer. This is an improvement of 87.5% in speed over scalar code execution for training the hidden layer. Although this is an example training scenario for a very basic neural network, and similar performance gains are anticipated on actual implementation.

In the proposed VeNNus, customized accelerator allows efficient parallel processing of execution which in-turn improves energy efficiency. In future, experimental analysis of energy efficiency will be included.

5 Conclusion and Future Scope

The need accelerator-centric architectures have led a rapid growth in the development of new research ideas in this field. Unfortunately, building such research prototypes is challenging due to licensing issues, but the emerging RISC-V open-source software and hardware ecosystem can partly address this challenge. RISC-V is simple and enables us to deliver world-class TeraFlop levels of computing without needing to resort to proprietary instruction sets, this in turn increases the availability of the software. Taking inspiration from existing AI accelerators such as Radeon Instinct MI25, NVIDIA Tesla V100, Google's TPU, Shakti I class processor and LACore, we were able to successfully design the architecture for our own VeNNus processor, which uses the RISC-V ISA. RISC-V ISA offers benefits such as flexibility, lower costs, and high efficiency. In order to increase the energy efficiency and performance of our AI accelerator, 16 custom vector instructions were added in extension to the RISC-V ISA. The VeNNus processor includes redundant Arithmetic-Logic Units (ALUs) and uses quantization of training weights to 8-bit integers to deliver better performance, driving higher throughput.

RISC-V community has been ideating on the inclusion of proposed vector instructions in the ISA; however, they haven't been included in the ISA yet. The RISC-V Vector Extension can have some interesting features. A vector register file can be deployed for different data types and different sizes, and it can provide an additional support for different data shapes in vector registers such as scalar, vector, and matrix.

As future work on the VeNNus processor, we would implement the design on an FPGA and boot it with Ubuntu LTS. Further, we would run multiple artificial intelligence applications to observe the training time reduction of the convolutional network. This would be a test for our processor and prove to be real indicator of the performance improvement. We have thought about two ways of scaling our architecture. First option is to have redundant FPGAs with the same architecture as VeNNus for parallel processing. Second option is to use Spark for distributing workloads. Thus, getting the architecture from the drawing board to the hardware and running AI applications on it would be the course of future work.

References

1. Asanovic K, Patterson D (2014) Instruction sets should be free: the case for RISC-V. Technical report UCB/EECS-2014–146, EECS Department, University of California, Berkeley
2. Balkind J, McKeown M, Fu Y, Nguyen T, Zhou Y, Lavrov A,Shahrad M, Fuchs A, Payne S, Liang X, Matl M, Wentzlaff D (2016) OpenPiton: an open source Manycore research framework. In: International conference on architectural support for programming languages and operating systems (ASPLOS)
3. Chen T, Du Z, Sun N, Wang J, Wu C, Chen Y, Temam O (2014) DianNao: a small-footprint high-throughput accelerator for ubiquitous machine-learning. In: International conference on architectural support for programming languages and operating systems (ASPLOS)

4. Chen Y, Lu S, Fu C, Blaauw D, Dreslinski R Jr, Kim TMH-S (2017) A programmable Galois field processor for the internet of things. In: International symposium on computer architecture (ISCA)
5. Chen Y-H, Krishna T, Emer J, Sze V (2016) Eyeriss: an energy-efficient reconfigurable accelerator for deep convolutional neural networks. In: International solid-state circuits conference (ISSCC)
6. Courbariaux M, Bengio Y, David J-P (2015) Binaryconnect: training deep neural networks with binary weights during propagations. In: Conference on neural information processing systems (NIPS)
7. Courbariaux M, Hubara I, Soudry D, El-Yaniv R, Bengio Y (2016) Binarized neural networks: training deep neural networks with weights and activations constrained to +1 or −1. arXiv: 1602.02830
8. Dreslinski RG, Fick D, Giridhar B, Kim G, Seo S, Fojtik M, Satpathy S, Lee Y, Kim D, Liu N, Wieckowski M, Chen G, Sylvester D, Blaauw D, Mudge T (2013) Centip3De: a 64-core, 3D stacked near-threshold system. IEEE Micro 33(2):8–16
9. Goulding N, Sampson J, Venkatesh G, Garcia S, Auricchio J, Babb J, Taylor M, Swanson S (2010) GreenDroid: a mobile application processor for a future of dark silicon. In: Symposium on high performance chips (hot chips)
10. Kim J, Jiang S, Torng C, Wang M, Srinath S, Ilbeyi B, Al-Hawaj K, Batten C (2017) Using intra-core loop-task accelerators to improve the productivity and performance of task-basedparallel programs. In: International symposium on microarchitecture (MICRO)
11. Krashinsky R, Batten C, Asanović K (2008) Implementing the scale vector-thread processor. ACM Trans Design Automat Electron Syst (TODAES) 13(3)
12. Magaki I, Khazraee M, Vega L, Taylor M (2016) ASIC clouds: specializing the datacenter. In: International symposium on computer architecture (ISCA)
13. Rastegari M, Ordonez V, Redmon J, Farhadi A (2016) XNOR-net: imagenet classification using binary convolutional neural networks. arXiv:1603.05279
14. RISC-V foundation. https://www.riscv.org. Accessed 15 Aug 2017
15. Srinath S, Ilbeyi B, Tan M, Liu G, Zhang Z, Batten C (2014) Architectural specialization for inter-iteration loop dependence patterns. In: International symposium on microarchitecture (MICRO)
16. Taylor MB (2012) Is dark silicon useful? Harnessing the Four Horsemen of the coming dark silicon Apocalypse. In: Design automation conference (DAC)
17. Taylor MB, Kim J, Miller J, Wentzlaff D, Ghodrat F, Greenwald B, Hoffmann H, Johnson P, Lee W, Saraf A, Shnidman N, Strumpen V, Amarasinghe S, Agarwal A (2003) A 16-issue multiple-program-counter microprocessor with Pointto-point scalar operand network. In: International solid-state circuits conference (ISSCC)
18. Venkatesh G, Sampson J, Goulding N, Garcia S, Bryksin V, Lugo-Martinez J, Swanson S, Taylor MB (2010) Conservation cores:reducing the energy of mature computations. In: International conference on architectural support for programming languages and operating systems (ASPLOS)
19. Whatmough PN, Lee SK, Lee H, Rama S, Brooks D, Wei G-Y (2017) A 28nm SoC with a 1.2GHz 568nJ/prediction sparse deep-neural-network engine with >0.1 timing error rate tolerance for IoT applications. In: International solid-state circuits conference (ISSCC)

DDoS Attack Detection in SDN Using CUSUM

P. V. Shalini, V. Radha, and Sriram G. Sanjeevi

Abstract Software Defined Networking (SDN) is a network paradigm which separates the control plane from data plane. Due to this separation, SDN gives the advantages of programmability, flexibility, and centralized control to the network. However, SDN requires communication between the data plane and control plane, which may create a bottleneck in the network due to limited bandwidth. In addition, there may be a possibility of an attack over centralized controller. Because of the abovementioned requirement and issue, SDN may be a victim of DoS/DDoS attack. In this paper, detection of DDoS attack is carried out by periodically monitoring TCP handshake packets. It is based on TCP protocol behavior. It applies the cumulative sum (CUSUM) algorithm to detect change point in number of half-open connections. It is implemented in the controller. We have compared our work with existing DDoS solutions with CUSUM and shown that our method gives better results.

Keywords SDN · DoS · DDoS · TCP

1 Introduction

Software Defined Networking (SDN) is a three layered architecture consisting of the data plane, the control plane, and application layer. The openflow protocol [4] allows communication between control plane and data plane and makes network manageable from centralized controller.

P. V. Shalini (✉) · S. G. Sanjeevi
Department of Computer Science and Engineering, National Institute of Technology, Warangal, India
e-mail: pvshalini@idrbt.ac.in

S. G. Sanjeevi
e-mail: sgs@nitw.ac.in

P. V. Shalini · V. Radha
Institute for Development and Research in Banking Technology, Hyderabad, India
e-mail: vradha@idrbt.ac.in

N. Chaki et al. (eds.), *Proceedings of International Conference on Computational Intelligence and Data Engineering*, Lecture Notes on Data Engineering and Communications Technologies 56, https://doi.org/10.1007/978-981-15-8767-2_26

The downside of SDN is that all the security threats which are present in traditional networks also exist in SDN. In addition to that communication requirement between switch and the controller could be exploited to launch a DoS attack by exhausting link bandwidth and memory at the switches and controller. One of the popular DoS attacks include TCP-based SYN flood, which exploits network resources. These attacks are a severe threat to SDN as they quickly exhaust the content addressable memory (CAM) of switches and bandwidth between switches and the controller. In this paper, we represent the three messages exchanged during TCP handshake [11] as SYN, SYN+ACK, and ACK. Also use $IPdst$, $IPsrc$, $Dstport$, $Srcport$, $seqno$, and $ackno$ to represent destination IP address, source IP address, destination port, source port, sequence number, and acknowledgement number in packet header, respectively. The client sends the SYN packet to initiate the connection with server. When server receives the SYN packet, it replies with a SYN+ACK packet. The $IPdst$ of SYN+ACK $=IPsrc$ of SYN, $Dstport$ of SYN+ACK $=Srcport$ of SYN. Also $ackno$ of SYN+ACK= $seqno$ +1 of SYN. When client receives SYN+ACK packet, it replies with ACK packet, to complete the connection establishment process. The $IPsrc, IPdst, Srcport, Dstport$ of SYN and ACK packets are same. The $seqno$ of ACK= $seqno + 1$ of SYN and $ackno$ of ACK= $seqno + 1$ of SYN+ACK. By using these properties, we distinguish the ACK from other acknowledgement packets. Also, for the SYN packet to find the corresponding SYN+ACK and ACK. In case, SYN packet does receive either SYN+ACK or ACK packet, such connections are incomplete or half open.

The statistical techniques detect the DDoS attack in very less time compared to other techniques like machine learning techniques, but still, there are some challenges in them like Chi-square test method is a hypothesis test that predicts a DoS attack. But it does not address the challenges like identify the host under attack [5]. The entropy techniques proposed by [12] are good at detecting DoS attack, only when a single host in the network is under attack. The CUSUM algorithm is used for DDoS detection. The advantage of CUSUM is that it can detect small shifts in the mean of the parameter used in the algorithm [13]. In this paper, we focus on SYN flood. Our method applies statistical cumulative sum (CUSUM) algorithm for early detection of a DoS attack. Our model considers all three messages of TCP three-way handshake, and also considers the RST and FIN packets. It is further explained in Sect. 3.1.

The rest of the paper is structured as follows: Sect. 2 describes related work. The proposed solution is described in Sect. 3. Section 4 presents the experimental setup. The evaluation results are given in Sect. 5. Section 6 gives the conclusions.

2 Related Work

We have reviewed the papers that used CUSUM algorithm for DDoS detection. The detection method in [6, 7, 10, 14] papers identifies the number of half-open connections during the given window time period, and applies CUSUM algorithm for it. Table 1 gives a comparison of the reviewed papers. In the papers [6–10], we have

Table 1 Comparison of reviewed papers for DDoS detection

References	Parameter for CUSUM	
	Packets considered at Server	Packets considered at Client
[6] 2004	SYN received, SYN+ACK sent, number of FIN (number of SYN - FIN)	SYN sent, SYN+ACK received, RST,FIN (number of SYN-FIN/RST)
[16] 2007	SYN received, FIN,RST (number of SYN - FIN/RST)	SYN, FIN, RST (number of SYN-FIN/RST)
[7] 2010	SYN received, SYN+ACK sent,RST (number of SYN+ACK-ACK)	SYN+ACK received, ACK sent (number of SYN+ACK-ACK)
[8] 2006	–	SYN sent, ACK sent (number of SYN - ACK)
[9] 2004 [17] 2017	Number of SYNs	–
[10] 2007	–	SYN sent, SYN+ACK received (number of SYN-SYN+ACK)
Proposed solution	SYN received, SYN+ACK sent, ACK received, RST, FIN (number of SYN-ACK/RST/FIN)	SYN sent, SYN+ACK received, ACK sent, RST, FIN (number of SYN-ACK/RST/FIN)

used CUSUM algorithm with parameter extracted from TCP handshake process, for DDoS detection. In our proposed method, we consider all three messages of TCP handshake, unlike only two messages considered in abovementioned reviewed papers. For every SYN packet, we consider corresponding SYN+ACK and its corresponding ACK packet of the third round of three-way handshake. The SYN+ACK or ACK packets which do not correspond to any SYN packet are not considered by our method.

3 Proposed Solution

In this method, we use database _DB($IPdst$,[$Dstport$],[$seqno$,$ackno$]) to store the $IPdst$, list of open port numbers for that destination IP, and list of $seqno$ and $ackno$. When a SYN packet arrives at the controller the $IPdst$, $Dstport$, $seqno$, $ackno$ are extracted, if $IPdst$ belongs to the given network then these entries are stored in the DB. When SYN+ACK packet arrives the $ackno$ in corresponding entry is updated. When the final ACK arrives the corresponding entry is deleted from the database. To get the final ACK packet at the controller we install the flow rule for SYN and SYN+ACK with low hard timeout value [18].

We use a sliding window protocol with t as time slice. Using these database entries, we calculate the number of half-open as well as full TCP connections during every t seconds. The number of half-open connections is calculated by using the number of entries in database DB. The number of full TCP connections is obtained by counting the number of deleted entries in the database in the current time period t.

In case the SYN packet is sent to server's wrong port number, the server replies with a RST packet. Also, a network connection may be terminated at any time by sending a FIN packet. In such cases, the RST and FIN are also considered to get the correct count of the number of half-open connections. If matching database entry is found for the RST/FIN, such entry is deleted and the count of the number of full connections is incremented. If matching entry is not found for RST/FIN, these packets are sent to close already established TCP connections and such a packet is ignored.

3.1 CUSUM for Detection of Change Point

We obtain the number of half-open and also full TCP connections from the database. Let C_t be number of half-open connections, $F(t)$ be the number of full connections, obtained from DB, during the time period t, where $\{t = 0, 1, \ldots\}$. The mean of C_t depends on number of hosts in the network, time of the day, etc. To make the algorithm general, the dependencies should be removed. To do this, we normalize C_t by the average number of full TCP connections, represented by $\overline{F}$. The average number of full TCP connections at time t is evaluated by assigning weights to average full TCP connections at time $t-1$ and number of full TCP connections at the current time period t. The $\overline{F}(t)$ is evaluated by assigning the weightage to $\overline{F}(t-1)$ and the number of full connections (F) at current time t.

$$\overline{F}(t) = \alpha\overline{F}(t-1) + (1-\alpha)F(t) \tag{1}$$

where $0 \leq \alpha \leq 1$

$$X_t = C_t/\overline{F} \tag{2}$$

Now X_t is an independent random variable. As the decision function in CUSUM algorithm compares the parameter X_t with the upper threshold. Negative drift for parameter before the change occurs and a positive drift after the change occurs is required. So, we choose a number to subtract X_t to make it negative, under normal attack free conditions. In general, the expectation (average) of X_t, $E(X_t) = v << 1$. The upper bound of v is chosen, say a, such that $a > v$, and we define,

$$\tilde{X}_t = X_t - a \tag{3}$$

to have a negative mean of $\tilde{X}_t$ for normal traffic and positive mean during SYN flood. For the traffic behavior, we define a stationary random process Y_t.

$$Y_t = (Y_{t-1} + \tilde{X}_t)^+, \tag{4}$$

$$Y_0 = 0, \tag{5}$$

where

$$X^+ = \begin{Bmatrix} X \ if \ X > 0, \\ 0 \ \text{otherwise} \end{Bmatrix} \tag{6}$$

The decision function,

$$d_{Th}(Y_t) = \begin{Bmatrix} 0 \ if \ Y_t \leq Th \\ 1 \ if \ Y_t > Th \end{Bmatrix} \tag{7}$$

where Th is the threshold value assigned.

The detection time depends on t. For every t, the CUSUM algorithm takes input as the number of half-open and full connections and gives output Y_n value, comparing Y_n with threshold, the result is either change or no change. Once the CUSUM algorithm detects the change point, it is treated as DoS attack.

4 Experimental Setup

The Mininet [1] network emulator with POX [2] controller is used to conduct the experiments. Scapy [3] is used to generate and send the network packets. The system properties are We used ubuntu 16.04 LTS operating system, with 15.6 GB memory, intel core i5 processor with 2.50 GHzx4 CPU, with 500 GB Hard Disk.

We have run the algorithms for detection and mitigation on various topologies. In this section, the results are displayed for tree topology with depth 3 and fanout 4 as shown in Fig. 1. In this topology we have 64 hosts,represented as H_i, where $i = 1, 2, 3 \ldots, 64$. 21 switches, represented as S_i, where $i = 1, 2, 3 \ldots 21$ and one controller represented as C_0. Our detection model considers only TCP SYN, SYN+ACK, ACK, RST, FIN packets. We conduct the experiments by generating 30, 40, 50 and 60% of traffic as SYN packets with no acknowledgement and remaining 70, 60, 50, 40% SYN packets with acknowledgements, respectively. For each time period t, the number of half-open and full connections are obtained from the DB, which are used to calculate the parameters (C_t) and (F) for CUSUM algorithm. The α value in Eq. 1 is taken as 0.6 and a in Eq. 3 value is taken as 0.8. The Y_n values are calculated for each time period to detect the attack. The threshold value for the decision function in Eq. 7 is taken as 0.3. The results of the experiments are displayed in the next section. We sent attack packets with an interval of 0.2 s, and sent the

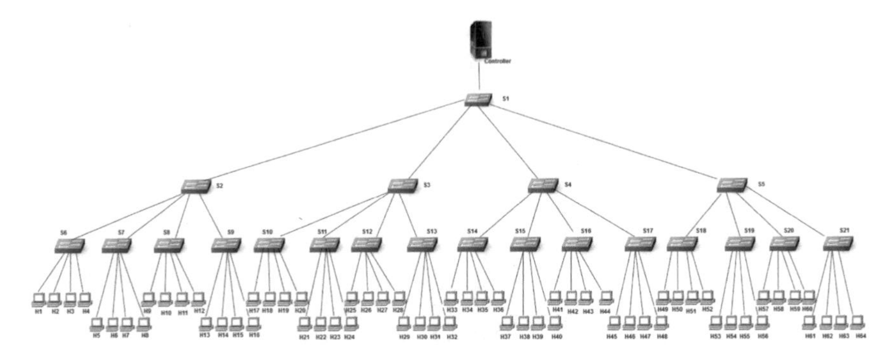

Fig. 1 Topology of Experimental network

normal packets with an interval of 0.5 s from each sender. We capture the packets at controller as PacketIn messages.

5 Evaluation Results

The experiment results show that our proposed model successfully detects DDoS attack for 30% or more attack traffic in the network. The graphs in Fig. 2 represent the evaluation results of DDoS detection. In this, each unit on X-axis displays the time periods used in CUSUM, Y-axis represent the value of Y_n, i.e., decision parameter for detection. In this, Fig. 2a–d represent graph for 30%, 40%, 50% and 60% attack traffic, respectively. We considered the methods in papers [7, 8] to compare with our detection method. The method by Zhang et al. in [7] is represented as Zhang et al.'s method. The method by Chen et al. in [8] is represented as Chen et al.'s method.

In Chen et al.'s method, RST packets are not considered for counting number of half-open connections. According to [15] 2.9% of SYN packets receive RST as reply. So, in this method, 2.9% of SYN are assigned RST as reply instead of SYN+ACK. Zhang et al.'s method represents the evaluation results of [7] in which final ACK packets are not considered at the server side. In case, a server is flooded with spoofed IP packets, then the acknowledgement is sent to another IP address, and the connection at server is still open. But [7] counts it as complete connection. So, attack is being ignored or unseen. In this method, we have taken 2.9% of ACK packets as missing and produced the results. This results are represented as Zhang et al.'s method in the graph in Fig. 2.

Comparing our proposed method with [7] method: As [7] uses SYN - SYN+ACK as the parameter for detecting full connection at the server side. However, when an ACK packet is missed, the connection is not complete, but the server side method in [7] counts it as a complete connection. In the case of the DoS attack, attacker sends the SYN packet to a server, and upon receiving SYN+ACK packet, the attacker does not send final ACK packet. In such cases, the attack is being undetected or being

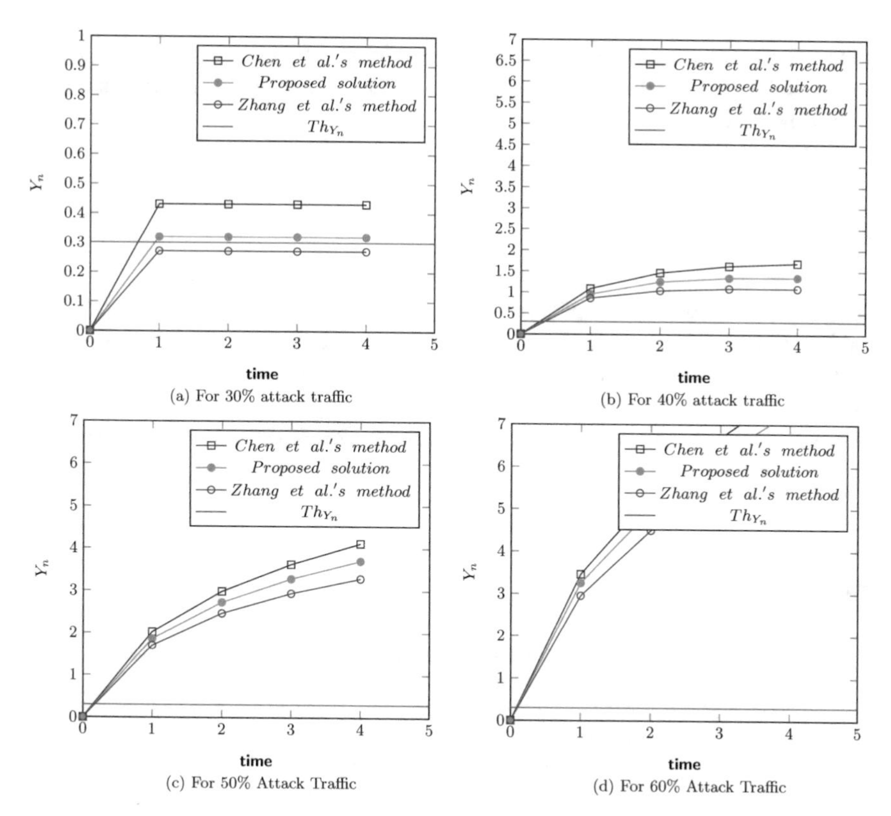

Fig. 2 Detection results for **(a)** 30% attack traffic (b) 40% attack traffic **(c)** 50% attack traffic **(d)** 60% attack traffic

unseen in this method. The results displayed in Fig. 2 shows that our proposed method detects the results when 2.9% of ACK packets are missing. Our proposed method detects the DoS attack for 30% low rate SYN attack traffic, whereas Zhang et al.'s method could not detect DoS attack with 30% SYN attack traffic.

Comparing our proposed method with [8] method: In [8], the RST packets are not considered for counting the number of full connections. When a SYN packet receives a RST as reply, the connection is terminated. However, it is still counted as a half-open connection in Chen et al.'s method. The Fig. 2 displays the results for Chen et al.'s method, for 2.9% of missing the RST packets. This graph is always above our proposed method's graph, i.e., Y_n value is always more than the actual value. Therefore, this method gives more false positive results compared to our proposed method.

The number of half-open connections counted by our proposed model is accurate or close to the actual value. Whereas the number of half-open connections counted by Zhang et al.'s method are lesser than actual value and the number of half-open connections counts by Chen et al.'s method are more actual value. Therefore, results produced by our model are more accurate than both Zhang et al.'s method and Chen et al.'s method.

6 Conclusion

Our proposed model is a simple and robust model. We have shown that our proposed model gives accurate results for DDoS detection. We have overcome some of the limitations of existing models against DDoS attacks which used CUSUM algorithm for detection. Our proposed model is successfully implemented and validated. Our detection model works independently of traffic patterns and sites and it is stateless as it applies CUSUM.

References

1. Mininet (2019) An instant virtual network on your laptop (or other PC)-Mininet. In: https://www.mininet.org/. Last accessed 9 Aug 2019
2. noxrepo/pox (2019) The POX network software platform: GitHub. https://github.com/noxrepo/pox. Last accessed 9 Aug 2019
3. Scapy (2019) https://scapy.net/. Last accessed 9 Aug 2019
4. Openflow-specifications (2019) https://www.opennetworking.org/2014/10/openflow-spec-v1.3.0.pdf. Last accessed 9 Aug 2019
5. Leu FY, Pai CC (2009) Detecting DoS and DDoS attacks using chi-square. In: 2009 fifth international conference on information assurance and security, vol 2. IEEE, China, pp 255–258
6. Wang H, Zhang D, Shin KG (2004) Change-point monitoring for the detection of DoS attacks. In: IEEE transactions on dependable and secure computing, vol 1, number 4. IEEE, USA, pp 193–208
7. Zhang Y, Liu Q, Zhao G (2010) A real-time DDoS attack detection and prevention system based on per-IP traffic behavioral analysis. In: 2010 3rd international conference on computer science and information technology, vol 2. IEEE, China, pp 163–167
8. Chen W, Yeung DY (2006) Defending against TCP SYN flooding attacks under different types of IP spoofing. In: International conference on networking, international conference on systems and international conference on mobile communications and learning technologies (ICNICONSMCL'06). IEEE, Mauritius, pp 38–38. https://doi.org/10.1109/ICNICONSMCL.2006.72
9. Siris VA, Papagalou (2004) Application of anomaly detection algorithms for detecting SYN flooding attacks. In: IEEE global telecommunications conference 2004 GLOBECOM'04, vol 4. IEEE, Dallas, pp 2050–2054
10. Thing V, Sloman M, Dulay N (2007) Enhanced TCP SYN attack detection. In: IEEE/IST workshop monitoring, attack detection mitigation (MonAM). Toulouse, France (2007)
11. Comer DE (1998) Internetworking with TCP IP vol 1 principles, protocols, and architectures fourth edition. In: Publishing house of electronics industry. Wellington, FL, U.S.A, pp 237–239
12. Kumar P, Tripathi M, Nehra A, Conti M, Lal C (2018) SAFETY: early detection and mitigation of TCP SYN flood utilizing entropy in SDN. In: IEEE transactions on network and service management, vol 15, number 4. IEEE, pp 1545–1559

13. Basseville, M, Nikiforov IV (1993) Detection of abrupt changes: theory and application, vol 104. Prentice Hall Englewood Cliffs, France, pp 35–40
14. Wang H, Zhang D, Shin KG (2002) Detecting SYN flooding attacks. In: Proceedings. Twenty-first annual joint conference of the IEEE computer and communications societies, vol 3. IEEE, Piscataway, NJ, USA, pp 1530–1539
15. Ohsita Y, Ata S, Murata M (2004) Detecting distributed Denial-of-Service attacks by analyzing TCP SYN packets statistically. In: IEEE global telecommunications conference 2004 GLOBECOM'04, vol 4. IEEE, Dallas, Texas, USA, pp 2043–2049
16. Ohsita Y, Ata S, Murata M (2004) Detecting distributed Denial-of-Service attacks by analyzing TCP SYN packets statistically. In: IEEE global telecommunications conference 2004 GLOBECOM'04, vol 4. IEEE, Dallas, Texas, USA, pp 2043–2049
17. Conti M, Gangwal A, Gaur MS (2017) A comprehensive and effective mechanism for DDoS detection in SDN. In: 2017 IEEE 13th international conference on wireless and mobile computing, networking and communications (WiMob). IEEE, Rome, Italy, pp. 1–8
18. Mohammadi R, Javidan R, Conti M (2017) Slicots: an sdn-based lightweight countermeasure for tcp syn flooding attacks. In: IEEE transactions on network and service management, vol 14, number 2. IEEE, pp 487–497

Comprehensive Analysis of Deep Learning Approaches for PM2.5 Forecasting

Sivaji Retta, Pavan Yarramsetti, and Sivalal Kethavath

Abstract Air pollution is causing massive damage to human health. PM2.5 in particular, has been shown to have a significant effect on human health. So, forecasting of PM2.5 is essential. Approaches like the ARIMA model used for time series forecasting. The invention of Deep Learning, especially the Recurrent Neural Networks, revolutionized the methods of forecasting the time series to achieve predictions that are more precise. Variants of RNN like LSTM, GRU which had long term dependencies unlike basic RNN gives more accurate predictions. Temporal Convolutional Network, which is a synthesis of 1D Fully Convolutional Network and Causal Convolutions, came into the picture during 2018 and also provided successful results in sequence learning and Forecasting time series. We compared deep learning approaches LSTM, GRU, CovLSTM and Temporal convolutional networks using three types of losses. After comprehensive analysis, our results proved that TCN also gives comparable results for time series forecasting as LSTM and, GRU.

Keywords Time series forecasting · Recurrent neural networks (RNN) · Gated recurrent unit (GRU) · Temporal convolutional networks (TCN) · Dilated convolutions

1 Introduction

This is an era of Liberalization, Privatization, and Globalization. These technological advancements transformed the villages to towns, towns to cities and cities to smart cities. The rise of smart cities have become the beacon of global development. However, the mushroom growth of industries gave rise to the unprecedented Air Pollution levels in all the major cities. This became a global problem fueling global warming.

S. Retta (✉) · P. Yarramsetti · S. Kethavath
RGUKT-AP IIIT, Nuzvid, India
e-mail: sivajiretta171@gmail.com

N. Chaki et al. (eds.), *Proceedings of International Conference on Computational Intelligence and Data Engineering*, Lecture Notes on Data Engineering and Communications Technologies 56, https://doi.org/10.1007/978-981-15-8767-2_27

Due to the advancement of IoT, many smart cities are using sensor nodes at different locations in the city for continuous monitoring of the pollution contaminants. Air quality is not constant, it will vary and the forecasting of air pollution contaminants is necessary for air quality management. There are different approaches to forecast the time series. Based on previous history of pollution, time series models will forecast the pollution levels in the near future. Mostly the models such as ARIMA (Auto Regression Integrated Moving Average) used in time series forecasting. By the evolution of recurrent neural networks, forecasting of time series became accurate. Variants of RNN like LSTM, GRU, ConvLSTM outperformed the general RNN by providing resistance to the vanishing gradient problem. Temporal Convolutional Networks, which is a mixture of 1D Fully Convolutional Network and Causal Convolutions, appeared during 2018 and provided successful results in sequence learning and forecasting time series.

We used the Keras, a high-level application programming interface (API) of neural network written in Python and able to construct deep learning model on top of Tensorflow. We evaluated models on the Open Source Beijing PM2.5 data set available in UCI Machine Learning Repository. We splitted the dataset and compared with different deep learning approaches LSTM, GRU, ConvLSTM and Temporal convolutional networks taking root Mean Absolute Error, Mean Square Error and R square error as metrics for performance measurement. After comprehensive analysis, our results proved that temporal convolutional networks also give comparable results for univariate time series forecasting as LSTM, GRU and ConvLSTM.

The subsection of the paper is organized as In Grover et al. [2], we have illustrated works related to it. In Feng et al. [3], We briefly discussed the four deep learning models. In Voyant et al. [4], we implemented the models on PM2.5 data. In Chen et al [5], we concluded the paper with comparative analysis of four deep learning models for PM2.5 forecasting and provided some directions about future scope.

2 Related Works

An extensive research from many years gave rise to several methods to forecast the pollution contents especially PM2.5 to know the quality of the air. In [1], Satellite remote sensing techniques with the PM2.5 sensors have been used. In [2], authors forecasted the various parameters like wind, dew point, temperature but not PM2.5 using a deep hybrid neural network. In [3] to improve the forecasting accuracy, authors combined wavelet transformation and air mass trajectory. In [4, 5] time series models like ARIMA and classifiers based on ANN are used to predict the pollution.

Exponential smoothing method was used for short term PM2.5 forecasting in [6]. In [7], authors compared the three methods Multilayer Network, A Linear Algorithm and a Clustering Algorithm to forecast PM2.5 in a large city. In [8] authors used 1D Convnets and Bidirectional GRU-based deep learning models to forecast parameters like temperature, wind direction, wind speed, rainfall etc. In [9], the Air Quality Index was predicted using RNN and LSTM for Delhi city. In [10], authors used RNN and LSTM to forecast the air pollution. In [11–14] also authors used LSTM, GRU, CNN based sliding window model as well as variants of LSTM for time series forecasting problems. In our paper, we used four deep learning models namely LSTM, GRU, ConvLSTM and Temporal Convolutional Networks (TCN) for PM2.5 forecasting and we tried to use the latest CNN based TCN for forecasting and our results proved that it is giving comparable results like LSTM and GRU.

3 Models

3.1 Lstm

LSTM network is a extended version of RNN, introduced by Seep Hochreiter and Jürgen Schmidhuber to rectify the problem of the vanishing gradient in RNN during optimization. It is more appropriate for modelling long-range dependencies, and used in many applications such as machine translation, audio detection, video recognition, and analysis of many time series. Compared with RNN's, LSTM architecture. The LSTM architecture includes memory blocks instead of hidden units, similar to RNN's. To lower parameters, the cells of a memory block modulated by nonlinear sigmoidal gates share the same gates. The LSTM cell shown in the figure below is a gated structure with three gates, namely input gate, output gate and forget gate, which regulates the cell and neural network information (Fig. 1).

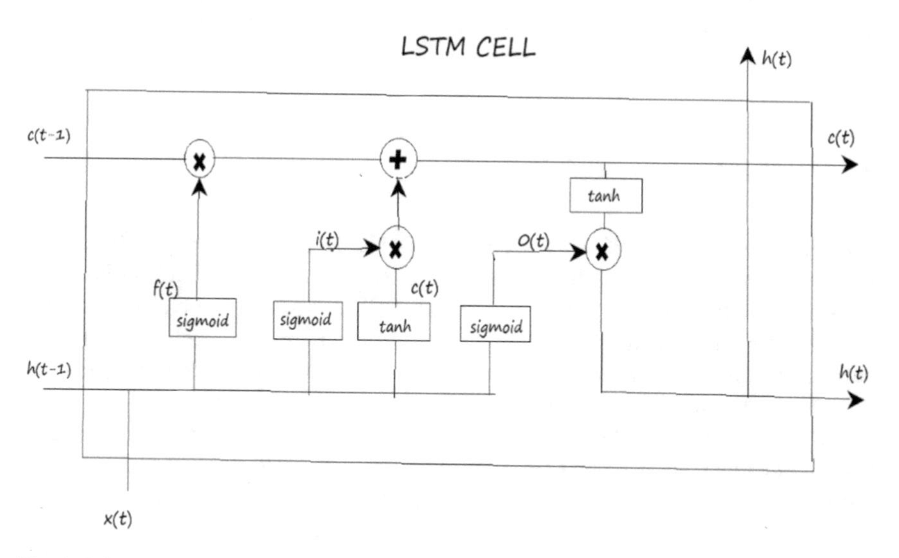

Fig. 1 LSTM structure

Attention weights is defined as: $w = (w1, w2, \ldots\ldots, wL)$. By the weights we can get sampling for the input data with $\tilde{X} = (x_t^1 \cdot w1, x_t^2 \cdot w2, \ldots.., x_t^L wL)$. Then, $(\tilde{X} = \tilde{X}_1, \tilde{X}_2 \ldots\ldots, \tilde{X}_T)$ is fed to LSTM networks. In addition, these formulations of the calculating method in LSTMs cells allow us to learn the nonlinear mapping function as follows:

Input gate layer: It consists of input data.

$\mathrm{i(t)} = \sigma(w_{xi}\, \tilde{X}_t + w_{hf} h(\mathrm{t}-1) + w_{ci} c(\mathrm{t}-1) + b_i)$ (1)

Forget gate layer: It determines which section of information is to be passed on.

$$f(t) = s(wxf \sim + w_{hf} h(t-1) + w_{cf} c(t-1) + b_f) \tag{2}$$

Cell State: Adds or removes the information across gateways by running throughout the network.

$$C(t) = f(t)c(t-1) + i(t)\tanh(w_{xc}X \sim t + w_{hc}h(t-1) + b_c) \tag{3}$$

Output gate: Consist of output produced by the LSTM

$$O(t) = s(w_{xo}\tilde{X} + w_{ho}h(t-1) + w_{co}c(t-1) + b_o) \cdot X_t \tag{4}$$

where $\sigma(\cdot)$ means the sigmoid activation function and tanh is the hyperbolic tangent function that produces a new vector that adds to the state and W matrixes with a double subscription of combination of weights between the two cells. The performance of the hidden layer in the current time step is: $ht = ot \tanh(ct)$. Finally, we can take output value ht as the forecasted value. It can be denoted as: $\tilde{y} = h_t$The final output value can be concatenated by the vector: $(\tilde{y}T = \tilde{y}^1, \tilde{y}^2, \ldots\ldots, \tilde{y}^T)$.

3.2 *Gated Recurrent Unit (GRU)*

Gated Recurrent Unit (GRU) is also a Recurrent Neural Networks variant. It was first introduced in 2014 by Kyunghyun Cho et al. RNN had a vanishing Gradient Problem. To overcome the vanishing gradient problem, GRU and LSTM were introduced but LSTM has complex architecture. GRU has nearly the same performance as LSTM with simple architecture thus fewer parameters needed to update the hidden state. So, It becomes simpler to train using GRU as compared to LSTM.

GRU has only two gates, which makes it simple in architecture namely Update Gate, and Reset Gate. Update gate determines how much old memory should transferred into the future and Reset gate dictates how much old memory should be lost.

Figure below is the basic architecture of GRU and the transition from one hidden state to the next hidden state is explained through mathematical equations as follows (Fig. 2).

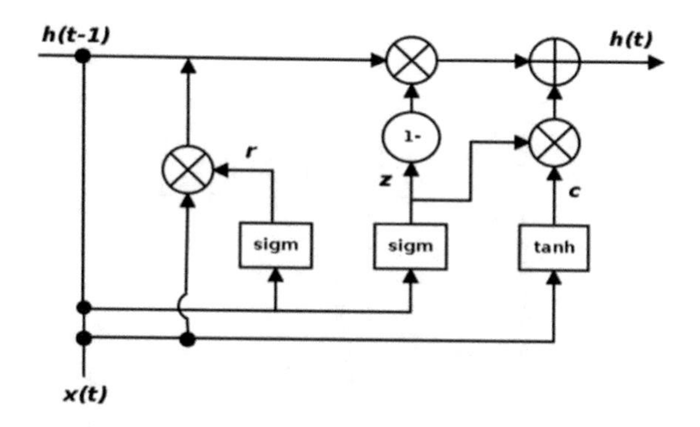

Fig. 2 GRU structure. *source* https://ieeexplore.ieee.org/document/8675076

$$z = \sigma(W_z h_t - 1 + U_z x_t) \quad (5)$$

$$r = \sigma(W_z h_t - 1 + U_z x_t) \quad (6)$$

$$c = \tanh(W_z(h_t - 1 \oplus r) + U_z x_t) \quad (7)$$

$$h_t = (z \oplus c) + ((1 - z) \oplus (h_t - 1)) \quad (8)$$

3.3 ConvLSTM

CNN attained better performance in image classification problems, some researchers explore 1D CNN on data regression problems. Convolutional LSTM (Convlstm) is one of the special variants of RNN. It has a capability to get rid of the long-term dependencies and is mainly designed for 3-d input data. It perform convolution operation in each gate of the lstm cell.

In convlstm, data collected in terms of time series and the The data flowing through the convlstm cells holds the input dimensions instead of being a single-dimensional vector with features. The input of convlstm is the set of data overtime. The output of convlstm is the combination of convolution and lstm output that is a set of data with shape.

The architecture in convlstm is based on the recurrent encoder and decoder. The first recurrent encoder of neural networks processes the input data sequentially and maintains the data's temporal correlation. The second recurrent neural network decoder initialized with the hidden state of the first recurrent network and then it is able to produce potential predictions one by one by unrolling the network (Fig. 3).

The following equations used at each stage in Convlstm:

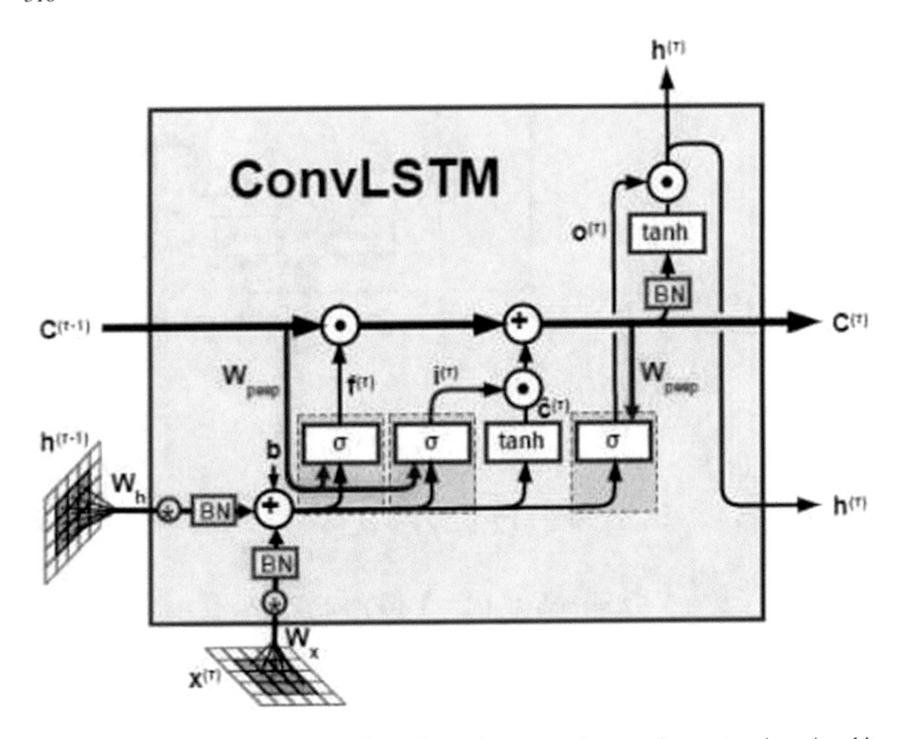

Fig. 3 Conv LSTM structure. *source* (https://www.jie-tao.com/category/computerscience/machinelearning/page/2/)

$$i_t = s(W_{xi} X_i + W_{hi} H_t - 1 + W_{ci} C_t - 1 + b_i) \tag{9}$$

$$f_t = s(W_{xf} X_t + W_{hf} H_t - 1 + W_{cf} C_t - 1 + b_f) \tag{10}$$

$$C_t = f_t(C_t - 1 + i_t \tanh(W_{xc} X_t + W_{hc} H_t - 1 + b_c) \tag{11}$$

$$o_t = s(W_{xo} X_t + W_{ho} H_t - 1 + W_{co} C_t + b_o) \tag{12}$$

$$H_t = o_t \tan_h(C_t) \tag{13}$$

ConvLSTM is used for maintaining the spatial structure of the data within the rnn cell. The data in features space flowing through the convlstm cell has a 3D form, instead of being a 1D vector only. It has a negative impact on training results and to prevent the negative effect we use scheduled sampling. The overall network comprises the Five main components,

1. Spatial encoder
2. Two-spatio-temporal encoder

3. Perpetual motivated loss layer.
4. The network should be trained end to end.

3.4 Temporal Convolutional Networks

Temporal Convolutional Network is a CNN based sequence model which can be used for different sequence modelling tasks. TCN is a blend of 1D fully convolutional Network and Causal convolutions. The two unique features of TCN are causal convolutions and the capability of its architecture to output the sequence of same length by accepting sequence of any length as input.

TCN 's have long memory without gating mechanisms. Length of each hidden layer of 1D FCN is same as length of input layer to get the second point. Causal convolutions are responsible for achieving the first point where only the elements from time t and the previous layer is convolved with the output at time "t". But, the problem with causal convolutions, they can only look at a background in the depth of the network with a linear scale. To accomplish this problem, Dilated convolutions came into picture with exponentially growing receptive field.

For an element s of a sequence, the dilated convolution F with an input sequence $tx \in Rn$ and a filter $f{:}\{0, k{-}1\} \rightarrow R$ is mathematically formulated as follows:

$$F(S) = (x *_d f)(s) = \sum_{i=0}^{k-1} f(i).x_{s-d.i} \tag{14}$$

where, d denotes the dilation factor, k is the filter size and $s{-}d.i$ determines direction of the past. Dilation is nothing but inclusion of fixed steps between every two adjacent filters.

As we can increase the depth, dilation factor and filter size to improve the receptive Field, stability of the network can achieved by replacing the Residual block by convolutional layer (Fig. 4).

Receptive field of a convolution network can be increased by increasing filter sizes (k) and also raising the dilation factor (d) and thus making flexible receptive

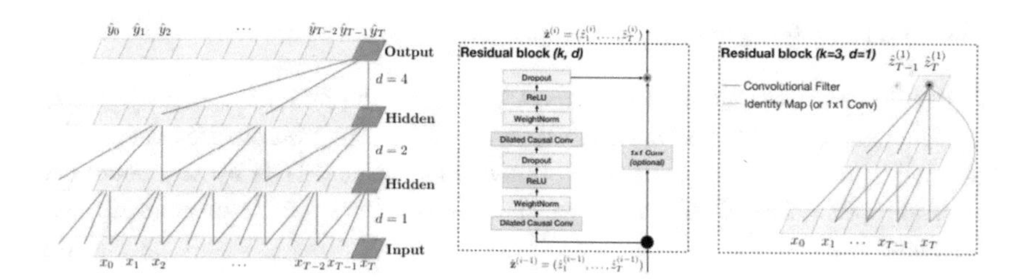

Fig. 4 TCN structure. *source* arxiv.org/pdf/1803.01271.pdf

field size. TCN has many advantages like parallelism, low memory requirements for training.

4 Implementation

4.1 Data and Data-Preprocessing

The dataset used in this paper is obtained from UCI Machine Learning Respiratory prepared by Song Xi Chen, Guanghua School of Management, Center for Statistical Science, Peking University. This dataset contains hourly data of PM2.5 concentration (ug/m^3), Dew Point, Temperature, Pressure, Wind direction and Speed, Cumulated hours of snow and rain of the US Embassy in Beijing. There are some missing values among the whole items. We removed them to avoid their influence on the model training. The dataset contains 13 attributes; there is no correlation between PM2.5 and other attributes. Therefore, we used univariate for the prediction of PM2.5 concentration. Later, we normalized the data using Min-Max strategy by mapping in the range zero to one and reshaped the data from 1D to 3D. The reshaped data split into 80% train and 20% validation set and some values for forecasting.

4.2 Model Design

4.2.1 Loss Function and Optimizer

In our proposed framework, we choose Mean Square Error (MSE) as the loss function and Adam optimizer to train the models. Adam is an optimization algorithm specially designed to train deep learning models. This includes the features of the algorithms Adaptive Gradient and Root Mean Square Propagation (RMSProp), which are capable of solving sparse gradients on noisy problems.

4.2.2 Hyper Parameters

We tuned hyper parameters of each model based on the individual keras package. For Convolutional LSTM after 1D CNN layer, we pass the result to the max pooling layer (with pool size = 2). For LSTM, Convlstm, GRU and TCN we set epochs and validation split as 20 and 0.2 respectively and all the other hyper parameters are fixed as default.

4.3 Model Evaluation Metrics

To test our prediction models, we follow three statistical metrics to calculate error. They are Mean absolute Error (MAE), Root Mean Square Error(RMSE) and R-square loss. If the error of MAE and RMSE is closer to 0 described as a better model and for R-square, closer to 1 is better. Additionally, we predicted the future rate $xt + 2$, $xt + 3$, $xt + 4$ and $xt + 5$.

$$\text{RMSE} = \sqrt{\frac{1}{n}\sum_{t=1}^{n}(y_t - \hat{y}_t)^2} \tag{15}$$

$$\text{MAE} = \frac{1}{n}\sum_{t}^{n}\left|y_t - \hat{y}_t\right| \tag{16}$$

$$\text{R}^2 = 1 - \frac{\sum_{t=0}^{n-1}(y_t - \hat{y}_t)^2}{\sum_{t=0}^{n-1}(y_t - \hat{y}_t)^2} \tag{17}$$

5 Results and Conclusion

In our paper, we compared the four deep learning approaches for PM2.5 forecasting. During the training process, the training loss and validation loss taking mean square error as metric are summarized in the below table (Table 1).

We have also evaluated our models on a testing set and the results are illustrated through the below figures Figs. 5, 6 and 7.

Table 1 Model performance based MSE loss

Model	No. of Epochs	Training loss	Validation loss
Conv LSTM	20	0.0014	0.0012
LSTM	20	5.7754e-0 4	4.6202e-04
GRU	20	5.7726e-0 4	4.5934e-04
TCN	20	5.8610e-04	4.6493e-04

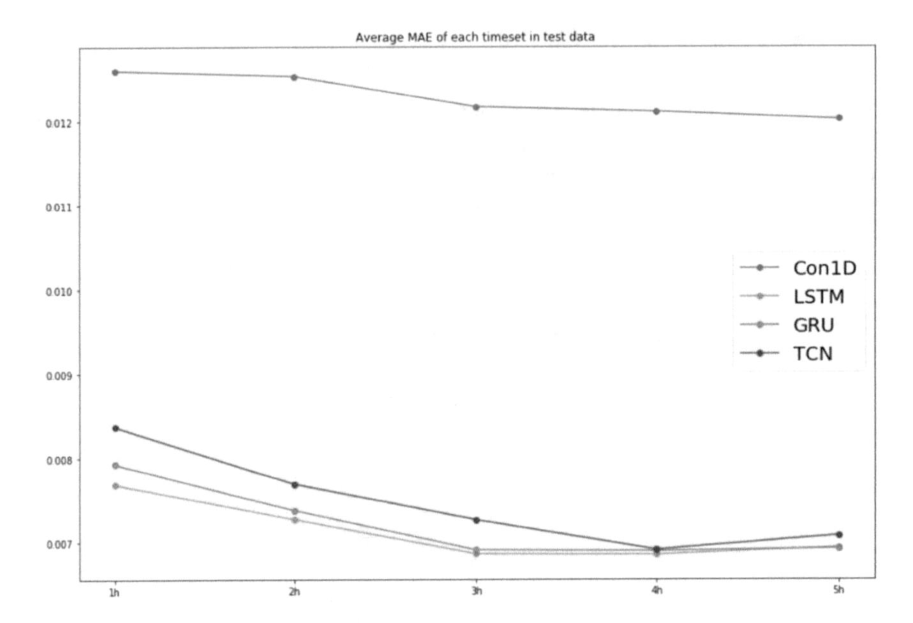

Fig. 5 Average MAE of each timeset in test data

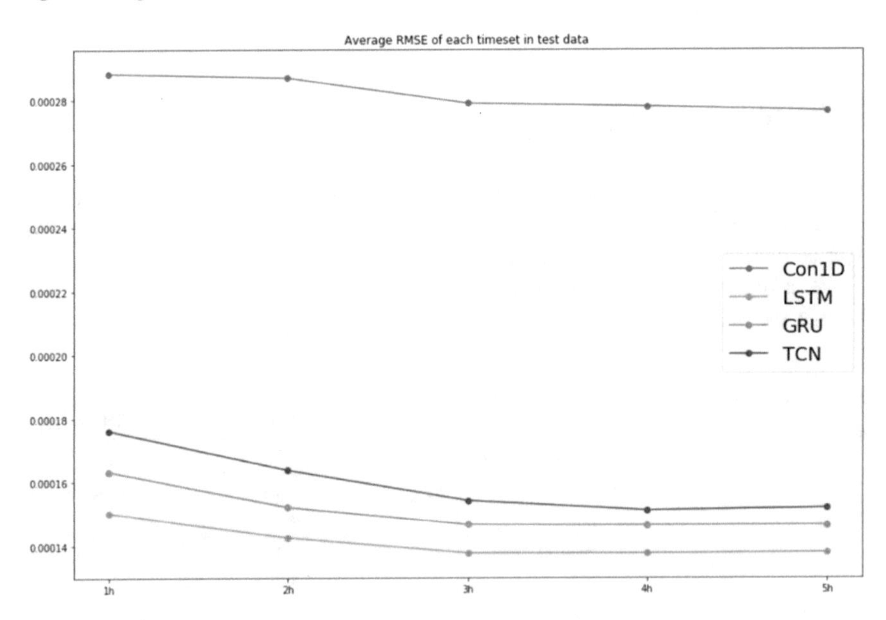

Fig. 6 Average RMSE of each timeset in test data

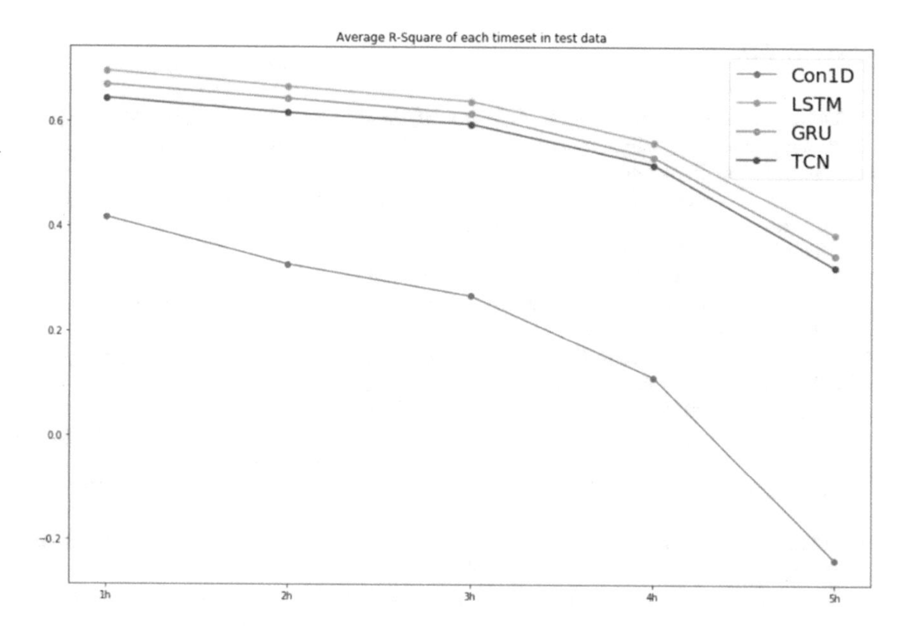

Fig. 7 Average R-square of each timeset in test data

6 Conclusion

In this paper, we proposed comparative analysis between four deep learning algorithms for univariate prediction of pm 2.5. We predicted the next five hours pollution level based on the last five hours samples and compared the performance of the model by the MSE and R-square loss. LSTM has got better performance compared other three models and ConvLSTM got least performance. Also, Temporal convolutional network, a novel deep learning approach for time series forecasting is also comparable results with LSTM and GRU.

7 Future Scope

Forecasting of PM2.5 is giving the best and accurate results with minimum loss. This enables the smarter and healthier mobility of citizens in our smart cities. We can show the "healthy path" between the two stations which has less exposure to the PM2.5 based on the parameters like distance taken and PM2.5 concentration. This creates a healthy movement on the polluted roads in the global cities like Beijing, New Delhi etc. by contributing to the global environment policies.

References

1. Di Nicolantonio W, Cacciari A, Tomasi C (2009) Particulate matter at surface: Northern Italy monitoring based on satellite remote sensing meteorological fields and in-situ samplings. IEEE J Sel Top Appl Earth Obs Remote Sens 2(4):284–292
2. Grover A, Kapoor A, Horvitz E (2015) A deep hybrid model for weather forecasting. In: Proceedings of the 21th ACM SIGKDD international conference on knowledge discovery and data mining, pp 379–386
3. Feng X, Li Q, Zhu Y, Hou J, Jin L, Wang J (2015) Artificial neural networks forecasting of PM2. 5 pollution using air mass trajectory based geographic model and wavelet transformation. Atmos Environ 107:118–128
4. Voyant C, Muselli M, Paoli C, Nivet ML (2012) Numerical weather prediction (NWP) and hybrid ARMA/ANN model to predict global radiation. Energy 39(1):341–355
5. Chen L, Lai X (2011) Comparison between ARIMA and ANN models used in short-term wind speed forecasting. In: Asia-Pacific power and energy engineering conference (APPEEC), pp 1–4
6. Mahajan S, Chen LJ, Tsai TC (2018) Sensors (Basel) 18(10):3223. Published online 2018 Sep 25. https://doi.org/10.3390/s18103223
7. PM2.5 forecasting in a large city: comparison of three methods Patricio Pereza, Giovanni Salini ab. https://doi.org/10.1016/j.atmosenv.2008.07.035
8. Tao, Q, Liu F, Li Y, Sidorov D (2019) Air pollution forecasting using a deep learning model based on 1D convnets and bidirectional GRU. IEEE Access 7:76690–76698
9. Bansal M, Aggarwal A, Verma, T, Sood A. Air quality index prediction of Delhi using LSTM. https://doi.org/10.13140/rg.2.2.26885.70884
10. Tsai YT, Zeng YR, Chang YS. Air pollution forecasting using RNN with LSTM. https://doi.org/10.1109/DASC/PiCom/DataCom/CyberSciTec.2018.00178
11. Mahajan S, Chen L, Tsai T (2017) An empirical study of PM2.5 forecasting using neural network. In: 2017 IEEE smartworld, ubiquitous intelligence and computing, advanced and trusted computed, scalable computing and communications, cloud and big data computing, internet of people and smart city innovation (SmartWorld/SCALCOM/UIC/ATC/CBDCom/IOP/SCI), pp 1–7. San Francisco, CA
12. Bai S., Kolter JZ, Koltun V (2018) An empirical evaluation of generic convolutional and recurrent networks for sequence modeling. arXiv preprint. arXiv:1803.01271
13. Selvin S, Vinayakumar R, Gopalakrishnan EA, Menon VK, Soman KP (2017) Stock price prediction using LSTM, RNN and CNN-sliding window model. In: 2017 International conference on advances in computing, communications and informatics (ICACCI), pp. 1643–1647. Udupi
14. Youru Li, Zhenfeng Zhu, Deqiang Kong, Hua Han, Yao Zhao () EA-LSTM: evolutionary attention-BASED LSTM for time Series prediction. Mach Learn (cs.LG); Neural Evol Comput (cs.NE); Mach Learn (stat.ML). https://arxiv.org/abs/1811.03760v1

Remotely Controlled Surveillance Vehicle with Live Video Streaming

Vinay Koneru, Irfan Shaik, Ch. Rupa, M. Krishnaveni, and R. Rachana

Abstract Technology is updating day by day so it also necessary to update according to it. These days anything can be done from home itself like meetings, online orders, and so on. So, the proposed project focusses on controlling the vehicle from anywhere remotely. This paper mainly concentrates on developing a surveillance vehicle that can be used to monitor dangerous areas, military bases, and other critical places using esp32 cameras. Our proposed design performs live streaming, including that it identifies the unauthorized person and stores it in SD card. Along with this, it also sends the image of that person to the authorized person's email. Here we use a mobile phone to control the car from anywhere through Blink Application. Blynk App is used to control the camera movements. The main strength of this paper is security features.

Keywords Esp32 camera · Blynk app · Pantilt · Servo motors · Google assistance

1 Introduction

The proposed system is used to control the vehicle from anywhere in the world. Here it gives the required output from the given input very quickly using Blynk application. It requires Internet connection to transfer the given input to the vehicle. It is a very sophisticated vehicle which cannot be hacked or malfunctioned. Here,

V. Koneru (✉) · I. Shaik · Ch. Rupa · M. Krishnaveni · R. Rachana
Velagapudi Ramakrishna Siddhartha Engineering College, Vijayawada, India
e-mail: vinay27.koneru@gmail.com

I. Shaik
e-mail: sk.irfan3616@gmail.com

N. Chaki et al. (eds.), *Proceedings of International Conference on Computational Intelligence and Data Engineering*, Lecture Notes on Data Engineering and Communications Technologies 56, https://doi.org/10.1007/978-981-15-8767-2_28

we use Blynk Application to control the vehicle remotely. First, the user needs to register in the Blynk Application by providing necessary details then it will produce a hash value to the email account which is a secure way of connecting to a vehicle. In the Blynk Application, we are using buttons, mobile tilting to control the vehicle. The control of the vehicle may become difficult from anywhere without seeing all the dimensions, so by adding an esp32 camera makes it easy to control the vehicle from anywhere in the world. This has live video streaming mode [1] which can be varied according to our desire. On executing the code in the esp32 camera, the link is generated in Arduino IDE. Now with the help of this link and the auth token from NGROK are executed in Python to generate a global link. The link generated using NGROK is opened in any browser to access the live video streaming of area around the vehicle. This live streaming mode can be accessed through mobile or pc.

Safety of the vehicle has become a major problem these days and the society is facing more car threats. Here the proposed design has an unauthorized alert system which takes the picture [2] of the intruder and stores it in the SD card and at the same time it is going to mail it to the authorized person's email account.

Surveillance [3] of the place is nothing but monitoring or carefully observing that place continuously without human intervention. Police patrolling has become negligible due to the inactive office hours of the police, but our proposed design makes this more efficient and simpler by patrolling this vehicle without physically present in the vehicle and simply controlling it from the station or control room.

Ultimately, this prevents the accidents and attacks that are happening during day and night times especially on highways. Nowadays delivery to the remote areas is not that much rapid and reliable as the delivery boys show reluctance in going there. So, by using our proposed system, the delivery to the remote areas can easily happen without any hassle.

2 Related Work

Ghute et al. [4] proposed a Design of Military Surveillance Robot. The designed system mainly focusses on controlling the robot and at the same time providing the live streaming capability to place this robot in unmanned areas. Here the author used Raspberry Pi, Camera, Ultrasonic sensor, Arduino Mega, and Google Assistance. The live video streaming is transmitted to the required destination through the Internet. Raspberry Pi stands as a heart of the device to control all the functionality. This is mainly used in the restricted areas. The drawbacks of this design are (i) The program logic is difficult to understand, (ii) It consumes more power, and (iii) Used Bluetooth which is limited to a certain range.

Chakraborty et al. [5] proposed an Autonomous Vehicle for Industrial Supervision Based on Google Assistant Services and IoT Analytics. It uses Arduino Uno, Node MCU, ultrasonic sensor, and some other sensors. The user gives voice commands as input to control the vehicle. Here the google assistance is used to convert the voice command into analog data and send it to the cloud. Later, the cloud is going to send the accepted data to the microcontroller to perform required operations. The sensors which are equipped to the vehicle senses the data and transmits it to the IOT platform, and then from this to the authorized user's device. The WLAN is used by this system to control the vehicle. It is mainly used for industrial inspection. The drawbacks of this design are (i) It cannot be controlled from longer distances and (ii) There might be a chance of misusing the data while transferring.

Raihan et al. [6] proposed a Design and Implementation of a Hand Movement Controlled Robotic Vehicle with Wireless Live Streaming Feature. The components used here are Accelerometer, Arduino Uno, RF module, Motor Driver, and Raspberry Pi. The input to the system is given through hand gestures where the values from the accelerometer are sent to the Arduino Uno. Later, the command is issued from Arduino to RF transmitter and from this it goes to RF receiver. Now from receiver it goes to the Arduino installed on the vehicle from there to motor driver, and now motor driver performs the operation upon the command received from the Arduino. Here Raspberry Pi helps in live video streaming. The limitations of this system are as follows: (i) Most expensive to implement and (ii) Gestures might result in incorrect readings at the receiver side.

Pawar et al. [7] proposed an IoT-Based Embedded System for Vehicle Security and Driver Surveillance. The components used here are Raspberry Pi, Raspbian, Camera, GPS module, GSM module, and vibration sensor. Whenever the user enters the car, the camera takes the image of the user and stores it in the memory, and it makes corresponding check with the database and if it matches with the one in database onlythen the access is given to the user. If not, then the corresponding image is mailed to the corresponding owner's account. It uses open CV to identify the person using camera. The vehicle starts only when the value received from the vibration sensor, i.e., when the person sits on the driver seat. The drawbacks of this design are (i) Power consumption is more and (ii) Taking more time to validate the image with the database.

3 Surveillance Vehicle Functionality

The main aim of the design is to access the vehicle remotely from anywhere through Internet based on live [8] video streaming using esp32 camera. The Blynk App is used to operate the vehicle through Internet. First the user needs to register in the

Blynk App, then it generates a unique auth token to provide the secure service from the cloud services from the cloud server. It provides cloud services to the user when the user requests for the service, then the libraries are called from the cloud server. Now the Blynk server checks the auth token whether it is online or not. If it is online then it sends a response to the vehicle in which the same auth token is detected.

The vehicle can be controlled remotely through buttons, mobile tilt using the inbuilt accelerometer in the mobile using Blynk App and voice commands. The pantilt servo module is used to get the required angle of the camera. Our proposed design can be controlled through Blynk Application which first takes the user registration. After validation, the user can control the servo motors using the sliders in the application.

Our design is used to solve the real-time problems like lock down where people are not allowed to go outside to bring their basic needs. This problem can be overcome by controlling of the vehicle remotely by the person to meet their basic requirements. The areas where people are unable to go and monitor [9], in such places our project is used to go and monitor with live video streaming in areas like military bases (Fig. 1).

In the above circuit diagram, different components are used to control the vehicle. The components used are ESP8266 Wi-Fi module (ESP is the abbreviation of its manufacturer, namely, "Expressif Systems"), it is used to transfer the response to the motor driver upon user request. L298N motor driver (dual channel H-bridge motor driver), it is used to control the motors. DC (Direct Current) motors are used to rotate the wheels attached to it. Power Bank, it is used as power source to the vehicle. Pan and tilt servo module is used for camera movement. ESP32 camera (ESP is the abbreviation of its manufacturer, namely, "Expressif Systems"), it is used for live streaming purpose. USB module (Universal Serial Bus), it is used for

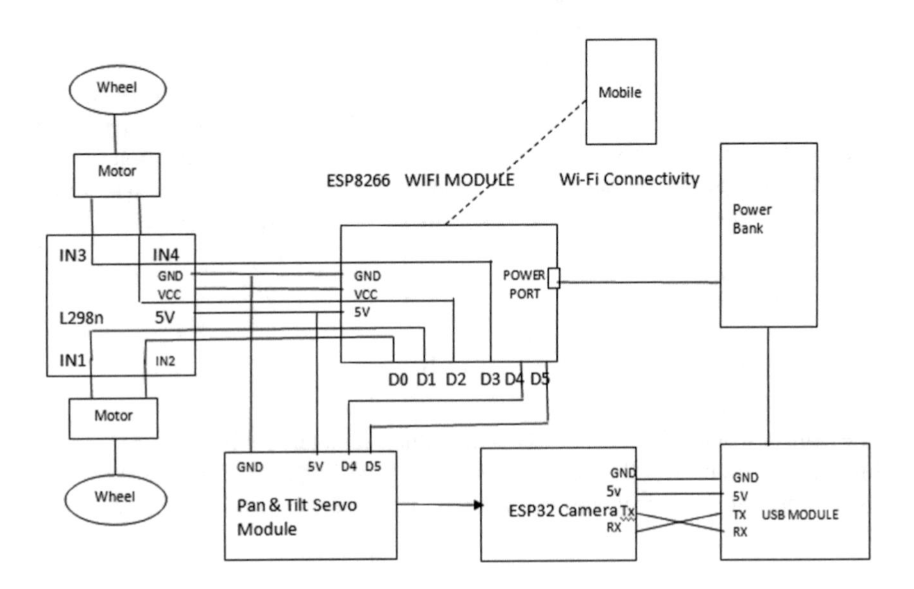

Fig. 1 Proposed system circuit diagram

Table 1 Controlling commands over voice

Virtual buttons	1	2	3	4	5
Input to vehicle	"FORWARD"	"LEFT"	"STOP"	"RIGHT"	"BACKWARD"
Vehicle movement	"Move Forward"	"Move Left"	"Stop"	"Move Right"	"Move Back"

programming the ESP32 camera. Mobile device is used to give input to vehicle upon user requirement.

The procedure is first the user gives the input through the mobile then it will transfer the request to the Wi-Fi module. The Wi-Fi module process the request and send it to the motor driver to perform the required task and it also sends the request to the pan tilt servo module. Now this is used for movement of the ESP32 camera which is connected to power bank through USB module. The live video streaming is achieved in the output screen. Our proposed project has three parts. Part 1 follows below steps (Table 1).

Part 1:

It consists of vehicle operations through buttons, mobile tilting, and voice commands.

```
Step 1: Take User registration details (Uid,pwd)
Step 2: evaluate hashvalue of the details
            V1 = hashvalue
Step 3: Storing default commands in IFTTT (IF This Then That)
            Di = defaultcommands
Step 4: Take User input through buttons or voice or tilting
            Ui=input
Step 5: Compare Ui with Di
            if(Ui==Di) perform operations
            else go Step 4
```

Part 2:

It consists of the work related to the live video streaming. Using pan tilt, we can get the total view of the ambience because the servo motors attached to it are controlled by the sliders. This can be achieved by following below steps.

Surveillance:

```
Step 1: Attach camera to the pan tilt mechanism.
Step 2: Now read the input from the Slider
            Xi=Slide (Ui)
Step 3: There are two Sliders, Horizontal(Hi) andVertical(Vi)(180 degrees each)
            if(Xi==Hi)
            camera moves horizontally
            else if(Xi==Vi)
                    camera moves vertically
            else
              go to Step 2
Step 4: Upload code into camera and run
Step 5: The link will be generated by running the code and use it for live video
        streaming
```

Part 3:

It consists of authorization where camera identifies the person whether he is authorized or not. If the user is authorized, then it will detect the user or else it will capture the image of the intruder and store it in the SD card, at the same time it will send it to the corresponding authorized person's email account.

Authorization:

```
Step 1: Upload code into camera and run
Step 2: The link will be generated by running the code and upload it in ngrok server
              L1=upload(Link) in ngrok
Step 3: Now using this link enroll the authorized person's image
              Auth=enroll(L1(image))
Step 4: Now validate the person
              If(User == Auth)
                  Detects the user
              Else
        Capture image of person and store it in sd card and send it to authorized gmail
              Mi=capture(User)
              Si=store(Mi);
              Gi=send(Si)
```

4 Results and Analysis

The technology used here is Arduino IDE where the program is written and executed, and then upload the program into esp32 camera. After execution, the camera generates a link which can be visible in serial monitor of Arduino IDE. The user needs to register in the NGROK before accessing its services, then it will generate a hashcode. Here, python shell is also used to generate the global link by executing the previous link generated in Arduino IDE and hashcode from NGROK in command prompt.

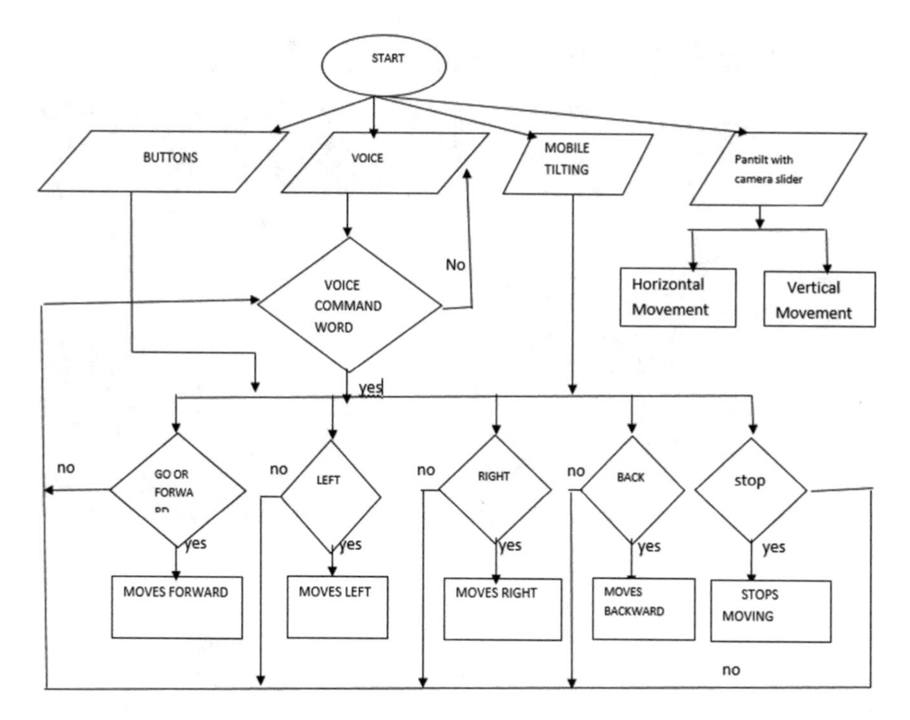

Fig. 2 Surveillance vehicle functionality

The proposed system has a capability of secure [10] live video streaming. The car can be controlled through mobile phone by tapping the buttons on the screen. The issues facing in the control of the vehicle can be overcome by mobile tilting feature. Here we need to tilt the mobile in the required direction to move the vehicle. Google assistance can be used to give users voice as an input. The IFTTT platform is going to validate user's voice input with default commands stored in it. The control of the car can also be done in our native language.

Wi-Fi module [11] is used as an integrated microcontroller which is going to handle the functionality that occurs. The Bluetooth module is not used in this system as it is only limited to short range. The esp32 [12] camera is used for video streaming. The video streaming is done through the link which will be generated when it runs the uploaded code in camera [13]. After loading the link in the browser then the browser asks for the enrollment of the authorized person's face (Fig. 2).

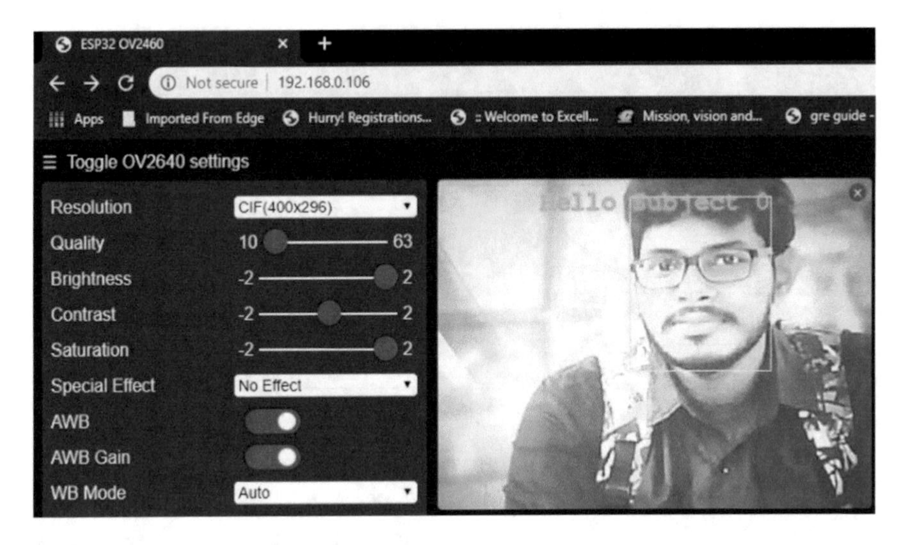

Fig. 3 Detecting the user with "Hello subject"

The previous generated link and the hash code of NGROK are executed in Python to generate the global access link. Here we are using NGROK server to generate a link through which global access from anywhere and from any device is possible.

For face detection, first the user needs to enroll their face. Later, the camera takes five samples of the user's face then it is going to recognize the face. Whenever the authorized person stays in front of the camera, it detects [14] the face and displays "Hello Subject" message (Fig. 3).

Whenever the unauthorized person comes in front of the camera then it displays "Intruder Alert" message (Fig. 4).

When there are more number of people in front of the camera then it recognizes both authorized and unauthorized user and displays the messages "Hello Subject" and "Intruder Alert," respectively (Figs. 5, 6 and 7).

Table 2 shows the comparison with the existing system by considering certain features like security services, live streaming, Google assistance, power consumption, and wireless network support.

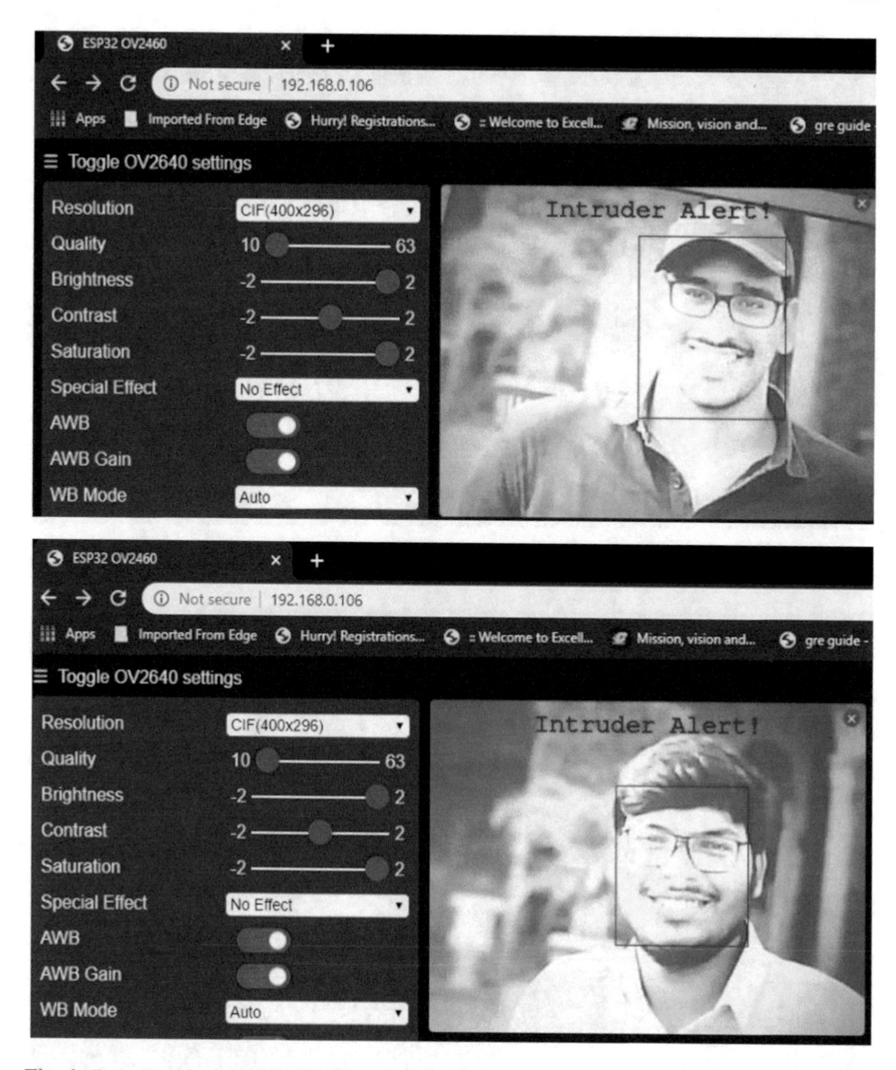

Fig. 4 Detecting the intruder with “Intruder alert”

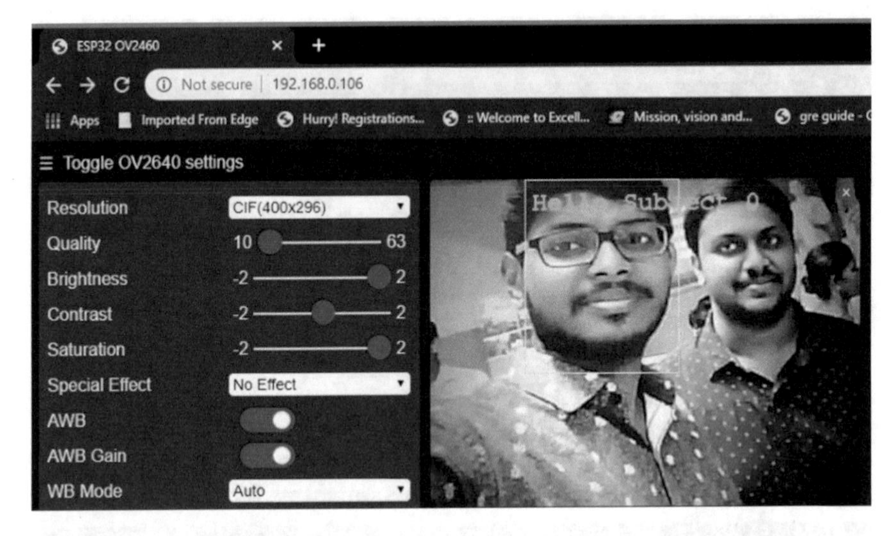

Fig. 5 Detection of an authorized user

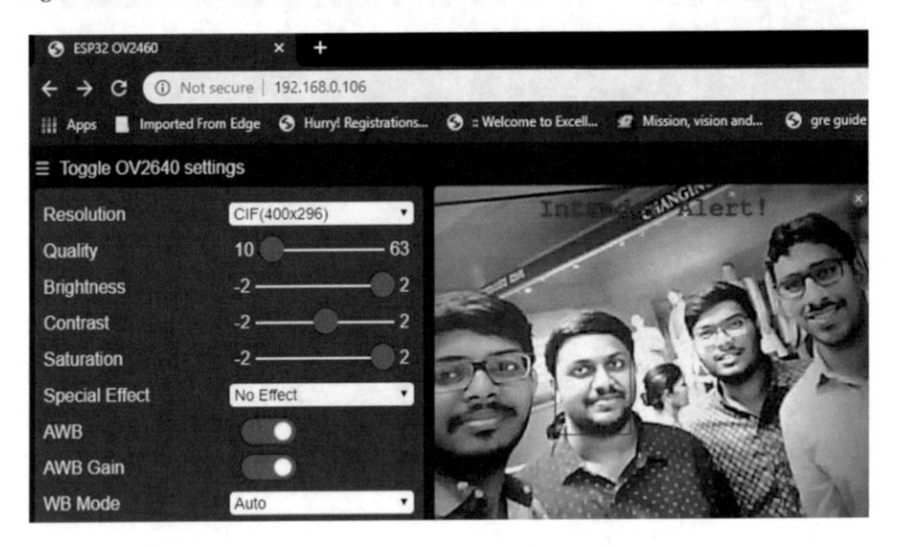

Fig. 6 Detection of an Intruder

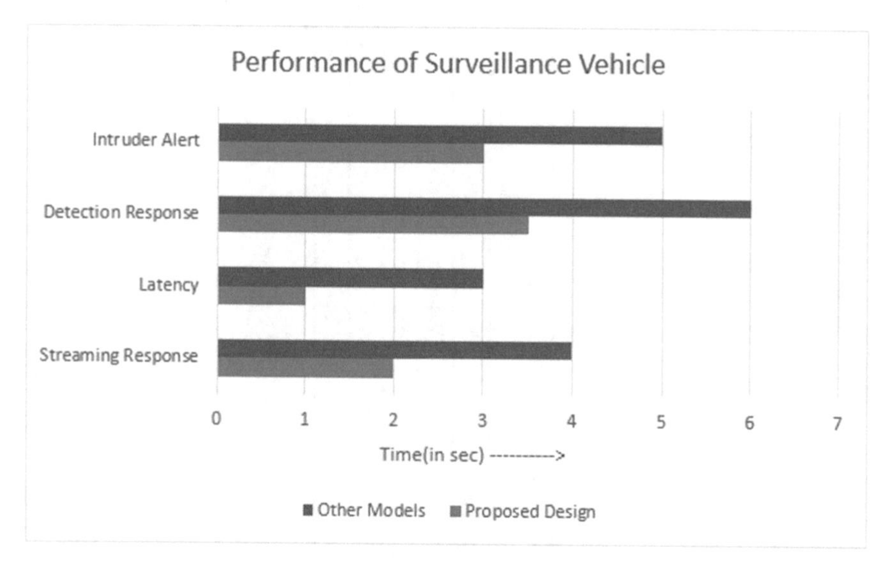

Fig. 7 Performance analysis of features of proposed system

5 Conclusion

In our proposed design, we use many techniques to control the vehicle remotely from anywhere. Here in our proposed design we wanted to develop a surveillance vehicle to reduce the risk involved in driving the vehicle in hazardous and dangerous places. The design is not complex to interact with the interface like enrolling face detection [15], face recognition [16], etc. The direction of the vehicle can be controlled through mobile phone easily. It makes the difficulty of police patrolling [17] easier in such a way that police can control their vehicle remotely from control rooms without being physically present in the vehicle. Here we can use live streaming from anywhere and from any device. This vehicle can also be controlled through buttons, voice and mobile phone tilting. Through live video streaming the user can have a sight of all the dimensions, so that the probability of accidents [18] will then be reduced.

Table 2 Feature comparison

Methods features	Bluetooth/Wi-Fi	Google assistant	Gestures	Authorization	Power consumptiion	Camera	Live streaming globally
Minal S	Yes/No	No	No	No	Mediium	Yes	No
Sandipan Chakraborty	No/Yes	Yes	Yes	No	High	Yes	No
S Nashif	No/Yes	No	Yes	No	High	Yes	No
Mahesh R	No/Yes	No	No	Yes	High	Yes	No
Proposed system	No/Yes	Yes	Yes	Yes	Low	Yes	No

References

1. Khalid H, Ahmed SU, Uddin R (2019) I-safe technology for smart and secure vehicles. In: International conference on computational intelligence and knowledge economy (ICCIKE)
2. Hsu Y-H, Chang S-Y, Guo J-I (2018) A multiple vehicle tracking and counting method and its realization on an embedded system with a surveillance camera. In: IEEE international conference on consumer electronics-Taiwan (ICCE-TW)
3. Geng H, Guan J, Pan H, Fu H (2018) Multiple vehicle detection with different scales in urban surveillance video. In: IEEE fourth international conference on multimedia big data (BigMM)
4. Ghute MS, Kamble KP, Korde M (2018) Design of military surveillance robot. In: First international conference on secure cyber computing and communication (ICSCCC)
5. Chakraborty S, Mukherjee S, Saha SK, Saha HN (2019) Autonomous vehicle for industrial supervision based on google assistant services & IoT analytics. In: IEEE 10th annual information technology, electronics and mobile communication conference (IEMCON)
6. Raihan MR, Hasan R, Arifin F, Nashif S, Haider MR (2019) Design and implementation of a hand movement controlled robotic vehicle with wireless live streaming feature. In: IEEE international conference on system, computation, automation and networking (ICSCAN)
7. Pawar MR, Rizvi I (2018) IoT based embedded system for vehicle security and driver surveillance. In: Second international conference on inventive communication and computational technologies (ICICCT)
8. Philo Stephy A, Preethi C, Mohana Prasad K (2019) Analysis of vehicle activities and live streaming using IOT. In: International conference on communication and signal processing (ICCSP)
9. Iqbal A, Saleem F (2018) Monitoring application for unmanned aircraft systems.In: Moscow workshop on electronic and networking technologies (MWENT)
10. Ayub MF, Ghawash F, Shabbir MA, Kamran M, Butt FA (2018) Next generation security and surveillance system using autonomous vehicles. In: Ubiquitous Positioning, Indoor Navigation and Location-Based Services (UPINLBS)
11. Dhikle MS, Shahnasser H (2018) An android application controlled video surveillance vehicle. In: 3rd international conference for convergence in technology (I2CT)
12. Rai P, Rehman M (2019) ESP32 based smart surveillance system. In: 2nd international conference on computing, mathematics and engineering technologies (iCoMET)
13. Zhang Q, Sun H, Wu X, Zhong H (2019) Edge video analytics for public safety: a review. Proc IEEE 107(8)
14. Imteaj A, Chowdhury MAIJ, Farshid M, Shahid AR (2019) RoboFI: autonomous path follower robot for human body detection and geolocalization for search and rescue missions using computer vision and IoT. In: 1st international conference on advances in science, engineering and robotics technology (ICASERT)
15. Mohanasundaram S, Krishnan V, Madhubala V (2019) Vehicle theft tracking, detecting and locking system using open CV. In: 5th international conference on advanced computing & communication systems (ICACCS)
16. Hu L, Ni Q (2018) IoT-driven automated object detection algorithm for urban surveillance systems in smart cities. IEEE Internet Things J 5(2)
17. Nintanavongsa P, Yaemvachi W, Pitimon I (2019) Performance analysis of perimeter surveillance unmanned aerial vehicles. In: 7th international electrical engineering congress (iEECON)
18. Desai G, Ambre V, Jakharia S, Sherkhane S (2018) Smart road surveillance using image processing. In: International conference on smart city and emerging technology (ICSCET)

Prediction of Biotic Stress in Paddy Crop Using Deep Convolutional Neural Networks

B. Leelavathy and Ram Mohan Rao Kovvur

Abstract The production of rice grain plays a major role in the global economy. In this context, there is a high demand to focus on yield production of staple food. In general, paddy crop fields undergo biotic stress which adversely affects stages of the growth of a plant which in turn leads to the low production of the crop. Conventional methods used to identify the biotic stress must rely on the experts manually through observations in the crop field, which isn't that reliable since there is a chance of error-prone observations that leads to the incorrect management of biotic stress management. We present a new approach to the development of the biotic stress prediction model, based on rice disease image classification, using deep convolutional networks. Preliminary steps required to perform the proposed method are the collection of datasets from different sources, deep learning framework to perform the training. The developed model can recognize different types of biotic stress out of diseased and healthy leaves which resulted in an accuracy of 98.33%.

Keywords Biotic stress · Deep learning · CNN · Image classification

1 Introduction

Paddy crop produces rice seeds that belong to the species of grass are named scientifically as Oryza sativa (in Asia) and Oryza glaberrima (in Africa). As a cereal, it is one of the widely consumed staple food by a huge part of the world's human population, specifically in Asian countries. It is the agricultural commodity with the third highest worldwide production (rice, 741.5 million tonnes in 2014), after sugarcane (1.9 billion tonnes) and maize (1.0 billion tonnes) [1]. Rice (Oryza sativa) production is merely disturbed by different kinds of abiotic and biotic stresses such

B. Leelavathy (✉) · R. M. Rao Kovvur
Department of Information Technology, Vasavi College of Engineering, Hyderabad, India
e-mail: leelapallava@staff.vce.ac.in

R. M. Rao Kovvur
e-mail: krmrao@staff.vce.ac.in

N. Chaki et al. (eds.), *Proceedings of International Conference on Computational Intelligence and Data Engineering*, Lecture Notes on Data Engineering and Communications Technologies 56, https://doi.org/10.1007/978-981-15-8767-2_29

as submergence, fungal, bacterial, drought, nematode, brown hopper, and stem borer. Few of the biotic stresses like bacterial leaf blight, rice blast, brown spot, leaf smut, gall midge, stem borer, brown plant hopper are responsible for decreasing the grain quality and it affects less grain productivity [2]. Biotic stresses include insect pests, fungus, bacteria, viruses, and herbicide toxicity. Plant response which is undergoing stress differs depending on the nature and severity of the stress involved, age of the plant, is capable of stress-tolerant nature. Paddy crops that are undergoing stress exhibit the change of symptoms in the color, shape, and texture of the leaves. It is obviously difficult to manually capture and quantify these micro-symptoms [3].

Taking the above context into consideration, a more improvised approach is needed that gives a dimension to the crop, resource management practices such as supplying sufficient nutrients, implementing irrigation regime, and many other factors when chosen by existing farmers as a practice which becomes a potential option for maximizing the crop yield without any certain gaps.

To have a model that automates the above-mentioned gaps that increase paddy crop production, one must implement various machine learning or artificial intelligence or deep learning [4, 5] technique. Because information technology has now changed, the realistic challenges of our day to day life with more augmented reality and automated applications. So, there should be some model that bridges the gap as agriculture specifically the staple food production is a major source for the world's global economy. There should be the best way of deploying the expected CNN model for the biotic stress prediction. The model is trained, tested, and validated with the given dataset. Hence, dataset collection is the basic and important step to be taken into concern.

The best way of collecting the dataset is to capture the images of leaves in the crop field that is infected at different growth stages of the field periodically and find out the severity of the stress is one of the simple ways to get rid of infection of leaves. The other best method is to deploy the cameras in the crop field and automate the capturing of images periodically on the specified time intervals like twice a day, once in a day, weekly or bi-weekly so that, at every growth stage of the crop, we can predict the severity of the damage that is likely to happen until it is going to be harvested. So, if we deploy the above-mentioned method where the system automates to detect the disease type from the images captured and also provide feedback in terms of pesticide selection would ease the farmers retrieving high yields than expected. The other method is to make use of available datasets from different sources like UCI Machine Learning Repository, Plant Village dataset [6].

We briefly propose an approach to apply the concepts of deep learning techniques to solve the problem of the conventional method and automate the biotic stress prediction of given rice leaf healthy, diseased images. We considered the most common biotic stresses that a paddy crop undergoes are Brown Spot, Leaf Smut, Bacterial Leaf Blight [7], built a model that automatically predicts the type of biotic stress it is possessing with. The dataset we used has both diseased and healthy images. The best-performing model achieves of accuracy score between 93 and 98.33%, by applying the technical aspects of our approach.

Bacterial Leaf Blight

Leaf Smut

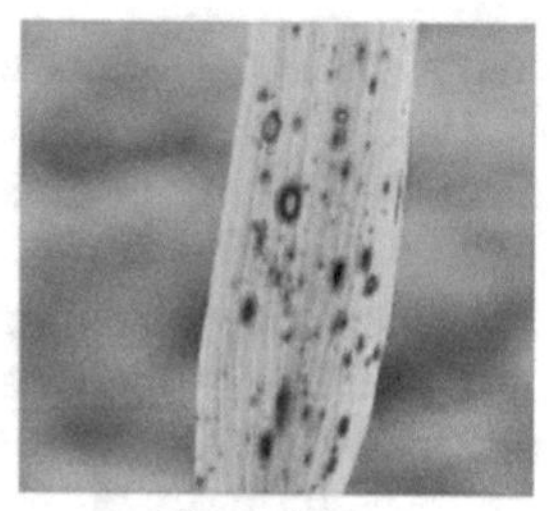

Brown Spot

Fig. 1 Types of diseases that come under Biotic Stress

2 Background

The biotic stress in paddy crops can be due to fungi or bacteria which can be identified at every growth of the plant. Following are the important types of rice plant diseases that are considered for the proposed methodology (see Fig. 1).

Bacterial Leaf Blight

It is caused by Xanthomonas oryzae pv. oryzae. It mainly affects the leaves of the plant causing yellowing of leaves, wilting of leaves and seedlings. During the seeding stage, infected leaves turn to grayish-green and roll-up. As it progresses the growth with a disease, the leaves turn from yellow to straw-colored and start wilting, leading whole seedlings to dry up and die [8].

Leaf Smut

Leaf Smut is caused by Entyloma oryzae. It commonly affects the leaves of plants causing small, slightly raised black spots or sori development on leaves. These small spots are somewhat rectangular, turning the leaf a non-uniform shape with reddish brown.

Brown Spot

Brown spot is a fungal disease that usually occurs on the host leaves and glume, as well as seedlings, sheaths, stems, and grains of the rice plant. The spots are gray-colored, oval, and fungi turn the leaves to reddish or dark brown color [8].

3 Related Work

For the development of image classifiers for the identification of biotic stress diagnosis, there is a need for large, verified set of images of both diseased and healthy plants [6]. There is no dataset existing so far, to address this problem, the Plant Village project has started collecting tens of thousands of images of healthy and diseased crop plants [9], which is an open-source dataset available. The dataset can perform the classification of 26 diseases in 14 crop species using 54,306 images with a CNN approach with the prediction of 38 possible classes and the trained model achieved the possible test results of accuracy about 99.35% [6]. In continuation, a rice blast feature extraction and disease classification method were proposed that is based on Deep-CNN, however, Quantitative analysis results indicate that CNN + SoftMax and CNN + SVM have performed similar, which is better than that of LBP + SVM and Haar-WT + SVM by a wide margin [10].

In a very few years, CNN despite its ability to extract features, it is in high extensive use under research in machine learning as well as in recognition of patterns [4, 11–16]. It is also stated that multi-layer NN has high learning capability, which learns features and expresses raw data ready for performing the classification [17]. CNN is one of the models to perform operations on images like recognition and classifications. The detection of objects, face recognition is some of the areas where CNN is widely used. Due to its high-end performance, we propose a method that uses CNN for predicting the biotic stress of a rice leaf image feature extraction and disease classification, and we ended up with good results by fine-tuning the necessary features of the model.

4 Proposed Work

To overcome the challenges, a computer vision-based approach [14, 18] has been introduced to classify and predict the biotic stresses in the paddy crop. The proposed method is categorized into the following stages, namely loading the datasets, data pre-processing, splitting of data in the training set and test set, data augmentation, recognition, and classification of stresses (see Fig. 2).

4.1 Proposed Algorithm

Step1: Input is a set of images with a dimension of 256 × 256
Step2: Input Pre-processing of an image to array representation
Step3: Further pre-processing to binary levels of [0, 1]
Step4: Data Splitting and data augmentation
Step5: Feeding into Conv2D first layer, ReLU, MaxPooling2D
Step6: Repeat step5 for another two Conv2D layers
Step7: Scale down to dense layer with Softmax Classifier at the last layer
Step8: Apply Adam optimizer, set the learning rate, compile the model
Step9: Calculate the model accuracy.

This experiment was executed on Intel® Core™ i7-8550U CPU@1.8 GHz, 2.00 GHz, Installed Memory (RAM) of 8 GB, 64-bit Operating System, ×64-based processor.

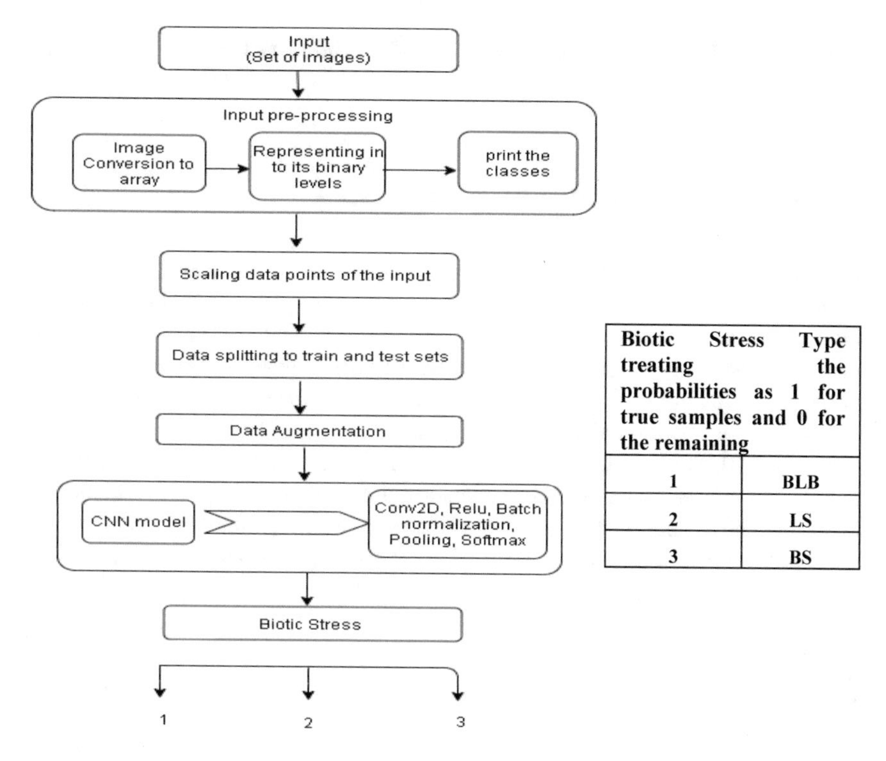

Fig. 2 Flow diagram illustrating all the stages of the proposed methodology

4.2 *Input*

A dataset consisting of different biotic stress like Bacterial Leaf Blight (BLB), Leaf Smut (LS), and Brown Spot (BS) and healthy images altogether 1000 images were taken as input. The dataset is a combination of different images taken from different sources. Loading of the dataset is done before with necessary packages like NumPy, Sklearn, Keras [19] that run on the top of TensorFlow [5, 20]. Also, Matplotlib must be imported.

4.3 *Input Pre-processing*

Pre-processing of input images is done so that the feature extraction becomes simple. Once the dataset is loaded, in this step every image from the dataset is converted to an array, later converting them to binary levels of 0 and 1, i.e., we have transformed the image labels using Scikit Learn's LabelBinarizer, printing the class labels that are assigned to the processed images. The images are further pre-processed by scaling the data points from 0 to 255 (the minimum and maximum) RGB values to the range of [0, 1].

4.4 Data Splitting and Data Augmentation

Splitting of the dataset into two disjoint subsets of optimal test size of 0.20 leading to the test-train split of 80–20% for training and testing purposes. Data augmentation consists of a suite of techniques and methods that increases the size and quality of the given training images/datasets which enables better deep learning model [21] to be built using them. This can be derived from simple transformations such as flipping horizontally, color, space augmentations, and cropping the image randomly [22]. We have used ImageDataGenerator method in Keras by passing the arguments rotation_range used for random rotations, width_shift_range specifies the horizontal shifts, height_shift_range specifies the vertical shifts, shear_range intensity of angles in counter-clockwise direction, zoom_range, horizontal_flip, fill_mode fills the points out of the boundary by nearest mode.

4.5 CNN Model

The deep learning CNN model is technically used to train and test, each input image which passes through a series of convolution layer with filters of size 32 with ReLU(Rectified Linear Unit), Maxpooling, convolution layer with filters of size 64, 128 followed by ReLU, Maxpooling, fully connected layers, applying a SoftMax function to classify an image that is undergoing biotic stress. For the complete illustration of CNN to process an input image and classification of the diseased images based on values (see Fig. 3).

Convolution Layer: When convolution operator must be applied on this layer, we can use dimensions depending on the model we build. There can be convolution with 1D, 2D, or 3D. This function takes certain arguments like input_dim which specifies the number of channels or dimensions does the input consists of, input_shape must be provided in conjuction with input_dim if we wish to use this as a first layer for

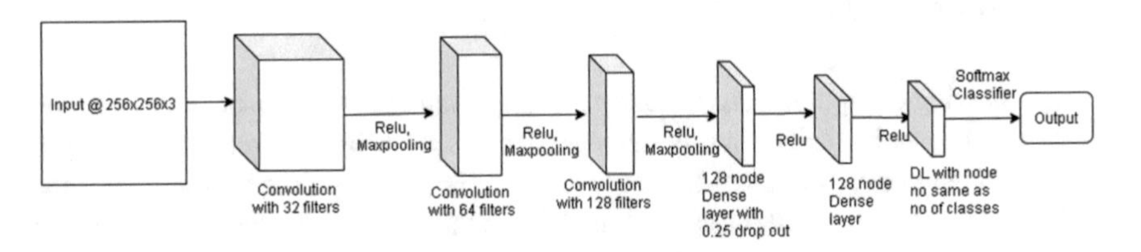

Fig. 3 CNN model flow architecture

Table 1 Findings observed during the execution

Biotic stress type	Number of images	Number of epochs	Accuracy (%)
BLB	1000 {healthy, infected}	50	93.33
LS		25	97.61
BS		25	98.33

the desired model. For our proposed method, we have used 2D convolution layers height and width of an input of size 256 × 256 with depth = 3 with 32, 64, 128 filters, following with activation function of argument "ReLU." Batch normalization, MaxPooling operation with 2D is applied which represents the input tensor shape to the mentioned output tensor of pool size = (3, 3) and (2, 2) downscaling the vector horizontally and vertically the size of the image, with a dropout of 0.25 followed by dense layer with the last layer using SoftMax classifier to generate the appropriate values as output.

Image matrix = (h*w*d), Filter = fw* fh* d, Stride of size = 2, ReLu is used for non-linear function whose output is obtained using f(x) = max(0, x). Maxpooling layer obtains the maximum value from the rectified feature map acquired. Fully connected layer flattens the features to build a model. The obtained model is given to the activation function called SoftMax to classify and predict the biotic stress levels.

5 Results and Observations

The model was implemented by importing Keras libraries, using TensorFlow in the background [5, 23, 24] with the above-mentioned methodology, by calculating the accuracy results of training and validation [3, 4]. Findings are mentioned in Table 1, the accuracy obtained for biotic stress prediction is 98.33% for biotic stress type BrownSpot and the model outputs are demonstrated that clearly state the predictions.

6 Metrics for Evaluation

To evaluate the performance of different methods in use, there are many statistical parameters used as metrics. Few of the quantitative measures are below.

6.1 Accuracy

The classification accuracy acts as the principal indicator; the higher it is, leading to a better performance by the specified classifier (see Fig. 5). The accuracy, in general,

```
print("[INFO] Calculating model accuracy")
scores = model.evaluate(x_test, y_test)
print(f"Test Accuracy: {scores[1]*100}")
```

Fig. 4 Code Snippet of the classifier that calculates the model accuracy for the number of correct predictions

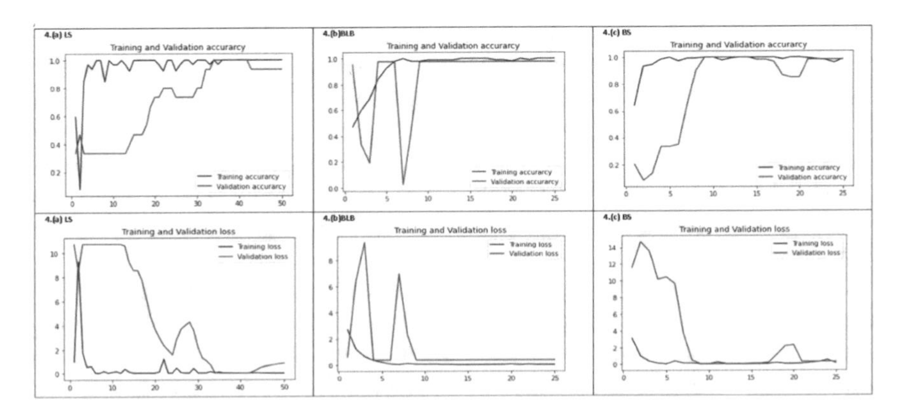

Fig. 5 Representing different patterns observed during the validation and training of the dataset based on the test set of 0.20. **a** LS **b** BLB **c** BS

can be calculated using Eq. (1) (Fig. 4)

$$\text{Accuracy} = \frac{\text{Number of correct predictions}}{\text{Total number of predictions performed}} \tag{1}$$

6.2 Receiver Operating Characteristic Curve (ROC)

The other important objective metrics in the task of image classification, which is defined by true positive rate and false positive rate; the larger the area under the ROC curve, the best classification it is. A graph plot that illustrates the binary classifier (see Fig. 5). Area under curve is the percentage of the area that lies under the ROC curve, with a range between 0 and 1 (Fig. 6).

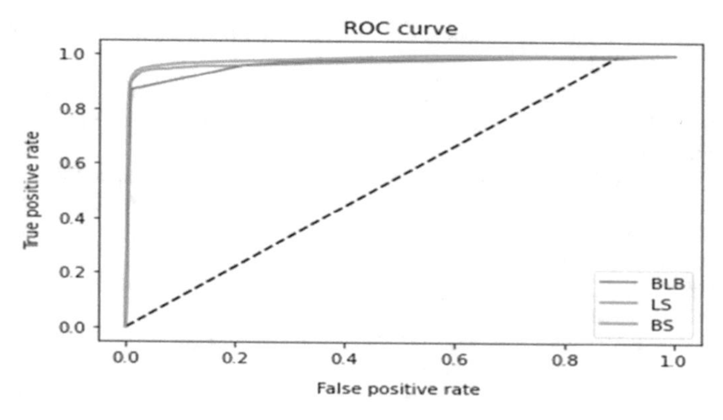

Fig. 6 ROC curve

7 Conclusion

The work has successfully classified the biotic stress prediction of paddy crop leaf disease images which automates the system of identification of biotic stress under different types of diseases the crop undergoes. It is quite typical to interpret the type of disease the crop is undergoing with manual intervention. The presented scenario of approach is flexible and robust enough to handle field crop images. To be generic, the model can be applied for autoprediction on different rice varieties/grain crops that are undergoing biotic stress which is a preliminary check that maximizes the period of potential growth of the paddy crop. However, prediction of food security and the yield of production after applying this model on the real-time paddy crop field can be done as a future scope.

References

1. Crops/Regions/World list/Production quantity (pick lists), rice (paddy), 2018. UN Food and Agriculture Organization, Corporate Statistical Database (FAOSTAT). 2020. Archived from the original on May 11, 2017. Accessed 11 Oct 2019
2. Singh P et al (2020) Biotic stress management in rice (Oryza sativa L.) through conventional and molecular approaches. In: Rakshit A, Singh H, Singh A, Singh U, Fraceto L (eds) New frontiers in stress management for durable agriculture. Springer, Singapore. https://doi.org/10.1007/978-981-15-1322-0_30
3. Anami BS, Malvade NN, Palaiah S. Classification of yield affecting biotic and abiotic paddy crop stresses using field images. Inf Process Agric. https://doi.org/10.1016/j.inpa.2019.08.005
4. Duan M, Li K, Yang C, Li K (2018) A hybrid deep learning CNN–ELM for age and gender classification. Neurocomputing 275:448–461
5. Chollet F (2015) Keras. GitHub Published online 21 November, 2015. https://github.com/fchollet/keras
6. Mohanty SP, Hughes DP, Salathé M (2016) Using deep learning for image-based plant disease detection. Front Plant Sci 7:1419. https://doi.org/10.3389/fpls.2016.01419

7. Rice production (peace crops) Chapter 14—Disease of rice [Online]. http://www.nzdl.org. Last accessed 23 Nov 2015
8. http://www.knowledgebank.irri.org/decision-tools/rice-doctor/rice-doctor-fact-sheets
9. Hughes DP, Salathé M (2015) An open access repository of images on plant health to enable the development of mobile disease diagnostics. arXiv:1511.08060
10. Liang W, Zhang H, Zhang G et al (2019) Rice blast disease recognition using a deep convolutional neural network. Sci Rep 9:2869. https://doi.org/10.1038/s41598-019-38966-0
11. Jiao ZC, Gao XB, Wang Y, Li J (2016) A deep feature based framework for breast masses classification. Neurocomputing 197:221–231
12. Ypsilantis PP et al (2015) Predicting response to neoadjuvant chemotherapy with PET imaging using convolutional neural networks. PLoS ONE 10
13. Liu ZY, Gao JF, Yang GG, Zhang H, He Y (2016) Localization and classification of paddy field pests using a saliency map and deep convolutional neural network. Sci Rep 6
14. Johnson J, Karpathy A, Fei-Fei L (2016) DenseCap: fully convolutional localization networks for dense captioning. In: 2016 IEEE Conference on computer vision and pattern recognition (CVPR), pp 4565–4574
15. Kang MJ, Kang JW (2016) Intrusion detection system using deep neural network for in-vehicle network security. PLoS ONE 11
16. LeCun Y, Bengio Y, Hinton G (2015) Deep learning. Nature 521:436–444
17. Hinton G, Salakhutdinov RR (2006) Reducing the dimensionality of data with neural networks. Science 313:504–507
18. CiresSan D et al (2012) Multi-column deep neural networks for image classification. In: 2012 IEEE Conference on computer vision and pattern recognition. IEEE, pp 3642–3649
19. CiresSan DC et al (2010) Deep, big, simple neural nets for handwritten digit recognition. Neural Comput 22:3207–3220
20. Krizhevsky A et al (2012) ImageNet classification with deep convolutional neural networks. In: Pereira F et al (eds) Proceedings of the 25th international conference on neural information processing systems, NIPS 2012, vol 1. Curran Associates Inc, pp 1097–1105
21. Simonyan K, Zisserman A (2014) Very deep convolutional networks for large-scale image recognition. CoRR abs/1409.1556
22. Shorten C, Khoshgoftaar TM (2019) A survey on image data augmentation for deep learning. J Big Data 6:60. https://doi.org/10.1186/s40537-019-0197-0
23. Srivastava N et al (2014) Dropout: a simple way to prevent neural networks from overfitting. J Mach Learn Res 15:1929–1958
24. Szegedy C et al (2016) Inception-v4, inception-ResNet and the impact of residual connections on learning. CoRR abs/1602.07261

High Average Utility Itemset Mining: A Survey

Mathe John Kenny Kumar and Dipti Rana

Abstract The past two decades has seen a great amount of research being done in the area of "Data Mining". "Frequent Itemset Mining" is one application of "Data Mining" used to mine "Frequent Patterns" from databases. Itemsets that occur frequently only extract patterns that are frequent. This type of mining has its disadvantage. Items that are frequent may or may not be profitable to an organization. For this type of disadvantage to be overcome "High Utility Itemset Mining" had been introduced. It is used to mine profitable patterns from databases. This can help organizations to market profitable patterns. "High Utility Itemset Mining" also has its own disadvantage, because while mining itemsets, the lengths of the itemsets are not taken into consideration. The bigger the length of the itemset the more the profit. This does not give a real representation of the value or profit of the itemset. To get over this problem "High Average Utility Itemset Mining" had been introduced.

Keywords High Average Utility Itemset Mining · Transaction utility · Transaction maximum utility

1 Introduction

The main idea of "Data Mining" [1, 2] is to discover knowledge in humongous amounts of data. Knowledge from data can be discovered using "Frequent Itemset Mining". The items that are frequent are used to mine association rules [3]. Itemets which occur frequently can be discovered using "Frequent Itemset Mining" algorithms. For example, in a super market setting, milk and bread are bought more frequently than milk and diapers; therefore, milk and bread is an itemset which is frequent as it exceeds minimum support [3]. Many contributions have been made till

M. J. Kenny Kumar (✉) · D. Rana
Sardar Vallabhbhai National Institute of Technology, Surat, India
e-mail: johnkennykumar@gmail.com

D. Rana
e-mail: dpr@coed.svnit.ac.in

N. Chaki et al. (eds.), *Proceedings of International Conference on Computational Intelligence and Data Engineering*, Lecture Notes on Data Engineering and Communications Technologies 56, https://doi.org/10.1007/978-981-15-8767-2_30

date for "Frequent Itemset Mining" [3–9]. A critical drawback of "Frequent Itemset Mining" is of the fact that it only mines "Frequent Itemsets" and not the itemsets that are profitable to an organization. For example, an itemset of Home automation system (1), Laptop (1) is more profitable than Bread (10), Milk (10). In this example, although home automation system and laptop have frequency of only 1, the itemset is more profitable than bread and milk which have a frequency of 10. To overcome this drawback "High Utility Oriented Itemset Mining" (HUIM) [10–13] was proposed. "Utility" represents the profit or weight of an itemset. Unlike "Frequent Itemset Mining", "High Utility Itemset Mining" (HUIM) does not adhere to the "Downward Closure Property". Developing upper bounds to eliminate itemets is a challenging task in *HUIM*. Applications of "Utility Itemset Mining" include analysis of users usage of a website [5, 14, 15], marketing [8, 16, 17], and bioinformatics. Various sub-fields of "Utility Pattern Mining" such as "High Utility Pattern Mining" [18], "High Utility Rules" [19, 20], "High Utility Sequential Patterns" [21, 22], and "High Utility Episodes" [23, 24] have also been proposed. Although "High Utility Itemset Mining" [18] is more effective than "Frequent Itemset Mining" [3, 4], there is a drawback with respect to the length of the itemsets. While discovering "High Utility Itemsets" the lengths of the itemsets are not taken into consideration. This is not correct as the profits of the set of items will be more for itemsets that have greater number of elements. This obvious limitation has been overcome by, "High Average Utility Itemset Mining" (HAUIM). This paper has the following contributions.

- We discuss various applications of "High Average Utility Itemset Mining" (HAUIM) and motivation to explore *HAUIM*.
- We review all the "High Average Utility Itemset Mining" algorithms from the year 2009–2019.
- We explain in detail about two algorithms, namely, HAUI-Miner [25] and EHAUPM [26] with a common example
- We discuss various features of "High Average Utility Itemset Mining" in terms of related work, approach, number of phases, limitations.

The structure of this paper is as follows. Section 2 describes various applications of "High Average Utility Itemset Mining". This section also provides the motivation to explore *HAUIM* algorithms. Section 3 describes various preliminaries and definitions related to "High Average Utility Itemset Mining". Section 4 describes various algorithms in detail from the year 2009–2019. Section 5 lists the various features of HAUIM algorithms and all the "real" as well "Synthetic Datasets" used in each HAUIM algorithm. Section 6 describes various Challenges and Research Opportunities with respect to finding "High Average Utility Itemsets".

2 Applications and Motivation of High Average Utility Itemset Mining

2.1 Applications

Market Basket Analysis In market analysis, when a customer buys a particular item it is seen that he/she buys other set of items along with the first item. Frequent Itemset Mining [8] approaches like the "Level-Wise" and FP-growth [27] were used to extract frequent patterns which give an idea into the pattern that is followed by the customer when he/she buys a particular item. This can help to market the products better through mail or through various social media networks. Rack arrangement according to "Frequent Patterns" can also be done to help customers pick up products along the way. The drawback of "Frequent Itemset mining" [8] is that it does not mine the utility of a pattern or the value of a pattern. By using "High Average Utility Itemset Mining" [25] techniques better market basket analysis can be done with respect to the profit of the itemset.

Web Mining Web data contains rich information. Web logs contain users information about the number of clicks in each session as well as any purchases that have been made. "Internal Utility" is the number of clicks of a user in a particular session. "External Utility" can be expressed in terms of the importance of each webpage depending upon the users preferences. By using "Utility Oriented Pattern Mining", users usage of the web can be mined easily. The results from mining web logs can be used to build personalized web pages, personalized search engines, recommendation of web pages and products.

E-Commerce IOT technologies that are used in mobile phones, wireless networks, and Global positioning devices are used to capture and store information about users behavior. Information about users behavior can be used to perform data analytics which can infer useful and interesting patterns. "Utility Oriented Pattern Mining" can be used to find user behavior which can be of great importance. Sequential Pattern Mining [21] for mobiles are used to discover "High Utility Sequence Patterns" from mobile data. It is used to mine "Association Rules" with respect to customers purchase behavior and his/her location. It can also act as a recommender system to recommend products based on locations.

Utility Patterns in Bioinformatics When experiments are done in biomedicine a row can be defined as a set of genes and their "Internal Utilities" as levels of expression. Each gene in a set of genes has some importance which can be represented by "External Utility". Useful associations between genes can be established using bio informatics. Gene regulation patterns were successfully discovered from a time course of comparative data of gene expressions in bioinformatics. High Average Utility Pattern discovery in gene expression data can help in discovery of new medicines.

Motivation: The motivation for mining "High Utility Patterns" is because of its various real-time applications like real-time monitoring of profits of supermarket, marketing various products to customers depending upon the utility of their previous transactions. Various other mining techniques like "Clustering" and "Classification" can also be used to group similar customers and products according to their buying behavior. "Utility Mining" can also be applied to group users based on web logs. This technique can be implemented to find new diseases based on protein sequencing and gene sequencing. The above real-time applications and many others motivate for further research of "Utility Oriented Pattern Mining" in terms of discovery of new algorithms and other real-time applications.

3 Basic Concepts of Utility Oriented Pattern Mining

This section introduces the basic notations that will be used throughout the paper. It also introduces various definitions related to "High Utility Itemset Mining" and "High Average Utility Itemset Mining" (Table 1).

3.1 Preliminaries for High Average Utility Itemset Mining

Preliminary 1 ("Frequent Pattern Mining" and "Association Rule Mining"): "Association Rule" of the form $P \rightarrow Q$, where $P \subset I$, $Q \subset I$, and $P \bigcap Q = \phi$. Given a database, a pattern is said to be "frequent" if its items occur frequently. The number of times an itemset *X* occurs in a database is represented as *sup*(X). An "Association Rule" is defined as $R : P- > Q$ where *P* and *Q* is disjoint and *Q* is non-empty. Association rules are constructed using frequent itemsets. If the "Confidence" of the itemset is no less than the minimum threshold (min_conf) the association rule is called a strong association rule. For example, $\{laptop\} \rightarrow \{antivirus\}[sup = 5\%, conf = 80\%]$, this means that laptop and antivirus are purchased together 5% of the times and 80% of the times customers who purchase laptop also purchase antivirus.

Prelimnary 2 ("High Utility Itemset", *HUI*): *HUI* gives the value or the profit of the item set instead of the frequency of the item set of a database. This paper adopts similar definitions as defined in [28]. This paper will use the below example Fig. 1 tables, namely, Transaction Utility and External Utilities through out to maintain consistency among different examples that will be explained in detail.

Prelimnary 3 ("High Average Utility Itemset", *HAUI*): "High Utility Itemset" is the profit of the itemset in a database whose profit exceeds the minimum threshold value. The longer the length of the transaction the greater is the utility. To get a fair or a real value of the itemset(X) the average of the itemsets are considered.If the utility of the itemset $\{AB\}$ is 25 the average utility of the itemset is $25/2 = 12.5$.

Table 1 Symbols and meaning

Symbols	Meaning
X	Set of unique items n distinct items $it = \{it_1, it_2, it_3, \ldots, it_n\}$
DB	Quantitative database $DB = \{Tr_1, Tr_2, Tr_3, \ldots, Tr_n\}$
TID	Each $Tr_n \in DB$ is a unique transaction in a database *DB*
X	m-itemset having m unique items $i = \{it_1, it_2, it_3, \ldots, it_m\}$
$sup(X)$	No of times *X* occurs in Database *DB*
$q(it_j, Tr_q)$	Number of items it_j in transaction Tr_q
$eut(it_j)$	External Utility of an item it_j
$ut(it_j, Tr_q)$	Profit of an item it_j in transaction Tr_q
$ut(X_j, Tr_q)$	Profit of a set of items X in transaction Tr_q
min_sup	Predefined Threshold for Support
min_conf	Predefined Threshold for confidence
$tut(tr_q)$	Summation of all the utilities of transaction Tr_q
δ	Minimum high average utility threshold
min_util	$\delta \times \sum(Tr_q, Tr_q \in \ DB)$
min_autil	$\delta \times \sum(Tr_q, Tr_q \in DB)$
$twut$	Transaction weighted utility of a pattern
$tmut$	The maximum utility of all the utilities in a transaction Tr_q
$htwui$	Itemset whose *twu* is greater than min_util
hui	Itemset whose utility is greater than min_util
$auub$	Summation of maximum profits of a set of items or an item X occuring in database DB
$hauub$	If $auub(X) > min\_autil$ minimum average utility threshold, then $auub$ is said to be $hauub$
$huim$	"High utility itemset" mining
$m\text{-}itemset$	Itemset which contains m number of items

Transaction Table

T_{id}	Items				*tu*	*tmu*
T_1	B<2>	C<8>	D<2>		52	24
T_2	A<4>	B<6>			44	36
T_3	B<3>	E<2>			38	20
T_4	A<2>	B<2>	C<7>		37	21
T_5	A<6>	B<5>	D<4>		74	32
T_6	A<4>	B<4>	D<5>	E<3>	102	40
T_7	D<3>	E<4>			64	40
T_8	D<2>				16	16

External Utilities

Item	Profit
A	2
B	6
C	3
D	8
E	10

Fig. 1 Transaction table and external utilities

Definitions

Definition 1 Consider $IT = \{it_1, it_2, \ldots, it_m\}$ be a group of items and DB be a database $\{Tr_1, Tr_2, \ldots Tr_n\}$ such that a transaction $Tr_i \in DB$ is a subset of *IT*.

Definition 2 The "Local Transaction Utility", $lut(it_p, Tr_q)$, is the number of items of an item it_p in transaction Tr_q.

From Fig. 1 $lut(B, Tr_1) = 2$.

Definition 3 "External Utility" $eut(it_p)$ is the profit of an item it_p.

From External Utilities of Fig. 1 $eut(D) = 8$.

Definition 4 Utility $ut(it_p, Tr_q)$, is the product of internal and external utility for item it_p in transaction Tr_q, given by

$$ut(it_p, Tr_q) = lut(it_p, Tr_q) \times eut(it_p) \tag{1}$$

$ut(D, Tr_1) = 2 \times 6 = 12$ from Transaction Table and External Utilities of Fig. 1.

Definition 5 The profit of an itemset X in transaction Tr_q, $ut(X, Tr_q)$ is given by

$$ut(X, Tr_q) = \sum ut(it_p, Tr_q) \tag{2}$$

$ut(BD, Tr_1) = 2 \times 6 + 2 \times 8 = 28$ from Transaction table and External Utilities of Fig. 1.

$$ut(X) = \sum_{Tr_q \in DB} \sum_{it_p \in X} ut(it_p, Tr_q) \tag{3}$$

$ut(AD) = ut(AD, Tr_5) + ut(AD, Tr_6) = 44 + 48 = 92$.

Definition 6 The "Transaction Utility" of particular transaction T_q is denoted as $tut(T_q)$. It specifies the total weight of all the items in a particular transaction

$$tut(Tr_q) = \sum_{it_p \in Tr_q} ut(it_p, Tr_q) \tag{4}$$

$tut(Tr_2) = ut(A, Tr_2) + ut(B, Tr_2) = 8 + 36 = 44$, from Transaction Table and External Utilities of Fig. 1.

Definition 7 The "Minimum High Average Utility Threshold" is the product of δ and the sum of the total utility values of all the transactions in database *DB*. In Transaction table and External utilities of Fig. 1 the sum of all the "transaction utility" values is 427. If δ is 0.12,then the "minimum utility" value can be defined as

$$min\_autil = \beta \times \sum tut(Tr_q), Tr_q \in DB \tag{5}$$

$min\_autil = 0.12 \times 427 = 51.24$ from Transaction table of Fig. 1

Definition 8 In *huim* "Downward Closure Property" cannot be maintained. From External utilities, "a" is a low utility item with utility of $32 < 51.24$ whereas utility of $\{A, B\}$ is 134 and $\{A, B, D\}$ is 146 which are both "high utility" item sets, therefore "Downward Closure Property" cannot be applied to mine "High Utility Itemsets".

Definition 9 As "Downward Closure Property" is not adhered to, "Downward Closure Property" is maintained using "Transaction Weighted Utilization". It is denoted by $twut(X)$, is the summation of the transaction utilities of all the transactions in the database DB.

$$twut(X) = \sum_{X \subseteq Tr_q \in DB} tut(Tr_q) \tag{6}$$

$twut(BC) = tut(T_1) + tut(T_4) = 52 + 37 = 89$, from Transaction Utility and External Utilities of Fig. 1. Here $min\_util = 51.24$. As $twut(BC) < min\_util$, any super pattern of "BC" cannot be a High twu itemset (candidate itemset) and obviously cannot be a "High Utility Itemset". X is a "High Transaction Weighted Utilization Itemset" (i.e., a candidate itemset) if $twut(X) \geq min\_util$.

Definition 10 The "average utility" $aut(X)$ of an itemset X with respect to a transaction database *DB* is the addition of profit values of X where X is contained in all the transactions of D divided by the number of items contained in X. It can be defined as

$$aut(X) = \sum_{Tr_q \in DB \bigwedge X \subseteq Tr_q} ut(X, Tr_q)/|X| \tag{7}$$

An itemset AB appears in transactions $Tr_2, Tr_4, Tr_5, and Tr_6$. Average Utility of AB is $aut(AB) = ut(BC, Tr_2) + ut(BC, Tr_4) + ut(BC, Tr_5) + ut(BC, Tr_6) = ((4 \times 2 + 6 \times 6) + (2 \times 2 + 2 \times 6) + (6 \times 2 + 5 \times 6) + (4 \times 2 + 4 \times 6))/2 = 134/2 = 67$, from Transaction Utility and External Utilities of Fig. 1

Definition 11 Let δ be a predefined "Minimum Average Utility Threshold". An itemset X is said to be a "High Average Utility Itemset" if it greater than the "Minimum Average Utility Threshold". For example, if the value of min_autil is 51.24 then the itemset the average utility of itemset AB, i.e., $aut\{AB\} = 67 > 51.24$ from Transaction table and External utilities of Fig. 1, so BC is a "High Average Utility Itemset" (*HAUI*).

Definition 12 "Frequent Pattern Mining" satisfies the "Downward Closure Property". But "High Average Utility Itemset Mining" does not, so this is a problem which is more difficult than mining "Frequent Itemsets". For example, $aut\{A\} = 32$ which is less than the minimum threshold, i.e., 51.24 whereas the average utility of $\{AB\}$, i.e., $aut\{AB\} = 67$, from Transaction table and External utilities of Fig. 1 which is greater than the minimum threshold. To overcome this Hong et al. proposed an $auub$ model to satisfy the "Downward Closure Property".

Definition 13 The maximum utility mu of a transaction is the largest value of a particular transaction Tr_q.

$$\max(Tr_q) = \max\{ut(it_{q1}, , Tr_q), ut(it_{q2}, Tr_q), ut(it_{qm}, Tr_q)\}, \tag{8}$$

where ut_{qj} is the utility of item it_{qj} in transaction Tr_q for $j = 1 \ldots m$

In Fig. 1 $\max(Tr_1) = 24$.

Definition 14 The "Average Utility Upper Bound" *auub* of an itemset *X* is the sum of all the maximum utilities of all transactions in a transaction database *DB* where *X* is found.

$$auub(X) = \sum_{Tr_q \in DB \bigwedge X \subseteq Tr_q} mut(Tr_q) \tag{9}$$

The itemset AB occurs in four transactions Tr_2, Tr_4, Tr_5, Tr_6 External Utilities of Fig. 1, so $auub(\{AB\}) = mu(Tr_2) + mu(Tr_4) + mu(Tr_5) + mu(Tr_6) = 36 + 21 + 32 + 40 = 129$.

Definition 15 An itemset *X* is called a "High Average Utility Upper bound" *HAUUB* itemset if its $auub(X) > min\_autil$, where min_autil is the "Minimum Average Utility Threshold".

For example, the itemset $\{AB\}$ is a *hauub* itemset because $auub(\{AB\}) = 129 > 51.25$ from Transaction Table and External Utilities.

4 High Average Utility Itemset Mining Algorithms

This section describes the various algorithms developed to mine "High Average Utility Patterns" with respect to various data structures used such as apriori, tree based, list based. This section also describes the various pruning techniques used to prune candidate itemsets.

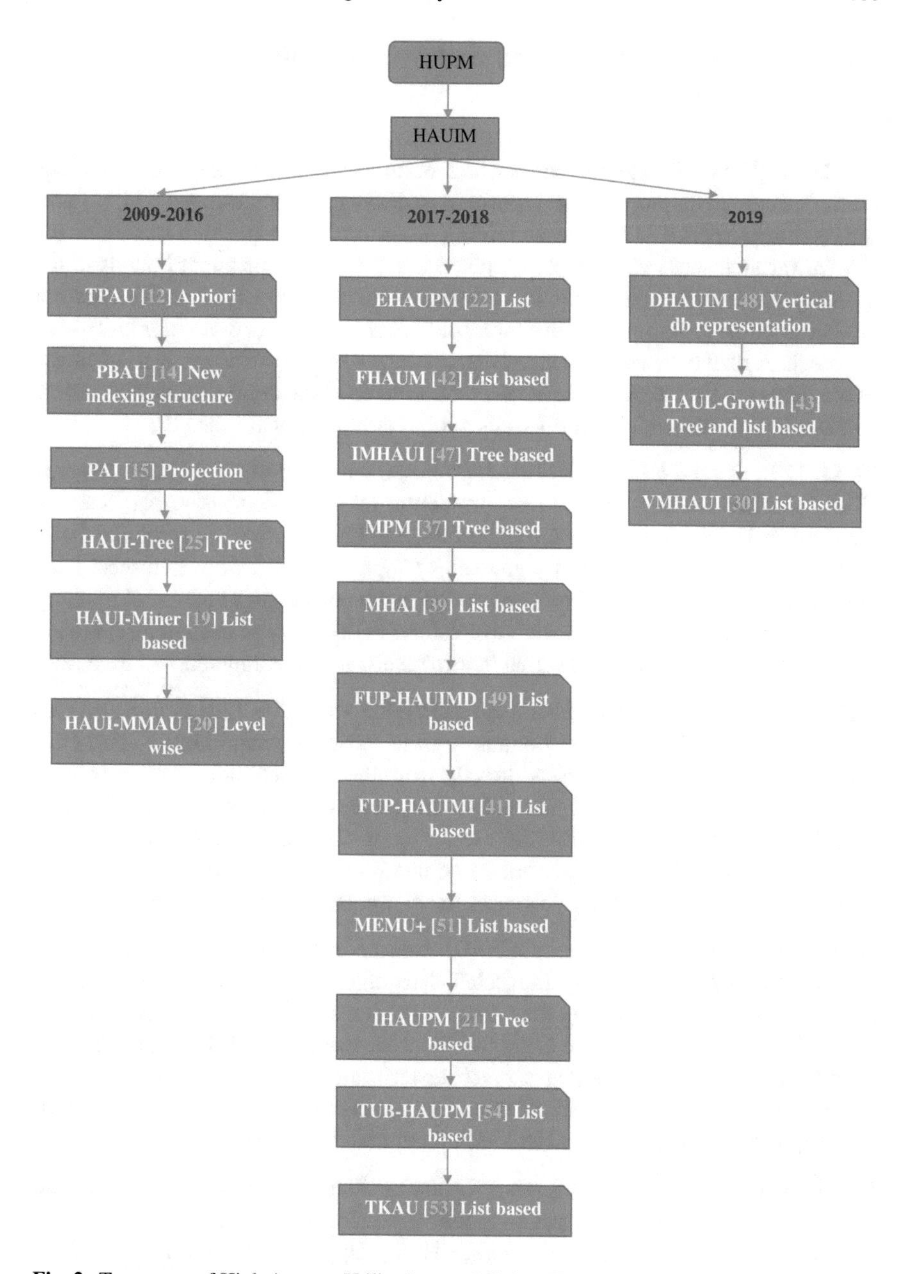

Fig. 2 Taxonomy of High Average Utility Itemset Mining (HAUIM) Algorithms

4.1 Algorithms with a Measure of Average Utility [2009–2016]

TPAU [10]: This was the first paper which outlined a new measure called "Average Utility". Hong et al. proposed an algorithm to discover "High Average Utility Itemsets". This paper is a "Two Phase" mining algorithm. The algorithm maintains the "Downward Closure Property" in the phase one by finding the upper bounds of the itemsets. This upper bounds are used to prune the itemsets whose utilities are not greater than the "minimum utility threshold". This process is done in a level-wise approach. A final database scan is required in the phase two to find the real "utilities" of the itemsets. Due to overestimation of the utilities of the itemsets many candidate itemsets are pruned which in turn saves a lot of computational time [10].

PBAU [29]: GUO-CHENG LAN et al. proposed a method to mine "High Average Utility Itemsets" which is based on projection of prefix databases. An indexing mechanism is used to identify the transactions to be processed from the database. The indexing mechanism directly generates "High Average Utility Itemsets" from the database. As the original database is not directly copied huge amount of memory is saved. The pruning strategy which overestimates the utility of each itemset is used to prune unpromising itemsets which in turn results in less computational time [29].

PAI [30]: PAI algorithm is an improved version of PBAU [29] approach. This approach uses improved upper bounds to mine "High Average Utility Itemsets". This algorithm mines the itemsets directly from the transactional database using a projection technique. There is no need to first generate the candidate itemsets and then mine the real average utility itemsets. Due to the use of projection technique the entire original database need not be copied thus reducing memory usage. The improved upper bound greatly decreases the computational time [30] (Fig. 2).

HAUI-Tree [31]: This paper uses the HAUI-Tree algorithm and a novel data structure to mine "High Average Utility Itemsets". This algorithm prunes itemsets that are unpromising by overestimating the value of the itemset. Only one database scan is required. Experiments are done using two datasets, namely, BMS-POS and CHESS. This method generates candidates much faster than other methods. For example using BMSPOS dataset at 0.8% of minimum average utility threshold HAUI-Tree [31] requires 0.66 min compared to 157 min required by PAI [30].

HAUI-Miner [25]: This paper uses the "Utility List" structure to mine "High Average Utility Itemsets". A database scan is required to calculate the "maximum utilities" of all transactions as well as the "Average Utility Upper Bound" (*auub*) of each 1-itemset. Database is scanned again to remove 1-High average utility itemsets whose utility is less than "Minimum Average Utility Threshold". The database is revised to arrange each item in ascending order of *auub*. Database is projected for each 1-*HAUUB* and the item whose *auub* is less than the "Minimum Average Utility Threshold" is pruned. HAUI-Miner [25] utilizes a "Utility List" to build lists for each 1-*HAUUB* itemset using the projected database. This algorithm discovers "High

Average Utility Itemsets" by using a depth first search approach. Items which are unpromising are pruned efficiently [25].

List data structure has been used extensively to mine "High Utility Itemsets". We provide a working example of HAUI-Miner by using Transaction Table and External Utilities of Fig. 1

Step 1: Calculate *tmut* for each transaction T_q.For example, the $ut(A) = (2 \times 6) = 12, ut(B) = (8 \times 3) = 24, ut(2 \times 8) = 16$. Therefore the $tmut(12, 24, 16) = 24$. If the "Minimum Average Utility Threshold" is 12%, so the value of "Minimum Average Utility Threshold" is $(.12) \times (52 + 44 + 38 + 37 + 74 + 102 + 64 + 16) = 51.24$.

Step 2: "High Utility Itemset Mining" does not satisfy the "Downward Closure Property". So the upper bound of each 1-itemset is found. For example, the $auub(A) = 36 + 21 + 32 + 32 = 129$.

Step 3: In the above example the value of C, i.e., $45 < 51.24$ so the item C can be pruned. The 1-itemsets can be arranged in ascending order $E \prec A \prec D \prec B$.

Step 4: The Revised Transaction Table as shown in Fig. 3 is constructed by pruning item c

Step 5: The projected sub-database is extracted from the Revised Transaction table as shown in Fig. 4

Step 6: Find the *auub* values of all 1-itemsets with respect to Projected Database of $\{E\}$ as shown in Fig. 4

Step 7: The *auub* value of $\{A\}$ is less than the min_autil, i.e., 51.24, therefore the item $\{A\}$ is pruned from the projected database of $\{E\}$

Step 8: Utility lists are built for 1-itemsets.

Step 9: 2-Item utility lists are constructed by the joining the 1-itemsets constructed in Step 8. If the $aut\{ED\} > min\_autil$ then the itemset is added to the "High Average Utility Itemset", else the *tmu* value is checked if it is greater than min_autil. In the below example the $aut\{ED\} = 70 + 64/2 = 67$, which is greater than min_autil which is 51.24, therefore the itemset $\{ED\}$ is added to the HAUI. In the below utility list of {EB}, $aut(\{EB\}) = 38 + 54/2 = 46 < 51.24$.

Step 10: 3-item "utility lists" are built by joining the 2-item "utility lists" (Figs. 5 and 6).

HAUI-MMAU [32]: This paper presents "High Average Itemset Mining" with "Multiple Minimum Average Utility Thresholds". HAUI-MMAU [32] is an algorithm which has two phases. The first step in this algorithm is to find the least value of all the minimum average utility thresholds and compare it with the "Average Utility Upper Bound" of the 1-itemset. The *auub* value can be calculated by adding the *tmu* values of all the transactions which has the 1-itemset. If the *auub* value of the 1-item is not greater than or equal to the the least minimum threshold value then the item is

Transaction Maximum Utilities

	A	B	C	D	E	*tmu*	*tu*
T_1	0	12	24	16	0	24	52
T_2	8	36	0	0	0	36	44
T_3	0	18	0	0	20	20	38
T_4	4	12	21	0	0	21	37
T_5	12	30	0	32	0	32	74
T_6	8	24	0	40	30	40	102
T_7	0	0	0	24	40	40	64
T_8	0	0	0	16	0	16	16

auub Values

Item	*auub*
A	129
B	173
C	45
D	152
E	100

Revised Transaction Table

T_{id}	E	A	D	B	*tmu*
T_1	-	-	16	12	16
T_2	-	8	-	36	36
T_3	20	-	-	18	20
T_4	-	4	-	12	12
T_5	-	12	32	30	32
T_6	30	8	40	24	40
T_7	40	-	24	-	40
T_8	-	-	16	-	16

Fig. 3 Transaction maximum utilities and revised transaction db

auub values for Revised db

Item	*auub*
E	100
A	120
D	144
B	156

Projected Database of {E}

T_{id}	E	A	D	B	*tmu*
T_3	20	-	-	18	20
T_6	30	8	40	24	40
T_7	40	-	24	-	40

auub values of 1-itemet

Item	*auub*
E	100
A	40
D	80
B	60

Fig. 4 auub values and projected database of $\{E\}$

Utility List of {E}

T_{id}	E	*tmu*
T_3	20	20
T_6	30	40
T_7	40	40

Utility List of {D}

T_{id}	D	*tmu*
T_6	40	40
T_7	24	40

Utility List of {B}

T_{id}	B	*tmu*
T_3	18	24
T_6	24	40

Fig. 5 Revised projected database of $\{E\}$ and utility lists

Utility List of {ED}

T_{id}	{ED}	*tmu*
T_6	70	40
T_7	64	40

Utility List of {EB}

T_{id}	{EB}	*tmu*
T_3	38	20
T_6	54	40

Utility List of {EDB}

T_{id}	{EDB}	*tmu*
T_6	94	40

Fig. 6 Utility lists of 2-itemsets and 3-itemsets

pruned else the item is not pruned. An "Improved Estimated Co-occurrence Pruning Strategy" (IEUCP) is used for pruning the search space [32]. The "Pruning before Calculation Strategy" (PBCS) is implemented at the beginning of the second phase to reduce the number of "High Average Utility Upper Bound Itemsets" [32].

Algorithms with a Measure of Average Utility [2017–2018]

EHAUPM [26]: Previous algorithms like HAUI-Miner [25] used *auub* upper bound to overestimate the utilities of itemsets. *auub* is based on transaction maximum utility. This algorithm provided two novel upper bounds to prune unpromising itemsets. These upper bounds prune more unpromising itemsets. An List structure called MAU was developed to keep the (a) utility values (iutil) (b) revised maximum utility (rmu) (c) remaining maximum utility (remu) [26]. Using rmu and remu two new upper bounds are developed, namely, lub and rtub [26]. The minimum of *auub, lub, rtub* is compared with the min_autil, if the upper bound is less than or equal to the min_autil the itemset is pruned, else its supersets are explored. Three techniques have also been developed to eliminate itemsets that are not promising.
All the pruning techniques in the EHAUPM [26] are explained clearly using Transaction Table and External Utilities of Fig. 1.

Step 1: Calculate "Average Utility Upper Bound" of each 1-item from the database DB.Transaction Table and External Utilities of Fig. 1 are used to find the "Average Utility Upper Bound" of each item.

Step 2: *auub* values of each 1-items have been calculated and illustrated in Fig. 3.

Step 3: From *auub* values of Fig. 3 $auub(C) = 45$ which is less than min_autil ie 51.24 therefore the item C is pruned.

Step 4: The transaction table is rearranged according to ascending order of *auub* as illustrated in the table called "Revised Transaction db" of Fig. 3 and its *auub* values are re-calculated as illustrated in Fig. 4.

Until now *mu* (maximum utility) has been used to calculate the upper bound, this algorithm proposes *rmu*(revised maximum utility) and *remu* (remaining maximum utility).

$$rmu(X, Tr_q) = \max\{ut(it_1, Tr_q), u(it_2, Tr_q), \ldots ut(it_{|Tr_q|}, Tr_q)\} \tag{10}$$

From Fig. 7 $rmu(E) = 20 + 40 + 40 = 100$

$$remu(X, Tr_q) = \max\{ut(it_m + 1), Tr_q), ut(it_m + 2), Tr_q), \ldots ut(it_N, Tr_q)\} \tag{11}$$

From Fig. 7 $remu(E) = 18 + 40 + 24 = 82$
Step 5: Now MAU Lists for all 1-items E, A, D, B are constructed.

MAU List for {E}

E	90	100	82
tid	*iutil*	*rmu*	*remu*
3	20	20	18
6	30	40	40
7	40	40	24

MAU List for {A}

A	32	120	120
tid	*iutil*	*rmu*	*remu*
2	8	36	36
4	4	12	12
5	12	32	32
6	8	40	40

MAU List for {D}

D	128	128	66
tid	*iutil*	*rmu*	*remu*
1	16	16	12
5	32	32	30
6	40	40	24
7	24	24	0
8	16	16	0

MAU List of {B}

B	32	120	120
tid	*iutil*	*rmu*	*remu*
1	12	12	0
2	36	36	0
3	18	18	0
4	12	12	0
5	30	30	0
6	24	24	0

Fig. 7 MAU lists

Step 6: Using *rmu*and *remu* in Step 5 two new upper bounds were proposed, namely, *lub* and *rtub*.

$$lub(X, Tr_q) = ut(X, Tr_q) + |X| \times remu(X, Tr_q)/|X| \tag{12}$$

$$lub(X) = \sum_{X \in Tr_q \bigwedge Tr_q \subseteq DB} lub(X, Tr_q) \tag{13}$$

$lub(AD, Tr_5) = 44 + |2| \times 30/2 = 52$
$lub(AD, Tr_6) = 48 + |2| \times 24/2 = 48$
$lub(AD) = 52 + 48 = 100.$

$$rtub(X) = \sum_{X \subseteq T\acute{r}_q \in D} rmu(X, T'r_q) \tag{14}$$

$rtub(AD) = rtub(AD, Tr_5) + rtub(AD, Tr_6) = 32 + 42 = 72. auub(AD) = 72$

The minimum of the three upper bounds *rmu*, *remu* and *auub* is considered as an upper bound for the item set. In the above example for the item set upper bound of $\{AD\} = min\{remu(AD), rmu(AD), aaub(AD)\} = 72$. As $72 > 51.24$ itemset AD is not pruned.

Step 7: Three more pruning strategies have been developed to prune unpromising itemsets early. The above is a running example of the working of EHAUPM [26] using Fig. 1.

FHAUM [33]: "High Average Utility Itemset Mining" is used to get a fairer representation of each utility in an itemset. In previous algorithms *auub* is used as an upper bound to overestimate the profit of an itemset. This algorithm proposes two novel upper bounds to prune unpromising itemsets

1. The second pruning technique *lubau*, "Lower Upper Bound Average Utility" is defined as follows:

$$lubau(X) = ut(it_1) + ut(it_2) + \cdots (it_m)\%m \tag{15}$$

2. The third pruning technique $tubau(X)$ is a better upper bound than $lubau(X)$, $tubau(X)$, "Tighter Upper Bound Average Utility" of an itemset X is defined as follows:

$$tubau(X) = ut(Y) + ut(it_m + 1)\%2, \tag{16}$$

where $X(= Y \bigcup it_{m+1})(= it_1 \bigcup i_2, \ldots it_m \bigcup it_{m+1})$

Two versions of FHAUM algorithms were compared, namely, D-FHAUPM (depth first search version) and B-FHAUPM (breadth first version). The datasets used were accidents, chess, foodmart, kosarak, pumsb, retail.

IMHAUI [34]: An algorithm to mine "High Average Utility Itmesets" algorithm in incremental databases was proposed. Many candidates were generated in previous algorithms due to multiple database scans. In IMHAUI [34] a tree-based data structure is used to store all the information of a given incremental database. Based on a path adjusting method this algorithm performs a restructuring process to preserve the compactness of the data structure. IMHAUI [34] outperforms other algorithms in terms of time taken, usage of memory and performance of the algorithm when huge number of transactions were added to the database.

MPM [35]: A new method called MPM was introduced which used the "High Average Utility Pattern Mining" approach and the "Damped Window" approach which has previously been used in "Frequent Pattern Mining" to extract "High Average Utility Itemsets" from data streams. New data structures, namely, DAT and TUL and a new pruning strategy had been implemented to make the mining process more effective [35]. Through this method users can get information about patterns which are useful for the discovery of symptoms which can be useful in linking them to certain diseases from the data. Various real datasets like "Breast Cancer Wisconsin", "Liver Disorders", "Heart Cleveland", and "Hepatitis" were used to compare the performance of MPM with other algorithms. MPM [35] performs better than UP-Growth [14], IMHAUI [36].

MHAI [37]: This paper introduces a new "High Average Utility Pattern Mining" algorithm called MHAI. This paper uses HAI-List to store all the information necessary to extract "High Average Utility Itemsets" without the use of candidate itemsets [37]. Database is scanned twice to build the HAI Lists of all 1-itemsets. By using a recursive approach all "High Average Utility Itemsets" can be found by constructing utility list of all items which are greater than one. A new strategy to prune itemsets based on maximum average utility is used to prune unpromising itemsets so that costly join operations of the utility lists can be avoided [37]. Four datasets, namely, Chess, Accidents, Retail, Chain-Store were used. MHAI [37] has better runtime than TPAU [10], PBAU [29], HAUI-Tree [31].

FUP-HAUIMD [38]: This paper presents an effective algorithm FUP-HAUIMD for updating the "High Average Utility Itemsets" when there is a deletion of transactions

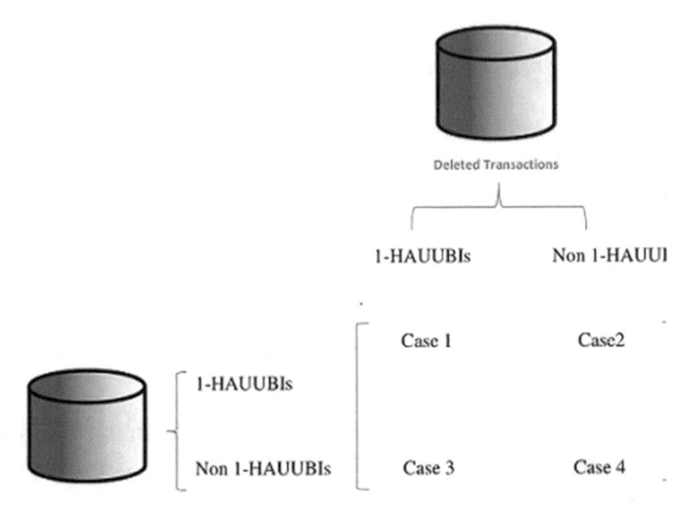

Fig. 8 MFUP model

from the database. For the algorithm to be correct and complete an MFUP(Modified Fast Updated) model was developed using FUP [39] model. This MFUP model decreases the number of times the database is scanned.There are four cases in MFUP concept they are

Case 1: An itemset could be a 1-"High Average Utility Upper Bound Itemset" or a non-"High Average Utility Upper Bound Itemset" if an itemset is a 1-"High Average Utility Itemset" in the primary database as well as in the transactions which have been deleted [38].

Case 2: An itemset is always a 1-"High Average Utility Upper bound itemset" in the database which has been updated if an itemset is 1-"High Average Utility Upper bound Itemset" in the primary database but a non 1-"High Average Utility Upper bound Itemset" in the transactions which have been deleted [38].

Case 3: The itemset is always a non 1-"High Average Utility Itemset" after the database has been updated if the itemset is a non 1-"High Average Utility Upper bound Itemset" in the primary database but a 1-"High Average Utility Upper Bound Itemset" in the transactions which have been deleted [38].

Case 4: In this case an other database scan is required to find the average utility of the itemsets and then the results have to be updated if an itemset is a non 1-"High Average Utility Upper Bound Itemset" in the primary database as well as the transactions which have been deleted [38] (Fig. 8).

Firstly, the transactions which have been deleted are scanned to obtain the Average Utility-Lists of 1-itemsets. Based on the MFUP model (each case is taken into consideration) the updated database utility lists of 1-HAUUBIs are obtained. min_autil of the updated database is given as $min\_autil = (TUT^{DB} - TUT^{db}) \times \delta$. If the average utility of the itemset is greater than min_autil then the itemset is added to the HAUI else the itemsets *tmut* value is used to check whether it is less than min_autil

[38]. If the condition is not satisfied the (k + 1) Average utility list is not constructed. A recursive depth first method is used to combine Average Utility lists of k-itemsets. If the updated Average Utility lists are greater than the min_autil it becomes a high average utility itemset [38]. The datasets used were accidents, kosarak, mushroom, retail, foodmart. FUP-HAUIMD [38] performs much better than HUI-miner [25] and EHAUPM [26].

FUP-HAUIMI [40]: This paper presents a way to discover high average utility itemsets without re-scanning the database repeatedly whenever the database is inserted with transactions. This paper uses the FUP method to develop an adapted FUP concept which categorizes the 1-HAUUBIs into four cases [40].
Case 1 An itemset is still a 1-"High Average Utility Upper bound itemset" after the database has been upadated if an itemset is a 1-"High Average Utility Upper bound Itemset" in the primary database and in the transaction that have been inserted [40].
Case 2 An itemset is said to be 1-"High Average Utility Upper bound itemset" or not a "High Average Utility Upper bound itemset" if an itemset is a 1-"High Average Utility Upper bound itemset" but not a 1-"High Average Utility Upper bound itemset" in the transactions that were inserted [40].
Case 3 An extra scan of the database is required to get the "Average Utility" of the itemset in the primary database and then update the database with the updated value of average utility if the itemset is a non-"High Average Utility Upper bound itemset" in the primary database but a 1-"High Average Utility Upper bound itemset" in the transaction that have been inserted [40].
Case 4 An itemset stays a non 1-"High Average Utility Upper bound itemset" after updating a database if an itemset is a non 1-"High Average Utility Upper bound itemset" in primary database and also in the transactions that have been inserted [40]. The first step is to build average utility lists of 1-itemsets from the inserted transactions. The second step is to merge the average utility lists of primary database and the transactions that have been inserted. By considering the merged transactions if the "Average Utility" of the itemset is not less than the "Minimum Average Utility", i.e., $min\_autil = (TUT^{DB} + TUT^{db}) \times \delta$ [40], then the itemset is said to be a "High Average Utility Itemset". An another database scan is required if the item does not exist in the primary database. The final step is to find the average utilities of the selected itemsets. For experiment purpose real datasets were used such as retail (sparse), kosarak(sparse), mushroom(dense), foodmart(sparse). FUP-HAUIMI [40] performances much better when compared to HAUI-Miner [25] and IHAUPM [17].
MEMU+ [41]: This algorithm mines "High Average Utility Itemsets" with "Multiple Minimum High Average Utility" Thresholds. List structure called MAU and sorted enumeration tree greatly reduces the search space [41]. An upper bound called *rtub* is used to remove unpromising itemsets. This upper bound is a better upper bound than the traditional auub upper bound. By using the rtub upper bound three new strategies to prune the search space have been proposed. [41]. The MEMU+ algorithm is one or two orders of magnitude faster than HAUIM-MMAU algorithm. When the runtimes of MEMU+ [41] and HAUI-MMAU algorithm are compared HAUI-MMAU algorithm is one or two order of magnitude slower than MEMU+ [41]

IHAUPM [42]: This paper presents an "Incremental High Average Utility Pattern Mining" algorithm to handle transaction insertion. The concept of FUP is changed to handle transaction insertion in dynamic databases. There are four cases which can occur when transactions are inserted.

Case 1 If an itemset X is a "High Average Utility Upper Bound Itemset" in the primary database as well as in the transactions that have been inserted then we can obtain [42]

$$auub(X)^U \geq (TUT^{DB} + TUT^{db}) \times \delta \tag{17}$$

Case 2 A couple of solutions can be inferred without scanning the original database D if an itemset is a "High Average Utility Upper bound Itemset" in the original database but a non-"High Average Utility Upper bound Itemset" in the transactions that have been inserted [42].

$$auub(X) \geq (TUT^{DB}) \times \delta if, (X) \geq TUT^{DB} \times \delta \tag{18}$$

$$auub(X) < (TUT^{DB}) \times \delta \, if, (X) < TUT^{DB} \times \delta \tag{19}$$

Case 3 When the primary database is scanned a couple of solutions can be obtained D if an itemset is non-"High Average Utility Upper bound Itemset" in the primary database but is a "High Average Utility Upper bound Itemset" is the transactions that have been inserted [42]:

$$auub(X)^{DB} + auub(X)^{db} = auub(X)^U \geq (TUT^{DB} + TUT^{db}) \times \delta \tag{20}$$

$$auub(X)^{DB} + auub(X)^{db} = auub(X)^U < (TUT^{DB} + TUT^{db}) \times \delta \tag{21}$$

Case 4 We can obtain the below solution for an itemset X if an itemset is a non-"High Average Utility Upper bound Itemset" in the primary database and the transactions that have been inserted [42].

$$auub(X)^U < (TUT^{DB} + TUT^{db}) \times \delta \tag{22}$$

HAUP-Tree stores all the 1-"High average utility upper bound itemsets". The database is not scanned except in case 3 when the itemset is a non-*HAUUBI* in the primary database but is a "High average utility upper bound itemsets" in the transactions that have been inserted [42] TPAU and PAI algorithms are one order of magnitude slower than IHAUPM [42] and the HAUI-Miner algorithm is 10 times slower than IHAUPM [42] algorithm. The HAUI-Tree algorithm is 10 times slower than IHAUPM [42]. The memory usage of PAI, TPAU, HAUI-Tree, HAUP growth algorithms are 7–8 times greater than IHAUPM [42].

TUB-HAUPM [43]: This paper proposes two new upper bounds to eliminate itemsets that are not promising and to keep the "Downward Closure property". The upper bounds *krtmuub* "Top k-revised transaction maximum utility upper bound" and *mfuub*(maximum following utility upper bound) are used to greatly reduce the search space. In this algorithm a recursive processing order prunes unnecessary items [43]. To reduce the evaluation time, transaction rival pruning strategy was formulated for giving a tight limitation for all the candidates [43]. In dense datasets like accidents, chess, and mushroom, the proposed upper bounds are precise [43]. The proposed algorithm eliminates huge number of branches in the search tree [43].The proposed upper bounds do not work well for sparse datasets. This algorithm works well for dense datasets with respect to runtime. TUB-HAUPM [43] greatly reduces the computational time when compared to EHAUPM [26].

TKAU [44]: This paper the problem of mining "Top k-High Average Utility Itemsets" is studied where k is the number of "High Average Utility Itemsets" that have to be mined. As "Top k High Average Utility Itemsets" need to be mined, an efficient AUO-List based algorithm TKAU [44] has been formulated to mine "Top k-High Average Utility Itemsets" without setting the "Minimum Average Utility Thresholds". This paper introduces five strategies namely, *EMUP* (Estimated Maximum Utility Pruning Strategy), *EA* (Early Abandoning), *RIU*, *CAD*, *EPBF* to eliminate itemsets that are unpromising from the search space and raise the "Minimum Average Utility Thresholds" in a more efficient manner [44].

Algorithms with a Measure of Average utility [2019]

DHAUIM [45]: This paper calculates values of utilities with a quantitative database which is vertical. This paper has proposed four upper bounds that are tighter than other upper bounds, namely, $\overline{aub_1}$, $\overline{aub}$, $\overline{iaub}$, $\overline{laub}$, and a novel structure to evaluate upper bounds in terms of pruning effects and three strategies to prune unpromising itemsets early [45]. A novel Tree structure called IDUL tree structure was used to calculate the "Average Utilities" and upper bounds of item sets using a process which is recursive in nature [45]. Experiments performed show that DHAUIM outperforms other "High Average Utility Itemset Mining" algorithms in terms of computational time. HUI-Miner [25], EHAUP [26], FHAUM [33] algorithms are one or two orders of magnitude slower than dHAUIM [45].

HAUL-Growth [46]: HAUL-Growth algorithms is used to mine "High Average Utility Itemsets". It uses a pattern growth approach called HAUL-Growth algorithm which proposed four new upper bounds called *eub*, *teub*, *bteub*, *max-reubk*. This paper also proposed two novel data structures called HAUL-Tree and IL [46]. The datasets used were chess, mushroom, pumsb, and bms.

VMHAUI [47]: This paper introduced four new vertical upper bounds, namely, $vmsub_1$, $vmsub$, $vmaub$, $vtopmaub$ which are much better than the upper bounds used by other state of the art "High Average Utility Itemset Mining" algorithms [47]. The upper bounds which have been proposed in this paper are more efficient than the existing upper bounds in terms of values and search space pruning. This paper also proposed two new pruning and tightening strategies based on new upper bounds to reduce unpromising itemsets early and a new list structure to compute average utilities and vertical WUBs of itemsets quickly based on reduced dataset and TPUT [47].

5 Features of High Average Utility Pattern Mining Algorithms

The Tables 2, 3, and 4 lists the features of varioius *HAUIM* algorithms with respect to 1. Related Work 2. Approach of the algorithm, i.e., the data structure used 3. No of phases used to mine HAUIS 4. Upper bounds used to prune unpromising itemsets and limitations of each algorithm. The real and synthetic datasets used in each algorithm are described in Tables 5, 6 and 7.

6 Research Opportunities and Challenges

This section describes various research opportunities using high average utility pattern mining. Research in *HAUIM* is open in the area of developing algorithms which perform better than the previous algorithms. Research is also open in the area of developing applications which use High average utility pattern mining.

Developing better algorithms Much work has been done in developing algorithms to mine "High Average Utility Itemsets". Some algorithms take huge amounts of time while some algorithms require huge amounts of memory. Tighter upper bounds can be developed so that unpromising itemsets can be pruned. Better pruning techniques can be developed so that the search space can be reduced. Research is open in *HAUIM* with respect to time consumed, memory required, number of candidates generated and scalability.

Application-Oriented Research Most of the research done in the field of *HAUIM* is related to the efficiency of the algorithms. Little work is done with respect to the application of *HAUIM*. Applications can be developed so that High Utility Patterns can be extracted from various social networks like Facebook, twitter, Instagram, etc.

Application-Oriented Research Most of the pattern mining algorithms use data that does not change with time. Different algorithms can be developed which can deal with complex data such as spatiotemporal data, Text, and streaming data. Most of the data that has been used is static data. In real-life applications data is dynamic in

Table 2 Features of HAUIM algorithms [2009–2016]

Algorithm and year	Related work	Approach	No of phases	Upper bounds	Limitations
TPAU [10] [2009]	TP [48]	Apriori based	Two	*auub*	Run time is very high due to multiple database scans
PBAU [29] [2012]	TPAU [10] [2009]	Depth first, new indexing structure	Two	*auub*, index based pruning	Runtime and memory usage can still be reduced using different data structures
PAI [30] [2012]	PBAU [29]	Projection based	Two	Improved upper bound	As databases are projected, memory usage is more
HAUI-Tree [31] [2014]	HUC-Tree [49], HUP-Tree [29]	Tree based	Two	*auub*	Huge amount of memory is used to store the tree structure
HAUI-Miner [25] [2016]	FHM [50], HAUI-Tree [31], PAI [30], PBAU [29], TPAU [10]	List based	One	*auub*	Huge number of costly join operations on lists are performed
HAUI-MMAU [32] [2016]	–	Level wise	Two	*auub*	Huge numbers of candidates are generated due to level-wise search

nature. For example data related to stocks are always dynamic. Efficient algorithms can be developed to mine "High Average Utility Itemsets" from dynamic data.

Developing various data structures Various data structures have been used like tree based, list based to mine "High Average Utility Itemsets". To make the algorithms more efficient various new data structures can be used such as graph data structures and modified list data structures.

Using distributive approach Huge amount of work has been done to mine "High Average Utility Itemsets" from transactional databases. A parallel approach can be used to mine HAUIs where runtime of the algorithms can be increased exponentially without compromising on the mining of HAUIs.

Table 3 Features of HAUIM algorithms [2017–2018]

Algorithm and year	Related work	Approach	No of phases	Upper bounds	Limitations
EHAUPM [26] [2017]	HAUI-Miner [25]	List based	One	*auub*, *lub*, *rtub*	Better pruning techniques can be developed
FHAUM [33] [2017]	HAUI-Miner [25]	List based	One	*auub*, *lubau*, *tubau*	Pruning techniques can be improved
IMHAUI [34] [2017]	TP [48], IHUP [14], UP-Growth [47], MU-Growth [51], HUPID [36]	Tree based	Two	*auub*	The tree structure occupies huge amounts of memory
MPM [35] [2017]	HAUI-Tree [31] IHUP [14] UP-Growth [52] MU-Growth [51], HUIPD [36]	Tree based	Two	*auub*	Memory requirement is huge as tree-based data structure is used
MHAI [37] [2017]	HAUI-Miner [25], FHM [50]	List based	one	*auub*	Costly join operations still need to be performed
FUP-HAUIMD [38] [2018]	FUP [39] Hong TP et al. [19]	List based	One	*auub*	Better pruning techniques can be developed to avoid costly join operations
FUP-HAUIMI [40] [2018]	FUP [39] Hong TP et al. [19]	List based	one	*auub*	Better pruning techniques can be developed to avoid costly join operations
MEMU+ [41] [2018]	HAUI-MMAU [32] HAUI-Miner [25]	List based	One	*auub*, *rtub*	Tighter upper bounds can be designed to further prune unpromising itemsets
IHAUPM [42] [2018]	FUP [39], Hong TP et al. [19], FUP-HAUIMD [38], FUP-HAUIMI [40]	Tree based	Two	*auub*	Huge amount of memory is requires to store the tree structure
TUB-HAUPM [43] [2018]	EHAUPM [26] FHAUM [33] HAUI-Miner [25]	List based	One	*auub*, *mfuub*, *krtmuub*	Efficient pruning strategies can be developed to prune unpromising itemsets
TKAU [44] [2018]	HAUI-Tree [31], HAUI-Miner [25], MHAI [37]	List based	One	*EMUP*, *EA*, *RIU*, *CAD*, *EPBF*	Memory usage can be improved

Table 4 Features HAUIM algorithms [2019]

Algorithm and year	Related work	Approach	No of phases	Upper bounds	Limitations
DHAUIM [45] [2019]	FHAUM [33] EHAUPM [26]	Vertical database representation based	One	$\overline{aub_1}, \overline{aub}, \overline{iaub}, \overline{laub}$	Tighter upper bounds can be designed
HAUL-Growth [46] [2019]	TUB-HAUPM [43] dHAUIM [45]	Tree based and list based	–	eub, teub, bteub, max-reubk, Tighter upper bounds can be designed to prune unpromising itemsets	
VMHAUI [47] [2019]	MHAI [37], TUB-HAUPM [43], dHAUIM [45]	List based	One	$\overline{vmsub}, \overline{msub}, \overline{vmaub}, \overline{vtopmaub}$	New data structures can be designed to improve the runtime as well as memory consumption

Table 5 Datasets used for algorithms [2009–2016]

Algorithm and year	Real datasets, thresholds shown for runtime	Synthetic datasets	Compared with
TPAU [10] [2009]	Major grocery store in America [0.008–0.012]	–	TP [48]
PBAU [29] [2012]	BMS-POS [0.20–1.00], chess [0.60–0.70]	T104N4K [0.1]	TPAU [10]
PAI [30] [2012]	BMS-POS [0.20–1.00], chess [0.60–0.70]	T104N4K [0.1]	PBAU [29]
HAUI-Tree [31] [2014]	BMS-POS [0.60–2.00], chess [4.0–6.0]	–	PAI [30]
HAUI-Miner [25] [2016]	Chess [4.8–6.4], mushroom [3.4–4.2], Sign [1.6–2.4], retail [0.01–0.05], kosarak [0.2–0.36]	T10I4D100K [0.3–0.7]	PAI [30], HAUI-Tree [31]
HAUI-MMAU [32] [2016]	Foodmart [Î² – (1–10) – (40–50), GLMAU = 1K Retail [Î² – (1–200) – (800–1000), GLMAU = 20K chess [Î² – (200–400) – (1000–1200), GLMAU = 168K	T40I10D100K [Î² – (6k–7k) – (10k–11k) , GLMAU = 2500K	$HAUI-MMAU_I EUCP$, $HAUI-MMAU\_PBCS$

Table 6 Datasets used for algorithms [2017–2018]

Algorithm and year	Real Datasets,Thresholds shown for runtime	Synthetic datasets	Compared with
EHAUPM [26] [2017]	Accidents [3.5–4.3] Chess [3–3.8] Kosarak [0.08–0.092] Mushroom [0.3–0.38] Retail [1.4e–3–2.2e–3]	T40I10D100K [0.2–0.28]	HAUI-Miner [25]
FHAUM [33] [2017]	Accidents [3.3–3.8] Chess [2.9–4.5] Foodmart [0.006–0.01] Kosarak [0.278–0.521] Pumsb [4.31–4.5] Retail [0.02–0.06]	T10I4N4KD|X|K [0.28%]	EHAUPM [25]
IMHAUI [34] [2017]	Chain-store [0.010–0.030] Foodmart [0.01–0.05] Mushroom [1.00–5.00] Breast-cancer Wisconsin [0.05–0.09]	T10I4D100K	ITPAU [53], HUPID [36]
MPM [35] [2017]	Breast-cancer Wisconsin [0.05–0.09] Liver disorders [0.05–0.09] Heart Cleveland [0.01–0.05]Hepatitis [1.6–2.0] Accidents [0.60–1.00] Connect [0.01–0.05]	–	UP-Growth [52] IMHAUI [51]
MHAI [37] [2017]	Chess [5.00–6.00] Accidents [2.50–3.00] Retail [0.010–0.020] Chain-store [0.0050–0.50]	–	HAUI-Tree [31] PBAU [29] TPAU [10]
FUP-HAUIMD [38] [2018]	Deletion ratio = 1% Accidents [5.5–5.9] Foodmart [5–9] Kosarak [0.5–0.9] Mushroom [2.4–2.8] Retail [0.02–0.024]	T10I4N4KD100k [0.02–0.028]	HAUI-Miner [25] EHAUPM [26]
FUP-HAUIMI [40] [2018]	Insertion ratio = 1% Retail [0.02–0.06] Kosarak [2–2.4] Mushroom [2–6] Foodmart [0.02–0.06] Insertionratio = 1%	T10I4D100K [0.02–0.06], T10I4N4KD100K [0.05–0.07]	HAUI-Miner [25] IHAUPM [42]
MEMU+ [41] [2018]	Runtimes for fixed glmau value and various glmau values: Kosarak, Chess, Mushroom, accidents, retail	T10I4D100K9	HAUM-MMAU [32]
IHAUPM [42] [2018]	Foodmart Kosarak mushroom	T10I4D100k, T40I10D100K	HAUI-Miner [25] HAUI-Tree [31] PAI [30] TPAU [10]
TUB-HAUPM [43] [2018]	Retail [0.0014–0.0022] Kosarak [0.006–0.012] Accidents [0.036–0.042] Chess [0.030–0.038] Mushroom [0.020–0.028]	–	EHAUPM [26]
TKAU [44] [2018]	Value of k [1–1000] Mushroom, Chess, Retail, Chain store, Kosarak, foodmart	T10I4DXK	HAUI-Tree [31] HAUI-Miner [25] MHAI [37]

Table 7 Datasets used for algorithms [2019]

Algorithm and year	Real datasets, Thresholds shown for runtime	Synthetic datasets	Compared with
dHAUIM [45] [2019]	Mushroom [0.3–0.56] Online retail [0.18–0.3] Chess [3–5.6]	T11I6N30D100K [1–2.2] T15I9N100D100K [0.42–0.9] T20I9N50D100K [1.2–2.3]	EHAUPM [26] MHAI [37] FHAUM [33] HAUI-Miner [25]
HAUL-Growth [46] [2019]	Chess [2.8–4.00] Mushroom [0.05–0.2] Accidents [2.75–4] Pumsb [2.25–3.0] BMS [0.105–0.132]	T20I9N50D100 [0.4–1.4]	TUB-HAUPM [43] dHAUPM [45]
VMHAUI [47] [2019]	Connect [3.5–5.5] Mushroom [0.05–2] Online-Retail [0.12–0.3]. BMS [0.113–0.15] Chess [2.60–5.0]	T11I4N100D40K [0.05–1.0]	MHAI [37], TUB-HAUPM [43], dHAUIM [45]

7 Conclusion

"High Utility Itemset Mining" has various applications in different fields. This paper presents the problem of "High Average Utility Itemset Mining", various notations, preliminaries and definitions. A detailed survey of all the "High Utility Itemset Mining" algorithms with a measure of average utility was presented in this paper. It also presents a detailed theoretical survey of all "High Utility Itemset Mining Algorithms". Finally the paper also presents the various challenges and Research opportunities related to "High Utility Itemset Mining" with a measure of average utility.

Conflict of interest

The authors declare that they have no conflict of interest.

References

1. Chen M-S et al (1996) Data mining: an overview from a database perspective. IEEE Trans Knowl Data Eng 8(6):866–883. https://doi.org/10.1109/69.553155
2. Data mining: concepts and techniques, 3rd edn. DATA Min. 560

3. Agrawal R et al (1993) Mining association rules between sets of items in large databases. In: Acm sigmod record, pp 207–216. ACM
4. Barber B, Hamilton HJ (2000) Algorithms for mining share frequent itemsets containing infrequent subsets. In: European conference on principles of data mining and knowledge discovery, pp 316–324. Springer
5. Barber B, Hamilton HJ (2003) Extracting share frequent itemsets with infrequent subsets. Data Min Knowl Discov 7(2):153–185
6. Berzal F et al (2001) TBAR: an efficient method for association rule mining in relational databases. Data Knowl Eng 37(1):47–64. https://doi.org/10.1016/S0169-023X(00)00055-0
7. Chang C-C, Lin C-Y (2005) Perfect hashing schemes for mining association rules. Comput J 48(2):168–179. https://doi.org/10.1093/comjnl/bxh074
8. Erwin A et al (2008) Efficient mining of high utility itemsets from large datasets. In: Pacific-Asia conference on knowledge discovery and data mining, pp 554–561. Springer
9. Dynamic itemset counting and implication rules for market basket data. https://dl.acm.org/citation.cfm?id=253325. Last accessed 23 December 2019
10. Hong T-P et al (2011) Effective utility mining with the measure of average utility. Expert Syst App 38(7):8259–8265
11. Raymond Chan et al (2003) Mining high utility itemsets. In: Third IEEE international conference on data mining, pp. 19–26. https://doi.org/10.1109/ICDM.2003.1250893
12. Yao H et al (2006) A unified framework for utility-based measures for mining itemsets. In: Proceedings of ACM SIGKDD 2nd workshop on utility-based data mining. Citeseer, pp 28–37
13. Yi-Dong Shen et al (2002) Objective-oriented utility-based association mining. In: Proceedings of 2002 IEEE international conference on data mining, 2002. pp 426–433. https://doi.org/10.1109/ICDM.2002.1183938
14. Ahmed CF et al (2009) Efficient tree structures for high utility pattern mining in incremental databases. IEEE Trans Knowl Data Eng 21(12):1708–1721
15. Shie BE et al (2010) Online mining of temporal maximal utility itemsets from data streams. In: Proceedings of the 2010 ACM symposium on applied computing. ACM, pp 1622–1626
16. Li Y-C et al (2005) Direct candidates generation: a novel algorithm for discovering complete share-frequent itemsets. In: International conference on fuzzy systems and knowledge discovery. Springer, pp 551–560
17. Li Y-C et al (2008) Isolated items discarding strategy for discovering high utility itemsets. Data Knowl Eng 64(1):198–217
18. Yao H, Hamilton HJ (2006) Mining itemset utilities from transaction databases. Data Knowl Eng 59(3):603–626
19. Hong T-P et al (2008) Incrementally fast updated frequent pattern trees. Expert Syst Appl 34(4):2424–2435
20. Lee D et al (2013) Utility-based association rule mining: a marketing solution for cross-selling. Expert Syst Appl 40(7):2715–2725
21. Ahmed CF et al (2010) A novel approach for mining high-utility sequential patterns in sequence databases. ETRI J 32(5):676–686
22. Yin J et al (2012) USpan: an efficient algorithm for mining high utility sequential patterns. In: Proceedings of the 18th ACM SIGKDD international conference on knowledge discovery and data mining. ACM, pp 660–668
23. Lin Y-F et al (2015) Discovering utility-based episode rules in complex event sequences. Expert Syst Appl 42(12):5303–5314
24. Wu C-W et al (2013) Mining high utility episodes in complex event sequences. In: Proceedings of the 19th ACM SIGKDD international conference on Knowledge discovery and data mining. ACM, pp 536–544
25. Lin JC-W et al (2016) An efficient algorithm to mine high average-utility itemsets. Adv Eng Inform 30(2):233–243
26. Lin JC-W et al (2017) EHAUPM: efficient high average-utility pattern mining with tighter upper bounds. IEEE Access 5:12927–12940. https://doi.org/10.1109/ACCESS.2017.2717438

27. Han J et al (2004) Mining frequent patterns without candidate generation: a frequent-pattern tree approach. Data Min Knowl Discov 8(1):53–87. https://doi.org/10.1023/B:DAMI.0000005258.31418.83
28. Peng AY et al (2017) mHUIMiner: a fast high utility itemset mining algorithm for sparse datasets. In: Pacific-Asia conference on knowledge discovery and data mining. Springer, pp 196–207
29. Lan G-C et al (2012) A projection-based approach for discovering high average-utility itemsets. J Inf Sci Eng 28(1):193–209
30. Lan G-C et al (2012) Efficiently mining high average-utility itemsets with an improved upper-bound strategy. Int J Inf Technol Decis Mak 11(05):1009–1030
31. Lu T et al (2015) A new method for mining high average utility itemsets. In: IFIP international conference on computer information systems and industrial management. Springer, pp 33–42
32. Lin JC-W et al (2016) Efficient mining of high-utility itemsets using multiple minimum utility thresholds. Knowl-Based Syst 113, 100–115 (2016). https://doi.org/10.1016/j.knosys.2016.09.013
33. A fast algorithm for mining high average-utility itemsets. SpringerLink. https://doi.org/10.1007/s10489-017-0896-1. Last accessed 23 December 2019
34. Efficient algorithm for mining high average-utility itemsets in incremental transaction databases SpringerLink. https://link.springer.com/article/10.1007/s10489-016-0890-z. Last accessed 23 December 2019
35. Yun U et al (2018) Damped window based high average utility pattern mining over data streams. Knowl-Based Syst 144, 188–205 (2018). https://doi.org/10.1016/j.knosys.2017.12.029
36. Yun U, Ryang H (2015) Incremental high utility pattern mining with static and dynamic databases. Appl Intell 42(2):323–352
37. Yun U, Kim D (2017) Mining of high average-utility itemsets using novel list structure and pruning strategy. Future Gener Comput Syst 68:346–360. https://doi.org/10.1016/j.future.2016.10.027
38. Maintenance algorithm for high average-utility itemsets with transaction deletion. SpringerLink. https://link.springer.com/article/10.1007/s10489-018-1180-8. Last accessed 23 December 2019
39. Maintenance of discovered association rules in large databases: an incremental updating technique. IEEE Conference Publication. https://ieeexplore.ieee.org/abstract/document/49209. Last accessed 23 December 2019
40. Zhang B et al (2018) Maintenance of discovered high average-utility itemsets in dynamic databases. Appl Sci 8(5), 769 (2018). https://doi.org/10.3390/app8050769
41. MEMU: More Efficient Algorithm to Mine High Average-Utility patterns with multiple minimum average-utility thresholds. IEEE J Mag. https://ieeexplore.ieee.org/abstract/document/8279384. Last accessed 23 December 2019
42. Lin JC-W et al (2018) Efficiently updating the discovered high average-utility itemsets with transaction insertion. Eng Appl Artif Intell 72:136–149. https://doi.org/10.1016/j.engappai.2018.03.021
43. TUB-HAUPM: tighter upper bound for mining high average-utility patterns. IEEE J Mag. https://ieeexplore.ieee.org/abstract/document/8329566. Last accessed 23 December 2019
44. Top-k high average-utility itemsets mining with effective pruning strategies | SpringerLink. https://link.springer.com/article/10.1007/s10489-018-1155-9. Last accessed 23 December 2019
45. Efficient vertical mining of high average-utility itemsets based on novel upper-bounds. IEEE J Mag. https://ieeexplore.ieee.org/abstract/document/8355591. Last accessed 23 December 2019
46. An efficient tree-based algorithm for mining high average-utility itemset. IEEE J Mag https://ieeexplore.ieee.org/abstract/document/8861288. Last accessed 23 December 2019
47. Truong T et al (2019) Efficient high average-utility itemset mining using novel vertical weak upper-bounds. Knowl-Based Syst 183, 104847 (2019). https://doi.org/10.1016/j.knosys.2019.07.018

48. Liu Y et al (2005) A two-phase algorithm for fast discovery of high utility itemsets. In: Pacific-Asia conference on knowledge discovery and data mining. Springer, pp 689–695
49. Ahmed CF et al (2011) HUC-Prune: an efficient candidate pruning technique to mine high utility patterns. Appl Intell 34(2):181–198
50. Fournier-Viger P et al (2014) FHM: faster high-utility itemset mining using estimated utility co-occurrence pruning. In: Andreasen T et al (eds) Foundations of intelligent systems. Springer International Publishing, Cham, pp 83–92
51. Yun U et al (2014) High utility itemset mining with techniques for reducing overestimated utilities and pruning candidates. Expert Syst Appl 41(8):3861–3878
52. Tseng VS et al (2012) Efficient algorithms for mining high utility itemsets from transactional databases. IEEE Trans Knowl Data Eng 25(8):1772–1786
53. an incremental mining algorithm for high average-utility itemsets. IEEE Conference Publication. https://ieeexplore.ieee.org/abstract/document/5381569. Last accessed 23 December 2019

Course Recommendation System for Post-graduate (Masters in Science) Aspirants

Shivani Pal, Anjali Shahi, Shubham Rai, and Reshma Gulwani

Abstract The choice of higher education program plays a major role in shaping up one's career. Hence, it is worth investing time in gathering the right information to make an informed decision. For this cause, to the extent of our knowledge there are no dedicated tools. Students tend to make decision either based on the peer review without giving much thought on it. The system aims to consider student's interest as sole parameter to recommend the course which will be best to pursue. The interest of the student will be gauged on subjects which the student has already undertaken as undergraduate courses. The proposed system provides the user with a Web application where the user can make its profile and input his/her interest based on the parameters. K-Nearest Neighbours Classification Algorithm is used to select the best-suited courses based on Euclidean distance. The proposed recommendation system is a content-based Recommendation System since it recommends items (courses in this case) based on content of items (the undergraduate courses) and user profile (the rating of courses based on interest). Our proposed system implements a recommendation prototype with a focus on Computer Science and Information Technology disciplines as the chosen field of knowledge.

Keywords Recommendation system · Content-based · Courses · Higher education

---

Supported by Ramrao Adik Institute of Technology.

S. Pal · A. Shahi (✉) · S. Rai · R. Gulwani
Ramrao Adik Institute of Technology, Navi Mumbai, Maharashtra, India
e-mail: anjalishahi848@gmail.com

S. Pal
e-mail: shiv040998@gmail.com

S. Rai
e-mail: srai28197@gmail.com

R. Gulwani
e-mail: reshma.gulwani@rait.ac.in

N. Chaki et al. (eds.), *Proceedings of International Conference on Computational Intelligence and Data Engineering*, Lecture Notes on Data Engineering and Communications Technologies 56, https://doi.org/10.1007/978-981-15-8767-2_31

1 Introduction

Recommendation systems have become a quotidian part of our lives, whether it is e-commerce websites suggesting products based on previous purchases or entertainment portals recommending highly rated shows. With a surge in information available on the Internet, searching for and making decisions has become difficult. This has resulted in recommendation systems becoming more of a necessity than a luxury. Recommendation System's main task is to seek to estimate and predict user content preference whether it be movies, games, music, books or course.

When opting for higher education in countries like USA, Canada, etc. selection of the right course is the most crucial step. Course selection at the Post-graduate or Master's level can be overwhelming due to the sheer bulk of courses offered by various universities in the Computer Science and Information Technology field itself. Students tend to choose courses in an aleatory manner without giving much thought. This may have ruinous consequences. The consequences may be that the student drops out of the university, under performing or just not happy with the course. Thus students should select courses that match their interest of study. Students get bombed with all kinds of suggestions about what they should study next, but they don't exactly know what to pursue. Students should select the courses that explicitly align with their attributes and attitude.

So the proposed Recommendation System aims to consider the interest of the student in particular to predict the course which best matches with the student's passion.

The proposed recommendation system will make use of content-based filtering. Content-based filtering recommends courses that have the characteristics that the student has given higher preference and rating. Context-based filtering works by suggesting course (item) by comparing the features with the user profile. The machine learning algorithm that will be used is K-Nearest Neighbour algorithm which is a supervised learning algorithm for classification.

This paper proposes predictions for student's course selection based on their preference and interest. The target audience for this study are engineering graduates who want to pursue Masters in Computer Science and IT related fields.

2 Recommender Systems

With the availability of a wide array of choices user's suffer with problems of selection Recommendation systems are dynamic information filtering systems that filter vital information from the huge volume of data available according to user's interests, preferences and observation [1, 2].
Recommendation system can be categorized into 3 types:

1. Collaborative filtering system
2. Content-based filtering system
3. Hybrid filtering system.

2.1 Collaborative Filtering

It is the most popularly used recommendation technique and is based completely on user's historical preference on a set of items. The conception that the users who agreed in their judgement of various items in the past are most likely to agree further again in the future is the ground idea for this system [3]. It can be further categorized into:

- Item-based CF—In this technique if the same user rates two items in a similar way, the items are considered similar. Based on this, prediction is made for a target user by evaluating weighted average of ratings on 'n' similar items from the same user.
- User-based CF—In this technique, the similarities between the target user and the other users is calculated. The top 'n' similar users are selected and the weighted average of ratings from 'n' users with similarities are taken as weights.

2.2 Content-Based Filtering

This system recommends items based on estimate of similarities between the user profile and the content of the items. Each item is represented as a set of features/attributes. The user profile is also represented with the same features or attribute chosen up by analysing the content of items. The engine becomes better in terms of accuracy as the user provides more inputs that is by taking actions on those recommendations. In this type of recommendation, recommendations acquired are specific to the user since the model does not involve use of any data from other users.

2.3 Hybrid Filtering

This system combines the advantages of both collaborative and content-based filtering. A hybrid approach combines the two types of techniques. The system combines collaborative and content-based filtering by making predictions based on a weighted average of the content-based recommendation and the collaborative recommendation. The highest recommendation is the one which receives highest ranks which is a measure for the weight.

Our proposed course recommendation system makes use of content-based filtering. The ratings given by students to individual undergraduate (UG) courses help create the students' profile. Each of the undergraduate courses are assigned descriptors and a large data-set is created. The rating of each of the UG courses is compared with the student profile and based on that the relevant Master's course is recommended.

3 Related Work

- Rao et al. [4]:
  Describes the progression of the unsupervised models that exploit the similarity of the members with the content of the course to supervised models that exploit the similarity of the members with the content of the course to supervised models and finally hyper-personalized mixed effect models with several million coefficients. The purpose of this course recommendation system is to predict the relevance of a course for a member. The actual classified list of recommendations for each member is calculated offline and stored in an online key-value store, accessed at the time of the request. This approach is in contrast to online rating, where the ranked list is calculated at the time of the request.
- Spanakis and Elatia [5]:
  The paper employs use of graduating attributes in the recommender system that uses input taken from students and recommend courses to them on the basis of their self-assessment, where graduating attributes are the qualities, skills and understandings that the student should develop during their time with the institution.
- Lin et al. [6]:
  In this paper sparse linear method is employed (SLIM). The technique here compares the accuracy of course recommendation with the ground facts that they collected using expert discussions. This paper too needs access to student registration data.
- Ng and Linn [7]:
  This paper overcomes the drawbacks posed by paper [4] and [7] by not relying on data from the university and recommends course for any student attending any university. It uses tag analysis and makes predictions based on course and professor ratings. It is mostly used in recommending courses beyond the student's area of study.

4 Algorithms

4.1 K-Nearest Neighbours

K-nearest neighbours (KNN) is a robust and versatile classifier. Despite its simplicity KNN can outperform most powerful classifiers. KNN classifies the new data based on similarity measure.

The logic behind which KNN works is to traverse through the neighbourhood, take up the test data point to be similar to the neighbours and extract the output. In K-Nearest Neighbours, we observe for 'k' neighbours and with respect to that prediction is made. K nearest neighbour essentially boils down to using similarity to define distance measure between two data points.

So, KNN involves two hyper-parameters:

1. K-value: it defines number of neighbours that will take part in the algorithm while considering the neighbours. K-value should be tuned properly for accurate prediction.
2. Distance function: it is used to calculate distance between any two data points. Euclidean distance, Manhattan distance, Hamming Distance, Minkowski distance are used to find similarities, euclidean being the most commonly opted choice.

Apart from the two parameters mentioned above, another factor that affects the performance of the algorithm is the approach to combine the class labels. Simple KNN takes the majority votes from the 'k' nearest neighbours. All the neighbours in this approach have a uniform weight to their vote. Simple KNN possesses a disadvantage in a situation where the neighbours vary widely in terms of their distance. In this case, the closest neighbour tends to indicate the class of the object better.

Weighted-KNN is used to overcome this disadvantage. Weighted-KNN is a modified version of KNN which gives more weight to the points which are close to the test point [8]. Weights are inversely proportional to the distance between the test point and the nearest neighbours.
The steps followed in weighted-KNN are as follows:

1. Unclassified data is taken as input.
2. Distance measure is calculated between the new data and all other classified data.
3. The value of K is defined.
4. Select k points based on the distance measure.
5. Predict the class of the query point, using distance-weighted voting.

4.2 Comparison with Other Algorithms

1. KNN versus Naive Bayes:
   Naive Bayes assumes that attributes are conditionally independent which make its decision boundaries (a decision boundary is a region in problem space where the output label is ambiguous) linear, parabolic or ecliptic and it looses flexibility [9].
2. KNN versus Decision tree:
   Decision tree is a tree-based algorithm where classification trees are used for dependent variable with discrete values. It seldom looses valuable information while handling continuous variables.
3. KNN versus Random Forest:
   Random Forest fetches the result by multiple decision trees. It's complex, hard to visualize the model. These tend to fail when the data is sparse.
4. KNN versus Neural networks:
   Neural networks unlike KNN require large training data to achieve required accuracy. NN requires a lot more hyperparameter tuning as compared to KNN.

5. KNN versus Logistic Regression:
   It works with algebraic calculation for best-fit curve for the complete population. In training, the curve is fitted to data points classifying the data points as per the classes. In terms of training k-nearest neighbours algorithm requires no training whereas Logistic regression requires some training. In terms of linearity, while logistic regression learns only a linear classifier KNN can learn non-linear boundaries as well.

So KNN is selected for our system based on the following reasons:

1. There is clear understanding about the input domain (The input of ratings based on interest).
2. Feasibly workable moderate data sample size (due to space and time constraints).
3. Colinearity and outliers are not a problem in our data-set.

5 Proposed System

5.1 Data-Set

A synthetic data-set is created and used for applying various models. It consists of more than 500 rows and 20 features. The advantage of using a synthetically data-set is that it contains no missing values. The data-set consists of ratings given by the aspirants based upon the interest towards various core-subjects from the student's Under Graduate curriculum as the features. The rating of their interest has been taken in the range of 1–10, hence ruling out the possibility of outliers. Also scaling is not needed since all the variables are in the specific scale of 1–10. The variable type of the features are numerical. It is a labelled data-set that is, it has both input and the output values.

The target values are the labels that are supposed to be predicted by the system after training. There are total 18 classes which are the courses which will be recommended to the user for Masters. They are categorical variables.

- Features: The core Under Graduate courses. Examples: 'Database Management', 'Java Programming', 'Python programming', 'Statistics', 'Data structure and analysis', etc.
- Labels: The Post-Graduate courses to be recommended. Example: 'Masters in Data science', 'Masters in Artificial Intelligence', 'Masters in Management in Information systems', etc.

'TARGET CLASS' is a variable consisting of string values with no logical order. We convert this variable into 'Category' type for memory optimization (Figs. 1 and 2).

Databases.	C++	analytics skills	python
7	1	8	10
7	2	8	9
7	3	8	9
8	0	6	7
8	1	7	8
8	2	8	9
8	3	9	9
7	2	8	9

Fig. 1 Sample data-set attributes

5.2 Execution

The Course Recommendation System for Post-Graduate aspirants works on concept of content-based filtering where the features of the labelled item-set is compared to the same features in the user profile. The data-set was split in the ratio of 80:20 as train and test set.

For training the model, 80% of the data is fed along with the output classes. Weighted K-Nearest Neighbours Algorithm is used to train the data. The rest 20% of the data which was not encountered by the system before is tested against the trained model to compare with the actual value and the accuracy.

Since the prediction is made based on labelled data, this is a Supervised Learning technique. Multi-class Classification is carried out as each data-point belongs to one of the ‘n’ different classes (where n is more than 2). Given a new data point, the goal of the algorithm is to predict the given new data-point to the correct class to which the data-point belongs.

Weighted K Nearest Neighbour algorithm here calculates the similarity between training data-set ratings based on the Under Graduate courses and student-profile ratings based on the same features. The result of which would be the course which is most similar in ratings to the student’s input of ratings.

The Euclidean distance method is used for calculating similarity between the ratings. The formula for calculating Euclidean distance is given by

$$\sqrt{\sum_{i=1}^{k} (x_i - y_i)^2}, \tag{1}$$

where, the value of ‘k’ defines the number of nearest neighbours to be considered. It is the most important parameter as choosing the right value of ‘k’ increases the accuracy.

Fig. 2 Sample target class

TARGET CLASS
Artificial intelligence
Artificial intelligence
Artificial intelligence
data science and analytics
data science and analytics
data science and analytics
data science and analytics
machine learning
robotics
human computer interaction
management information system
applied research and innovation
network management
cloud computing
operation research
enterprise architecture management
software engineering
cyber security
computer vision
videogame programming
distributed system an web technology

In our system, the value of 'k' is determined by plotting the 'Mean error versus k-value' graph (different values of k ranging from 1 to 40) (Fig. 3).

The algorithm selects 'k' points based on the similarity measure. The class, i.e. the course recommended would be selected based on the distance-weighted voting. This implies that the point which is the nearest (least distance) will have a higher weight (Fig. 4).

The control flow of the recommendation system is as follows:

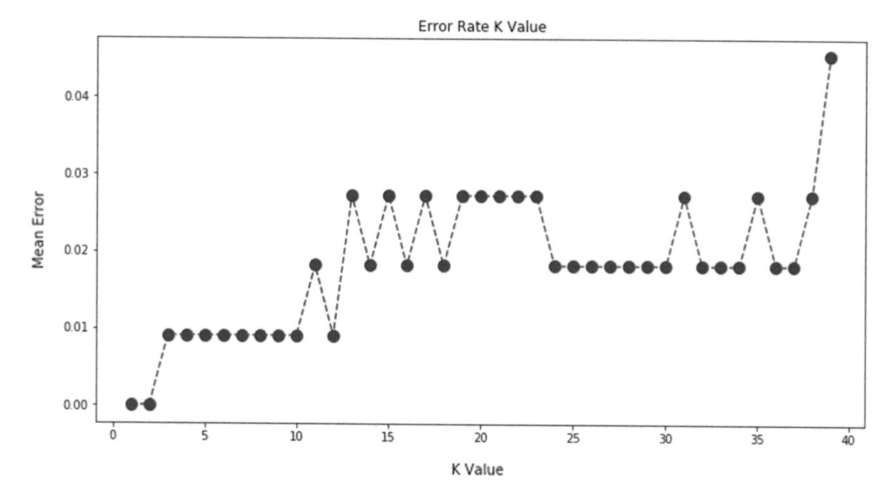

Fig. 3 Mean error versus K-value

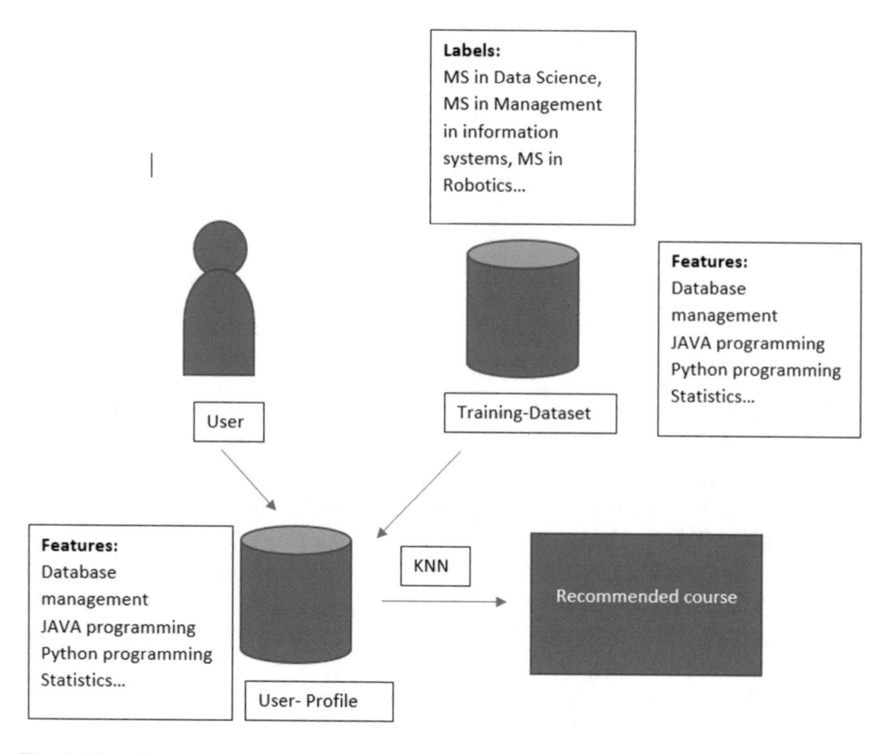

Fig. 4 Flow diagram

```
          data science and analytics       1.00      0.91      0.95        11
distributed system an web technology       1.00      1.00      1.00         7
                    machine learning       0.88      1.00      0.93         7
```

Fig. 5 Algorithm metrics and recommended course

6 Results

The terms precision, recall and f1 score are often associated with data classification. Precision and recall are statistical metrics used in classification and are defined using terms 'true positive' and 'true negative'.

Data points that are correctly classified or classified as positive by the model are called 'True positives' while those data points that wrongly identified as negative are called 'False negatives'.

- Precision: is the frequency of true positives (TP) divided by the sum of frequency of true positives and false positives (FP).

$$N_{TP}/N_{TP} + N_{FP} \quad (2)$$

- Recall: is the frequency of true positives divided by the sum of frequency of true positives and number of false negatives (FN).

$$N_{TP}/N_{TP} + N_{FN} \quad (3)$$

- F1 score: combining the value of precision and recall gives the f1 score. In mathematical terms, it is the harmonic mean of precision and recall.

Figure 5 indicates the precision, recall, f-score and support on application of KNN algorithm on the data-set. The first column indicates precision value followed by recall, f1 score and support.

The confusion matrix given below represents the performance of the classification model on the test data. There 18 classes. Thus the confusion matrix is an 18 × 18 Matrix. The left axis denotes the 'true class' and the top axis shows denotes 'class assigned' to an item with respect to that true class. Every element (i, j) of the matrix is the number of items with true class 'i' that was classified as being in class 'j' (Fig. 6).

Cohen kappa—a statistic that measures inter-annotator agreement is also calculated. The kappa statistic is often used as a measure of reliability between two raters. Regardless, columns correspond to one 'rater' while rows correspond to another 'rater'. In supervised machine learning, one 'rater' reflects ground truth (the actual values of each instance to be classified), obtained from labeled data, and the other 'rater' is the machine learning classifier used to perform the classification. Ultimately it doesn't matter which is to compute the kappa statistic, but for clarity's sake lets

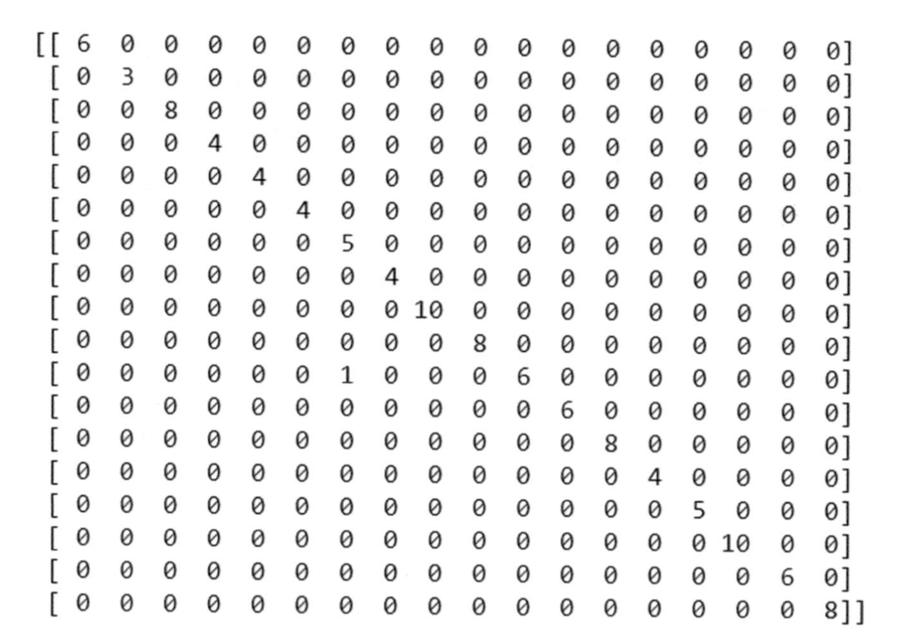

```
[[ 6  0  0  0  0  0  0  0  0  0  0  0  0  0  0  0  0  0]
 [ 0  3  0  0  0  0  0  0  0  0  0  0  0  0  0  0  0  0]
 [ 0  0  8  0  0  0  0  0  0  0  0  0  0  0  0  0  0  0]
 [ 0  0  0  4  0  0  0  0  0  0  0  0  0  0  0  0  0  0]
 [ 0  0  0  0  4  0  0  0  0  0  0  0  0  0  0  0  0  0]
 [ 0  0  0  0  0  4  0  0  0  0  0  0  0  0  0  0  0  0]
 [ 0  0  0  0  0  0  5  0  0  0  0  0  0  0  0  0  0  0]
 [ 0  0  0  0  0  0  0  4  0  0  0  0  0  0  0  0  0  0]
 [ 0  0  0  0  0  0  0  0 10  0  0  0  0  0  0  0  0  0]
 [ 0  0  0  0  0  0  0  0  0  8  0  0  0  0  0  0  0  0]
 [ 0  0  0  0  0  0  1  0  0  0  6  0  0  0  0  0  0  0]
 [ 0  0  0  0  0  0  0  0  0  0  0  6  0  0  0  0  0  0]
 [ 0  0  0  0  0  0  0  0  0  0  0  0  8  0  0  0  0  0]
 [ 0  0  0  0  0  0  0  0  0  0  0  0  0  4  0  0  0  0]
 [ 0  0  0  0  0  0  0  0  0  0  0  0  0  0  5  0  0  0]
 [ 0  0  0  0  0  0  0  0  0  0  0  0  0  0  0 10  0  0]
 [ 0  0  0  0  0  0  0  0  0  0  0  0  0  0  0  0  6  0]
 [ 0  0  0  0  0  0  0  0  0  0  0  0  0  0  0  0  0  8]]
```

Fig. 6 Confusion matrix

say that the columns reflect ground truth and the rows reflect the machine learning classifier classifications. The Cohen kappa score for KNN is found to be 0.99028.

Based on the precision score, f1 score, recall value and cohen kappa score we conclude that weighted-KNN is the optimum model for our Course Recommendation system. KNN gives the most accurate result for this particular dataset.

References

1. Nasaramma K, Bangaru Lakshmi M, Prasanna Priya G, HimaBindu G (2019) Recommendation system for student e-learning courses. Int J Appl Eng Res 14
2. Ndiyae NM, Chaabi Y, Lekdioui K, Lishou C (2019) Recommending system for digital educational resources based on learning analysis. In: 2019 proceedings of the new challenges in data sciences: acts of the second conference of the Moroccan Classification Society, New York, USA
3. Obeidat R, Duwairi R, Al-Aiad A (2019) A collaborative recommendation system for online courses recommendations. In: 2019 international conference on deep learning and machine learning in emerging applications (Deep-ML), Istanbul, Turkey
4. Rao S, Konstantin S, Polatkan G, Joshi M, Chaudhari S, Tcheprasov V, Gee J, Kumar D (2019) Learning to be relevant: evolution of a course recommendation system. In: 2019 Association for Computing Machinery
5. Spanakis G, Elatia S (2017) A course recommender system based on graduating attributes
6. Lin J, Pu H, Li Y, Lian J (2018) Intelligent recommendation system for course selection in smart education. Proc Comput Sci

7. Ng Y-K, Linn J (2017) CrsRecs: a personalized course recommendation system for college students. In: 8th international conference on information, intelligence, systems & applications (IISA), p 2017. IEEE
8. Chiang T-H, Lo H-Y, Lin S-D (2012) A ranking-based KNN approach for multi-label classification. In: Asian conference on machine learning
9. Jadhav SD, Channe HP (2016) Comparative study of K-NN, Naive Bayes and decision tree classification techniques. Int J Sci Res (IJSR) 5(1):1842–1845

Government Document Approval and Management Using Blockchain

N. S. Akhilesh, M. N. Aniruddha, Ghosh Anirban, and M. Dakshayini

Abstract Blockchain is a distributed database in the form of an immutable decentralized ledger spread across a peer-to-peer network. Blockchain contains several characteristics which are incredibly beneficial to any organization that requires a secure alternative to centralized database solutions. Because Blockchain is distributed, it enables essential data to be easily available to users even when offline, and with the help of cryptography, it can even ensure privacy and security. With all these benefits, one must wonder why Blockchain is not being used commonly among organizations that strongly necessitate such benefits. Organizations like the Government provide excellent use cases for Blockchain yet its implementation in this field is scarce. In this research paper, we hope to explain why this is the case and demonstrate how Blockchain could be implemented within the Government as a practical working database solution, namely, in the area of Civilian Document Management

Keywords Blockchain · Government document management · Distributed database

Supported by BMS College of Engineering https://bmsce.ac.in/home/Information-Science-and-Engineering-About.

N. S. Akhilesh (✉) · M. N. Aniruddha · G. Anirban · M. Dakshayini
BMS College of Engineering, Bangalore, India
e-mail: 1bm16is009@bmsce.ac.in

M. N. Aniruddha
e-mail: 1bm16is015@bmsce.ac.in

G. Anirban
e-mail: 1bm16is016@bmsce.ac.in

M. Dakshayini
e-mail: dakshayini.ise@bmsce.ac.in

N. Chaki et al. (eds.), *Proceedings of International Conference on Computational Intelligence and Data Engineering*, Lecture Notes on Data Engineering and Communications Technologies 56, https://doi.org/10.1007/978-981-15-8767-2_32

1 Introduction

Blockchain is a fascinating new technology since it is an alternative to conventional database systems. Conventional database systems work by storing data on a centralized server. User Machines can then make requests for this data and send them to the server when they need to, using a set of standard protocols (HTTP/S), the machines that do this are known as client machines or just clients. This system of storing and sharing data is known as the client-server architecture and is now ubiquitous in the technical industry. While this form of storing data has served us well and will continue to do so, it is not without its disadvantages.

In a Client-Server-based Architecture, all data is stored on a single server or a set of servers all of which are either stored in a single location or spread globally for better data redundancy and fault tolerance. In this system, the security of the data is entirely dependent on the security of the servers (or server) in which they are stored. If one of the servers or server were to either be hacked or stopped working for some reason, the data inside that server is compromised. Therefore, a great deal of care is put into ensuring that the servers are secure and fault tolerant since they represent a single point of failure. This is where Blockchain highlights its advantages. Unlike conventional database systems (Client-Server Architecture-based systems), Blockchain is a highly distributed database and has no single point of failure. We will explain the features of Blockchain below but before that, we must emphasize, that we call Blockchain an alternate to conventional databases, not a replacement. Conventional databases perform better than Blockchain in a large number of usecases and are also generally more flexible, easier to implement, and easier to manage. So conventional databases will continue to exist and be widely used for the foreseeable future.

Blockchain is often described as a distributed immutable ledger. The word "Blockchain" itself represents how data in a Blockchain is stored, in the form of blocks which are linked together like a chain thereby forming a chain of blocks or a "Blockchain". In computer science, one may think of it as a type of linked list. There are two properties in particular that make Blockchain special:

1. The first is that the Blockchain is immutable, meaning data can only be read from and written into the chain. Once in the chain, the data cannot be manipulated or removed.
2. The second is that the Blockchain is distributed or decentralized meaning that the actual chain of blocks is not stored in a single computer but rather spread across a large network of computers while being constantly synchronized between them

These two properties come together to ensure that any data being stored in the chain is secure. Since in order to compromise the data in the chain, one would need to attack every computer in the network that has a copy of the chain and replace the chains that is present in those computers with a malicious copy. (And doing this in a large network is almost impossible).

True Immutability in the world of software is impossible to implement, hence real-world Blockchains (like the one in Bitcoin) can only possess a form of pseudo-immutability by using a combination of hashing and networking. Hashing is a cryptographic function that takes any piece of information (of any form) and returns a unique fixed-length hash (id) based on the information passed. A well-designed hash function will always generate two entirely different and unique hashes for two different pieces of information passed to it even when these pieces of information differ slightly. Additionally, a hash cannot be used to retrieve the information used to generate it. The most widely used hashing function, as of the time of writing this paper, is the SHA-256 algorithm.

In Blockchain, each block is composed of three parts, any type of information that must be stored in that block, the hash of the contents of the previous block in the chain (if there is no previous block, then a random hash is taken) and the hash of the current block's contents (including the hash of the previous block). This leads to a form of cryptographic link that connects all the blocks in the chain linearly. And if someone tries to manipulate a block in the chain, all the blocks that come after that block would automatically break (since they used the hashes of the previous blocks to generate their own hashes), one could then simply check the hashes to ensure that the chain hasn't been tampered with. But this alone is insufficient to make the Blockchain immutable, since a machine if powerful enough can update a block in the middle of the chain and re-generate the hashes for the rest of the blocks in the chain. And this is where the distributed nature of Blockchain comes in.

As mentioned above, Blockchain is not stored on any one machine but on a large number of machines spread across a network known as a peer-to-peer network. A p2p network is a network in which all entities of the network are directly (using a link) or indirectly (via a peer/s) connected to each other. Blockchain is typically stored in every node in a p2p network and is constantly being synchronized within this network. This makes certain that if one of the nodes in the network (and by extension it's Blockchain) is compromised, the rest of the network can immediately identify and fix that node based on a majority. Additionally, whenever a new Block is pushed onto the chain, the new Block is immediately spread across the network and validated. These features make it increasingly harder for a hacker/s to hack the chain as the network grows larger, since it's necessary to compromise atleast 51% of the network before the chain can be manipulated. The actual process of making sure that the Blockchain is synchronized across the network and new blocks are validated is done using something known as a consensus protocol.

All the above mentioned features should highlight that Blockchain possesses a great deal of security compared to conventional databases, so why is it then that Blockchains are not used ubiquitously? As the technology would be incredibly useful in a great deal of areas such as Banks, Administrative Organizations, etc., where data security and safety is vital. In this paper, we address all of the above with respect to one specific usecase, Government Document Approval and Management.

2 Literature Review

2.1 Proof of Work and Bitcoin

The Proof-of-Work (or PoW) consensus protocol is the protocol used by Bitcoin and Ethereum [2], two of the biggest real-world Blockchain applications in the world. It first came to light in the famous Bitcoin paper by Satoshi Nakamoto 2008. In the paper, Satoshi Nakamoto provides an in depth explanation of how Blockchain can be used for a real-world usecase (in this case, for creating a digital currency). In the paper it is described that a real-world Blockchain application should have the following [8]:

An Immutable Ledger: It was the Bitcoin paper that suggested the blocks of a Blockchain be linked to each other cryptographically using hashing (as explained in the introduction). The intricacies involved in the creation of this hash was also used to create the process known as mining [8].

A Peer-to-Peer Network: A large scale p2p network (where each node carries a copy of the chain) spread across the entire web is essential to the security of Blockchain. Initially, Bitcoin used IRC seeding to create this network but later shifted to DNS Seeding. DNS Seeding works by having interested participants contact a public client who will then spread the participant's information across the rest of the network [8].

Network Consensus: In a network consisting of a massive number of nodes, it is vital that the nodes are constantly communicating with each other and ensuring that there is consensus between them (and the copies of the chain that they carry). In Bitcoin, new blocks are instilled into the network using the Gossip Protocol and every node only accepts a chain that is bigger than the one it possesses. In case of any conflicts, the network chooses the chain which gets a new block first (and this typically happens in the more powerful side of the network). These measures ensure that as long as atleast 50% of the network is not malicious, the chain is secure [8].

Bitcoin also uses a process known as mining to further fortify it's consensus protocol. Mining is a process in which nodes compete to solve a cryptographic puzzle, one which is hard to solve but easy to validate (in Bitcoin, this puzzle is to simply generate a hash with a predefined number of zeroes at the beginning). The solution of this puzzle is used to generate a new block; therefore, the node that finds the solution first gets to generate (or mine) the new block and is rewarded by the network for doing so. Using mining, Bitcoin has established two things [5]:

1. A controlled rate at which new blocks are added to the chain. The rate is manipulated by changing the difficulty of the puzzle ensuring that the chain is always synchronized across the network before a new block is created [5].
2. A competition for generating new blocks in the network. And since this competition is open for all and only one block can be added at a time, any user who wished to hack the chain would need to compete with an incredibly large number of gen-

uine users to do so (and this would require an unfathomable amount of computer resources to do since the network favors nodes with better performance) [5].

The protocol is called Proof-of-Work because every miner has to put in a great deal of work to generate a single block and that block then becomes proof of this. An issue with this system, however, is that for every new block generated, only the user that first generated the block is rewarded and the millions of other nodes that tried to generate this block are forced to simply move on to the next block. This leads to a great deal of resources being wasted. And this sole reason, makes PoW incompatible with the presented usecase [5].

2.2 *Proof-of-Stake*

In view of the massive power consumption attributed to the widely used Proof-of-Work protocol, many alternate protocols (ones which didn't consume as much power) were designed to try and take its place. The proof-of-stake protocol is one such protocol and perhaps the most popular [4].

Proof-of-Stake uses a process known as minting to replace the mining used by Proof-of-Work. Like PoW, PoS is also trying to create a competitive environment for generating new blocks that deters malicious users. Unlike PoW, PoS does not do this by making the competition based on computational power but rather based on a stake placed by each node in the network. Minting in PoS, works in a way that is similar to an Auction and the user who gets to generate the next new block is chosen based on three factors [4, 7]:

1. A certain predefined amount of randomness [4].
2. A monetary stake put up by that specific node [4].
3. And the amount of time the node has put up that monetary stake [4].

Thus a user who possesses the highest stake (or bid) and has spent more time with that stake than any other node has the best chance of being selected as the next new minter but even this doesn't guarantee that the node becomes the next minter since a certain amount of randomness also comes into play (this ensures that the network isn't dominated by nodes which are simply rich). But using the above methodology, PoS just like PoW can also [4]:

1. Control the rate at which new blocks are added to the chain [4].
2. Create a competitive environment that makes it increasingly hard for a fraudulent user to inject something malicious as the network grows [4].

While this system does provide the same level of security as PoW without the massive power consumption, it is not without its issues. PoS is hard to establish during the early stages of a Blockchain application when the application is relatively small. In PoS, the coins of the network are used as the staking value and when starting out, these coins don't possess a lot of value so acquiring them would not be difficult.

This becomes harder as the value of the coins increase, so in a large network such as Bitcoin, the rarity of the coins makes the system feasible [4].

Another problem is that unlike PoW, which has no barrier for entry to participation, in PoS, one needs a monetary stake to participate in the network. The stake is only used as an evaluation criteria (and security measure) and is returned to the user but it doesn't change the fact that a user would need to buy into the network in order to be a part of it [4].

2.3 Proof-of-Reputation

The Proof-of-Reputation consensus protocol is a protocol derived from the Proof-of-Authority protocol. In PoA, there exists multiple types of users. This is in contrast to PoW, where every user has equal power (as even a miner can be a regular user and vice versa) and the identity of each user remains anonymous. In PoA however, each user has an associated identity and this identity is used to define a role for that user. A user can either be a regular user that simply reads from the Blockchain or be a validator who validates blocks and puts them on the chain. This type of user authorization is especially useful in private organizations where the organization would want to manage user permissions and data access [3].

In Proof-of-Reputation, the users who take the role of validators are usually large corporations instead of single individuals (as it is in PoA). The reason behind this is because large organizations unlike individuals have a reputation to maintain and this reputation could potentially take damage if the organization were to do something malicious. This is why the protocol is called Proof-of-Reputation since the organizations who act as validators are effectively using their reputation as proof of the block's validity every time they put it on the chain [3].

As such, in order for the protocol to be as secure as possible, the validators must be organizations with a great deal of reputation to put at stake. Large organizations such as Google, Microsoft, etc., become ideal candidates to be validators in such a protocol. Once a list of validators is established within the protocol, this list is maintained within the Blockchain [3].

3 Problem Statement and Explanation

Government Organizations could greatly benefit from a technology such as Blockchain yet its implementation in this area is scarce, why is this the case? We will ignore political and social reasons for this (such as the largely mixed public view of Blockchain right now due to Bitcoin and other such crytpocurrencies) as of now and focus only on the technical limitations at play here. There are two main reasons we believe that contribute to Blockchain not being a mainstream technology and not being used by organizations such as the Government. The first is due to the nature of current Blockchain Architectures and the second is due to the state of existing consensus protocols being used.

There are predominantly two main types of Blockchain Architectures, the first is the public Blockchain and the second is the private Blockchain. In a public Blockchain, every user is anonymous. The best example of a public Blockchain is the Bitcoin network. A bitcoin wallet has little to no connection to its user's identity (this is one of the reasons why bitcoin is often used by hackers and malicious organizations). A private Blockchain is one where everyone's identity is known and only authorized personel are allowed to interact with the Blockchain. Private Blockchains are often used in companies, where only employees whose identities are known are allowed to participate in the network [6].

There is a stark difference in how the two Blockchain architectures is able to attain security. Incase of the public Blockchain, where everyone is anonymous and nobody knows anybody, there is no trust in the network. If someone were to create a new block in a public Blockchain, there is no way to verify the Block's genuineness based on the user alone, since the user's identity is not known. Hence in a public Blockchain, trust is handled instead by the consensus protocol used by that network. This leads to some incredibly complex consensus protocols such as the Proof-of-Work (being used right now by Bitcoin) and Proof-of-Stake. In a private Blockchain, however, every user's identity is known, so if one were to create a malicious block, that user could easily be pinpointed and punished. Thus there is trust in the network and so there is no need for a complex consensus protocol to be used. Instead private Blockchains often use protocols such as the much simpler Raft protocol to maintain consensus [1].

Unfortunately, neither of the existing architectures and consensus protocols are suitable for Government document management. Government documents such as Aadhaar[1] cards and PAN[2] cards are often jointly managed and owned by both the civilians to whom they correspond to and the government organizations in charge of managing them. So in a private Blockchain, where everybody knows everyone else, all users would have information about each other's ownership of documents which is not ideal. But in a public Blockchain, there is anonymity even between the civilians and the government organization who are directly responsible with managing their documents.

All of this isn't even mentioning that protocols such as Proof-of-Work are hard to setup since they facilitate the need for Mining, a process which has been proven to be incredibly inefficient (As of 2020, according to Digiconomist,[3] It has been estimated that the amount of electrical energy consumed by the bitcoin network is comparable

---

[1] A 12-digit unique identity number that can be obtained voluntarily by residents or passport holders of India, based on their biometric and demographic data.

[2] A permanent account number (PAN) is a ten-character alphanumeric identifier, issued in the form of a laminated PAN card, by the Indian Income Tax Department, to any person who applies for it or to whom the department allots the number without an application.

[3] https://digiconomist.net/bitcoin-energy-consumption.

to that of the country of Chile). And no Government would want to waste that much electricity.

4 Solution

To tackle the abovementioned problems, we propose a hybrid Blockchain architecture coupled with an alternate consensus protocol, the proof-of-reputation consensus protocol.

Instead of using a strictly public or private Blockchain, we propose a type of architecture which consists of two types of users: an Approver and a User. Based on these two types of users, we then define the following rules:

1. Anyone can become a User and the process for becoming one is relatively straightforward.
2. Only government approved offices and organizations can become Approvers. The process to become an Approver is extremely rigid and requires the provision of a verifiable and traceable proof of identity. (This is so that the approver can be punished for any acts of fraudulence in the future).
3. Users are anonymous to each other. But each User is assigned to an Approver/s and that/those Approver/s is/are allowed access to the User's identity.
4. All Approvers' identities are known to all (Both to each other and to all the Users).
5. Only Approvers are allowed to add blocks to the chain. Users have read-only access to the Blockchain.
6. Only Users are allowed to create the data (in this case Documents) that can be put into Blocks.

With these rules in place, the only way to add a Block to the chain is to have a User create the content of the Block and send it to an Approver. The Approver then examines and validates the content of the Block, before adding it to a Block and then adding the Block to the Blockchain. This is similar in many ways to how a normal Government document request is performed. For example, consider a request for an Aadhaar card, first a civilian (User) gets an Aadhaar application form and fills it (content) and then submits it to a government office (Approver), the office then checks the form, validates it, and then finally accepts it (adding a block to the Blockchain) thereby finally issuing the Aadhaar card.

Since the Approvers are the only ones who are allowed to put blocks into the chain, in the event of a fraudulent block being placed in the chain, the block can be traced back to the Approver who added it to the chain (Since the Approver's identity is known publicly) and the Approver in turn has access to the identity of the User/s involved in that Block. This effectively creates a sub-network of trust within the main Blockchain network that has the benefits of both a private Blockchain (where mining is not required since user authorization becomes the source of trust and simpler consensus protocols can be used) as well as that of a public Blockchain (where it is easy to participate in the network and there is privacy/confidentiality/anonymity between the members of the network).

5 Implementation

First, in order to establish a basic Blockchain, we need to setup the following:-

1. A Cryptographically Linked Ledger: In programming, this can be represented in the form of an Array or a Linked List, so just about any programming language can be used to create this. For the actual hashes, it is recommended to use the SHA-256 (since it is an industry standard) algorithm but that is not a strict condition.
2. A Peer-Peer Network: There are a number of ways to create a P2P network in the real world. The world wide web by itself is already a massive P2P network. In programming, one could create a P2P network using Websockets, IPFS, LAN, etc. In Bitcoin, this is achieved by using DNS Seeding, essentially every node in the network contains the addresses of a number of other nodes which inturn possess the addresses of other nodes and so on. To join the network, a new node needs to contact an existing node in the network for its address list and the information about the new node is then spread around the network.
   Additionally there exists certain software which ease the process of creating a P2P network such as Graph Databases like GunDB or Distributed Systems Softare like Docker Swarm and Kubernetes, etc
   (For the purpose of the experiment given below, we have just used LAN).

With these two in place, we can now move on to implementing the working of the different types of users in the network and the consensus protocol. To do this, we must establish a way to define two different types of users to the network: the Approver and the User. This can be achieved using Asymmetric Encryption.

Every User in the network is given two encryption keypairs: one for signing the block and the other for encrypting its content (to ensure privacy) and the Approver is given a single encryption keypair for signing the block. Every User and Approver then exposes a public key (the one that will be used for verifying their signatures) to the network where a list of these public keys is passed around the network so that it may be used for validation. Additionally, an Approver will also expose its public key through a custom portal (For example: Aadhaar Portal) and it may also provide the public keys of the Users it manages if requested.

For the Approver, the custom portal becomes a source of truth to its Users (much like Government Websites and Web portals are to its citizens) which provides Users with the Approver's public key as well as access to the Approver's information (such as the Approver's proof of identity) and the Approver's services (Aadhaar, PAN, etc).

So for a normal User to join the network, he need only provide his public key and an encrypted unique username which can be decrypted using the public key thus proving he is the owner of that username. While in the case of the Approver, a custom verifiable and reliable portal will also need to be setup that handles distribution of the

Approver's public key, information and services. And the Approver will also need to provide its public key and an encrypted link that points to it's portal to the network in order to join it.

Now before a User can access any of the services of the Approver, he will also need to securely transfer a public key to the Approver (or alternatively this can be done by just using a Public Key Infrastructure with Public Key Certificates issued by trusted authorities) that will be used to ensure the privacy of the data being exchanged between the User and the Approver. This can be done by either a physical exchange (A civilian visiting a Government office) or through secure electronic/digital means (Ex: SSL or through the Approver's secure portal). This public key is different from the public key used by the network to validate the User's signatures. Instead this public key (that is shared only between the User and the Approver) is used to ensure that the User's private information (which is stored inside the Block) can only be read by the User and the Approver.

(So to re-illustrate, the User possesses two keypairs, one is for signing the block and the public key for this keypair is shared with the entire network and the other is for encrypting the private User-specific information that exists within the block and the public key for this keypair is shared only with those Approvers whose services the User is using).

Now when a User wants to make a request for an Approver's service (such as a request for an Aadhaar card):

1. The User would first create a document containing all the necessary information for the (in this case) Aadhaar Card (this includes an id proof and address proof).
2. The User first encrypts all the information in the document that he/she does not wish to share publicly using his privacy keypair's private key and if he/she hasn't already, privately and securely shares the privacy keypair's public key to the Approver using one of the methods mentioned above.
3. The User then hashes the document and encrypts the hash using his/her private key (The one used for signing the block).
4. Finally the User sends this document along with its signature to the Approver.
5. The Approver upon receiving the document and its signature uses the User's public key to verify the User's signature on the document.
6. The Approver then uses the User's privacy keypair's public key to decrypt all the content of the document and then goes through the document to approve the User's request.
7. After approving the document, the Approver then adds the document to a Block, signs the Block (in a similar fashion to how the User signed his/her document) and finally adds the Block to the Blockchain.

When reading the above steps, one might wonder the need for a separate privacy keypair, since the user could just create a symmetric key, use that to encrypt the document, then encrypt the key using the user's main private key and share the encrypted symmetric key to the approver as is traditionally done in digital envelope technology. But in this approach, the Approver could also potentially use the symmetric key to re-encrypt and change the document that the user is trying to add to the blockchain.

To avoid this, the user is required to create a separate keypair and share the public key of that keypair instead since then, the approver would only be able to decrypt but not encrypt the document that the user is trying to sign.

Additionally, keep in mind that the Approver does not immediately add the Block to the Blockchain. And the reason for this is to ensure a constant block rate. Block rate is (as we define it) the rate at which new Blocks are added to the Blockchain, in the case of Bitcoin, the Block rate is approximately 10 min which is to say that in Bitcoin, a new Block is added once every 10 min. Ensuring a constant block rate is important especially in large networks since new blocks need time to spread across the network. Hence a large enough and fixed block rate such as Bitcoin's 10 min gives enough time for the network to have reached overall consensus with existing Blocks and Blockchains before a new Block is added.

In Bitcoin, a constant Block rate of 10 minutes is maintained by manipulating the difficulty of the cryptographic puzzle that needs to be solved by the miners in order to add new Blocks to the chain (See Literature Survey). But the concept of mining doesn't exist in the system we are using, so in order to achieve a constant block rate, we developed a technique known as time blocks.

A time block is a special kind of block which holds all the blocks generated by a number of peers within a specific time frame. Since when working with multiple different peers, it can be quite difficult to ensure a constant block rate, instead we allow the creation of blocks at a constant rate to be done automatically. To understand the process, consider a network spread across a country, each state has a single approver and so let's assume this country has 20 approvers and each citizen is a regular user. Then:

1. First, we define a block rate say 3 min. Additionally, we also make a predefined order for the approvers' blocks within the 3 min time period. So something as simple as an alphabetical order should suffice.
2. With the block rate set to three minutes, we now define a rule that states that every Approver in the network can only add a block at the end of a 3 min cycle.
3. Also of importance, is to define a central time source since locations around a large landmass can have different timezones, we define a central timezone for the blockchain that all participants of the chain must adhere to. (So if two participants, A and B are in two different timezones and A's timezone is the central timezone, then participant B must also use A's time when working with the Blockchain. The actual time itself doesn't matter in this system, what matters is ensuring that the value of time in the Blockchain between each set of blocks is always 3 min hence why a single timezone is needed).
4. Now when a user makes a document request to an approver, the Approver after approving it adds it to a block. And like this keeps adding requests to the block until reaching the end of a three-minute cycle at which point the approver then uploads the block, then creating a new block that will collect approved requests for the next three minutes.
5. With this, we now ensure that every 3 min (as per a central and globally accepted timezone), 20 new blocks (because there are 20 Approvers in this case) is added to the Blockchain.

6. The participants then group all of these blocks (within this specific 3 min cycle) and then put them inside another block (a sort of hyperblock) which is essentially the timeblock of that 3 min cycle. A hash is generated and included within this timeblock, before it is then made a part of the Blockchain. The blocks inside the timeblock are arranged in the predefined order of the network.
7. Thus the Blockchain, in this case, is composed of timeblocks which inturn contain all the blocks created by the network in 3 min (since that is the block rate in this case) intervals and these blocks inturn contain all the approved request collected by that specific block's approver in that specific 3 min cycle.

While this method does ensure a constant block rate within the chain. It also inherits a set of disadvantages:

1. Each timeblock is likely going to be massive since it contains a large culmination of blocks which also contain a large culmination of document requests.
   (This could be fixed by compressing blocks and also by not storing any images or documents directly on the chain but rather storing links to these images and documents)
2. It constrains each Approver to only upload Blocks at specific times and the timegap between them needs to be large enough for the network to circulate new timeblocks meaning that it is rare for any new blocks to be added instantaneously.
   (Though this hardly matters for this specific usecase since government document approvals are not a quick process anyway)

With all of these setup, we can implement a working solution that uses Blockchain for the management and approval of government documents

6 Proof of Concept and Design

A basic implementation of the solution explained above can be found at https://github.com/AkhileshNS/pba-demo, it is written in JavaScript and NodeJS. An official Proof of Concept website is also under the process of being built. The PoC website is built using the following technologies:

1. React: a JavaScript library used for building the user interface of the website.
2. MobX: A central state management library used to manage the state of the website.
3. NodeJS: A JavaScript runtime used for setting a backend REST Api where user registration and cryptography can be done.
4. Elliptic: A library that provides various encryption-based functionality based on elliptic cryptography. The program uses the ECDSA encryption algorithm provided by the library. Note. This implementation of the algorithm uses the secp256k1 curve which is relatively safe but it is highly recommended to not use ECDSA at all in real-world applications (even though Bitcoin uses it) and to instead use EdDSA wherever possible. The library does also provide implementation for the same as well.

5. Express: A library to create http-servers or REST Apis. The program uses this to create a REST Api through which users can interact with the program.
6. Crypto-js: A library that provides various cryptographic functions. The program makes use of the SHA-256 hashing function provided by this library.
7. Firebase: For storing user information (Note. While firebase is a central server-based service that works against the idea, it is only being used here for the proof-of-concept and can easily be replaced by a custom authentication and user management solution built into the Blockchain or by using something like Hyperledger Fabric).
8. GUNjs: A JavaScript Graph Database Library that provides a distributed graph database across a P2P network where the Blockchain for this application is going to be hosted.

Note. This is only a proof of concept and must not be treated as a real-world program.

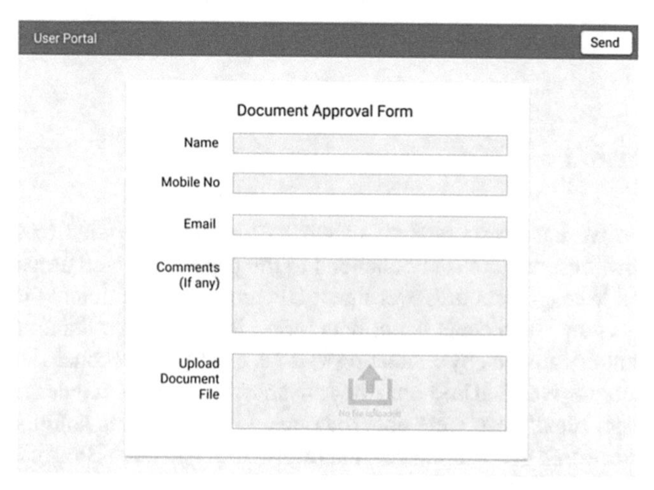

Portal where a civilian user can upload the document he/she wished to get approved

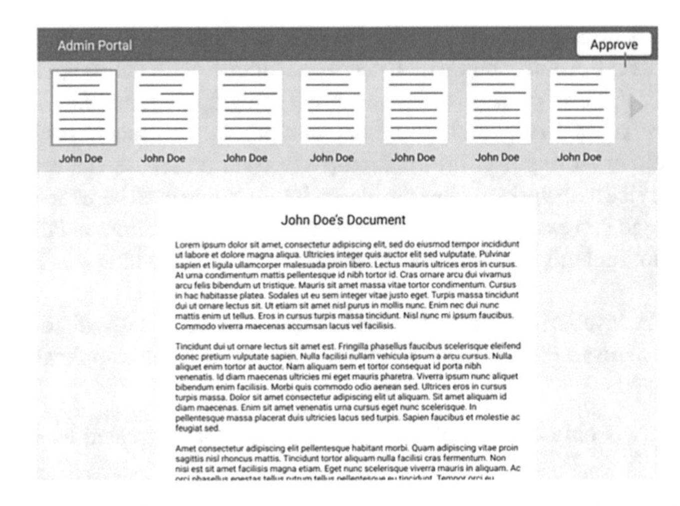

Portal where a government official can view documents and approve them.

7 Conclusion

In this paper, we have discussed about what Blockchain is, what features it provides and how these features are beneficial to the paper's covered usecase (namely, Approval and Management of Government Documents). Additionally we have also tried to assess why Blockchain is not being used for the approval and management of government documents, by evaluating various existing Blockchain solutions like Bitcoin, Ethereum, etc., and how well they are able to satisfy the requirements needed to do the same. Finally, we were able to conclude why existing solutions were not able to fit the covered usecase and use that information to craft a Blockchain solution which does.

Acknowledgements We would also like to thank our respected Principal, Dr. B. V. Ravi Shankar, for giving us this opportunity to work on this paper. We would like to express our deepest appreciation to all those who provided us the support to complete this paper. A special gratitude to our guide, Dr. Dakshayini M., whose contribution in stimulating suggestions and encouragement helped us through every step of writing this paper.

References

1. Androulaki E, Barger A, Bortnikov V, Cachin C, Christidis K, De Caro A, Enyeart D, Ferris C, Laventman G, Manevich Y, Muralidharan S, Murthy C, Nguyen B, Sethi M, Singh G, Smith K, Sorniotti A, Stathakopoulou C, Vukolic M, Weed Cocco S, Yellick J (2020) Hyperledger/fabric.

[online] GitHub. Available at: https://github.com/hyperledger/fabric/blob/master/docs/source/whatis.md
2. Ethereum A next-generation smart contract and decentralized application platform. Vitalik Buterin https://www.weusecoins.com/assets/pdf/library/Ethereum_white_paper-a_next_generation_smart_contract_and_decentralized_application_platform-vitalik-buterin.pdf Ethereum
3. Gai F, Wang B, Deng(B) W, Peng W Proof of reputation: a reputation-based consensus protocol for peer-to-peer network. https://www.researchgate.net/publication/325097454_Proof_of_Reputation_A_Reputation-Based_Consensus_Protocol_for_Peer-to-Peer_Network Proof-of-Reputation
4. Hooda P Proof of Stake (PoS) in blockchain. GeekForGeeks, [Online]. Available: https://www.geeksforgeeks.org/proof-of-stake-pos-in-blockchain/
5. Hooda P Proof of Work (PoW) Consensu. GeekForGeeks, [Online]. Available: https://www.geeksforgeeks.org/proof-of-work-pow-consensus/
6. Javeri P (2019) Smart contracts and blockchain. [Online]. Available: https://medium.com/@prashunjaveri/smart-contracts-and-blockchains-c24538418bf6
7. Moindrot O, Bournhonesque C Proof of stake made simple with casper. https://www.scs.stanford.edu/17au-cs244b/labs/projects/moindrot_bournhonesque.pdf
8. Nakamoto (2008) Bitcoin: A peer-to-peer electronic cash system. Bitcoin. Available: https://bitcoin.org/bitcoin.pdf

Quality Assessment of Orange Fruit Images Using Convolutional Neural Networks

B. Leelavathy, Y. S. S. Sri Datta, and Yerram Sai Rachana

Abstract For sustainability of life and rural development, assessment of fruits in a non-destructive manner is required. The fruits which are available in the market must fulfill buyer needs. Orange fruit analysis is generally done by visual examination and by observing the size. For large volumes, we cannot assess by human graders, so picture preparation is done on quantitative, solid, and predictable data. This research contains a division of orange images based on fresh and rotten criteria, the images are of different kind based on the rotation of images. Then the classification of images is done based on CNN, binary cross-entropy loss function, along with accuracy is calculated and resultant graphs are illustrated with an accuracy of 78.57%.

Keywords Rotten images · Classification · CNN · Adam optimization

1 Introduction

India is the country where there is a higher rate of exports as well as imports are done in case of vegetables, fruits, and it's a known fact. Among the fruits, our country ranks first in production of Bananas, Papayas, and Mangoes. The people living in India are more dependent on an agricultural domain, statistics say that 61.5% are dependent. Despite the ranking of India in the area of fruits, fruits have defects, for example, apple has defects like rot, blotch, and scab. In this case, here we will discuss orange fruit, Orange is a kind of citric fruit. The production of oranges per year is estimated to be 54.23 million tonnes and the highest producing country is Brazil which is 35.6

B. Leelavathy (✉) · Y. S. S. Sri Datta · Y. S. Rachana
Department of Information Technology, Vasavi College of Engineering, Hyderabad, India
e-mail: leelapallava@staff.vce.ac.in

Y. S. S. Sri Datta
e-mail: sridatta179@gmail.com

Y. S. Rachana
e-mail: yerramsairachana@gmail.com

N. Chaki et al. (eds.), *Proceedings of International Conference on Computational Intelligence and Data Engineering*, Lecture Notes on Data Engineering and Communications Technologies 56, https://doi.org/10.1007/978-981-15-8767-2_33

Fig. 1 Types of oranges—fresh and Rotten

million tonnes [1]. The classification of oranges is done based on Normal ones and Defected ones (see Fig. 1). The defects on oranges include insect damage, copper burn, phytotoxicity, etc.

2 Related Work

Ahmed M. Abdelsalam and Mohammed S. Sayed (2016) built a computer vision system that identifies the outdoor layered defects of orange fruits using a sensor called multi-spectral images. In this model [2], they have implemented an algorithm that captures orange fruit images by only a Near-Infra Red component. After taking the respective images, preprocessing steps on these images was done, which detects the seven different color components of orange which is a thresholding technique that was implemented. After all these above steps a process of voting is done on all the obtained seven different threshold color images for detecting the defected and defect-free orange fruits.

Hardik Patel, Rashmin Prajapati, and Milin Patel developed a model with a method of image preparation which is done in a computer, where the fruits are estimated on different factors like solid and quantitative data and mainly for avoiding human grading system. This method shows Bacteria spot defects and also they proposed a framework for reviewing. The factors like texture, color, and size are required for the early appraisal of the fruit orange. The side view of oranges and some characteristics of the fruits are removed by identifying some sort of calculations. By observing these characteristics they figured out the reviewing process [3]. In this reviewing a very decent Orange quality and evaluating pf zones is done by considering the advantages and disadvantages, the classification of different features was discussed.

Chen et al. [4] proposed a method on automatic grading detection of orange which is based on computer vision. In this method, the orange fruit images are gathered and preprocessed, an image segmentation and edge detection were applied to the images of orange fruits and are segmented. Based on the segmentation of images, main features of oranges are extracted, they are fruit color, fruit size, orange surface defect and shape of the fruit, all of these features were learnt by using neural network which yields automatic detection of grade first oranges. The accuracy of grading obtained was 94.38%, accuracy of classification of the grade first is 100%. When it was compared with artificial mode, excellent real-time identification performance rate was observed.

M. Recce, J. Taylor, A. Piebe, and G. Tropiano developed a novel system for orange grading into quality bands of three with respect to the characteristics of the surface [5]. This method is the only non-automated processing operation in the family of citrus. This system also handles a huge variety of defect marking and surface coloration, shape(highly eccentric to spherical) and also the size of fruit which is from 55 to 100 mm. In order to distinguish the defects, the point of attachment of stem which is nothing but calyx must also be recognized. For identifying the radial color variation they implemented a neural network classifier in the area of rotation invariant transformations, which is also called as Zernike moments. This method requires a complex pattern recognition and also a high throughput which needs 5–10 oranges per second. Grading of these fruits is achieved by imaging the fruit from all orthogonal directions simultaneously by passing them through the chamber of inspection. The first stage contains the histograms of all the views of orange fruit which are analyzed by using classifiers of neural networks. From these views, the one which contains defect is further analyzed by utilizing five independent masks combined with a classifier of neural network. The expensive process of stem detection is applied to a slight part of the images. From all these steps results are presented with performance analysis.

Satpute and Jagdale [6] proposed a system which is based on automatic inspection of fruit quality for tomato defect detection, grading, and sorting of tomato. In this method as a primary step, they segmented the tomato-based on the algorithm called OTSU. After the segmentation of tomato, extracting of features like color detection, size detection was done. Methods like Erosion and Dilation were used for size detection, these features where in turn utilized for size detection like large, medium, and small. In order to extract colors like green, yellow, and red, color detection was used for tomato sorting.

A model was proposed for the classification of citrus fruit using the parameter GLCM. They performed the conversion of RGB image into the gray scale image, extraction of features was implemented using the GLCM feature extraction [7]. The feature of GLCM is four types like Energy, Correlation, Contrast, and Homogeneity, all these features are used for feature extraction. A model was constructed for grading and sorting of agricultural product. Primarily RGB images are converted into scale GRAY image which is nothing but preprocessing. The secondary step is the extraction of feature, here in this method extraction is based on shape feature, based on fuzzy logic and Support vector machine classification of fruits was performed.

A model for processing images based on Dates Maturity Status and dates classification [8] was constructed by T. Najeeb and M. Safar. The primary step proposed was to resize the images, i.e., preprocessing of input data. The second step was to perform segmentation within the threshold. Post Segmentation they have processed the image measurement labeling. The final step was, based on color detection and size detection through which extraction of a feature of the fruit was proposed.

A model for olive fruit detection of defects using an automatic method. The following procedure was proposed with firstly preprocessing the images from RGB to GRAY on the given olive fruit. Segmentation was done based on the threshold, following the feature extraction of the olive fruit [9]. They have implemented the SICA which is the Special Image Convolution Algorithm,(THMT) which is a Texture Homogeneity Measuring Technique in the proposed method.

3 Proposed Method

Data preprocessing is a technique of preparing the data by cleaning and organizing it. In the proposed method we have used several libraries/modules like sklearn.preprocessing, keras.preprocessing to preprocess our data, i.e., images. LabelBinarizer from sklearn.preprocessing which is used to convert multi-class labels to binary label, i.e., belongs/doesn't belong to the class. The img_to_array method of Keras library is used to convert the image instances to NumPy array. ImageDataGenerator of keras.preprocessing.image is a great tool for data augmentation of our images and to generate batch samples for our model/network. If the data used numbers as classes, then **to_categorical** method was used to transform these numbers to vector representation, making it suitable for our model.

The proposed model was built with the help of **Convolutional Neural Networks (CNN)**. Convolutional Neural Networks are used effectively in the areas of image recognition and classification. As the name suggests, it has a connection with biology or neuroscience. CNN's take the biological inspiration from the visual cortex. Image Classification deals with taking an image as an input and classifying it into a class or probability of classes that describe it as the best. This task is done using a series of convolutional layers by understanding the low-level features of the images like intensity, edges, curves, etc. The input image is passed through a set of convolutional, pooling, and fully connected layers to get an output.

The **test_train_split** function of sklearn.model_selection is a method used to split the dataset into two disjoint subsets for training and testing purposes. The optimal test_size is 0.20 referring to the train-test split as 80–20%. The model is built/trained using a training dataset while tested against the testing dataset.

The **loss function** used is binary cross-entropy loss, which is a default loss function to be used for binary classification. It is always intended to use with binary classification problems as the target values always belong to the set $\{0, 1\}$.

The **Cross-Entropy Loss** is defined in Eq. (1) as

$$CE = -\sum_{i}^{c} t_i \log(s_i), \tag{1}$$

where t_i and s_i are the ground truth and the CNN score for each class $_i$ in C. As usually an activation function (Sigmoid/Softmax) is applied to the scores before the CE Loss computation, we write $f(s_i)$ to refer to the activations.

In a binary classification problem, where $C' = 2$, the Cross-Entropy Loss can be defined Eq. (2) as

$$CE = -\sum_{i=1}^{C'=2} t_i \log(s_i) = -t_1(\log s_1) - (1 - t_1)\log(1 - s_1), \tag{2}$$

where it's assumed that there are two classes: C_1 and C_2. t_1 [0, 1] and s_1 are the ground truth and the score for C_1 and $t_2 = 1 - t_1$ and $s_2 = 1 - s_1$ are the ground truth and the score for C_2.

The **activation function** used is "**ReLU**"—Rectified Linear Unit. It is of the nonlinear activation function. The nonlinear activation functions always help the model to generalize or adapt to different data. The advantage of the ReLU activation function over other activation functions is that it doesn't activate all the neurons at the same time. The function and its derivative are monotonic in nature. ReLU is the default loss function for CNN as it allows the model to learn faster and perform better. Mathematically, ReLU is defined as: **y = max(0, x)**. The Softmax function makes the outputs of each unit to be between 0 and 1. Also, it divides each output such that the total sum of the outputs is equal to 1. Mathematically, it is represented as in Eq. (3)

$$\sigma(Z)j = e^z j\sigma\left(z_j\right) = \frac{e^z j}{\Sigma_{k=1}^{k_{e^{zk}}}}. \tag{3}$$

The Optimization used here is **Adam optimization**. It is used to iteratively update the network weights in training data. It is used instead of the classical stochastic gradient descent algorithm. Adam's algorithm has an advantage of the Adaptive Gradient Algorithm and Root Mean Square Propagation. The Adaptive Gradient Algorithm maintains multiple learning rates per parameter to enhance the performance. **Root Mean Square Propagation** helps to maintain per-parameter learning rates that are adapted based on the average of recent magnitudes of the gradients for the weight.

This experiment was executed on Intel® Core™ i7-8550U CPU@1.8 GHz, 1.99 GHz., Installed Memory (RAM) of 12 GB, 64-bit Operating System, × 64-based processor.

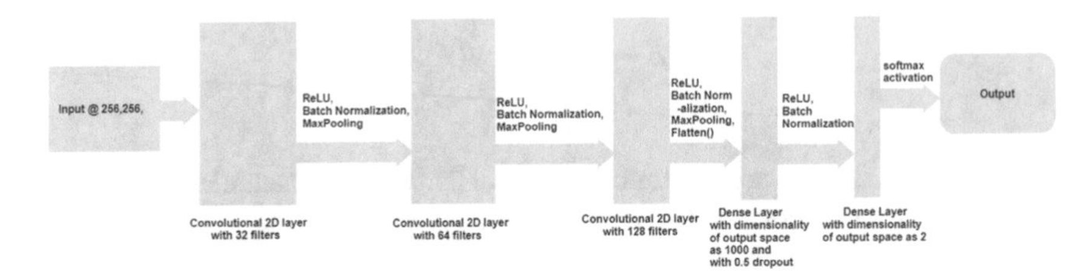

Fig. 2 CNN model flow architecture

3.1 Proposed Algorithm

Step 1: Import the dataset which contains the images of fresh and rotten oranges.

Step 2: Perform data preprocessing on the input images using the functions img_to_array, to_categorical, LabelBinarizer.

Step 3: Using train_test_split function, split the dataset into training and testing subsets with test_size=0.20.

Step 4: Using the ImageDataGenerator function, which augments the images and generates batch samples to the model.

Step 5: Create a Sequential Model (see Fig. 2).

Step 6: Compile the model.

Step 7: Apply fit_generator using on the model and store it in history variable/object.

Step 8: Assign the values into acc, val_acc, loss, val_loss from history dict based on keys and initialize epochs=range (1, len(acc),+1).

Step 9: Plot the graphs.

Step 10: Calculate the model accuracy.

3.2 Working of the Model

In the proposed model we used a dataset that is obtained from an internet source, which contains the images of oranges of *Fresh* and *Rotten* categories. Uploading the images to the Google Colab platform was performed. There are some API's which are needed to be imported so that the functions of them can be used for our model.

The number of epochs defines how many distinct times you consider the training data set. We initialized 25 epochs and then we initialize the variables height and width of the input image as 256 and depth is 3. Then, preprocessing steps are performed which are conversion of image to an array by using img_to_array which converts the image into NumPy array. A list is created with contents as array objects of the images. Using LabelBinarizer, we convert multi-class labels to binary labels.

The traditional train_test_split is applied to the data. The train size and test size are taken as 80 and 20. The sequential model is generated for which a series of steps are done which are adding a convolutional 2D layer to the sequential model with the activation function of ReLU, Batch Normalization is done to improve the stability of the previous added step Max pooling is done for reducing the unnecessary dimensions of the given images from the previous step. The dropout layer is added.

This series of steps is repeated for a couple of times by changing the convolutional filters to 64 and 128, respectively. After these steps, the output is flattened in order to do the required classification. Then, a 2D dense layer with the activation function of "Softmax" is added. After these steps, Adam Optimization is used as an optimizer and then we use fit_generator() to the model and store that in history object/variable. By using the history variable and aug which is an object of ImageDataGenerator, the generation of epochs is done which is the "penultimate" step that gives the insight of the entire process (whether or not the grading is done in a proper and accountable manner or not). The final step is to generate the graphs of training based on accuracy and loss. Finally, the accuracy is calculated which justifies the model.

Sample Pseudocode of the experiment (see Figs. 3, 4, 5).

```
history = model.fit_generator(
    aug.flow(x_train, y_train, batch_size=BS),
    validation_data=(x_test, y_test),
    steps_per_epoch=len(x_train) // BS,
    epochs=EPOCHS, verbose=1
    )
```

Fig. 3 Code snippet of training the CNN model

```
print("[INFO] Calculating model accuracy")
scores = model.evaluate(x_test, y_test)
print("Test Accuracy: {scores[1]*100}")
```

Fig. 4 Code snippet of calculating the accuracy

```
opt = Adam(lr=INIT_LR, decay=INIT_LR / EPOCHS)
# distribution
model.compile(loss="binary_crossentropy", optimizer=opt,metrics=["accuracy"])
```

Fig. 5 Code snippet of Adam optimization method

4 Findings and Observations

Using various values produced during model.fit_generator() like the accuracy of the training, accuracy of the validation, loss during training and loss during validation, we can plot various graphs to understand the model (see Fig. 6).

Summary related to the analysis of our proposed model is given in Table 1.

Hence, the model was compiled with two different number of epochs as 20 and 25. When compiled with train and test split of 80–20 we achieved a highest accuracy of 78.57%.

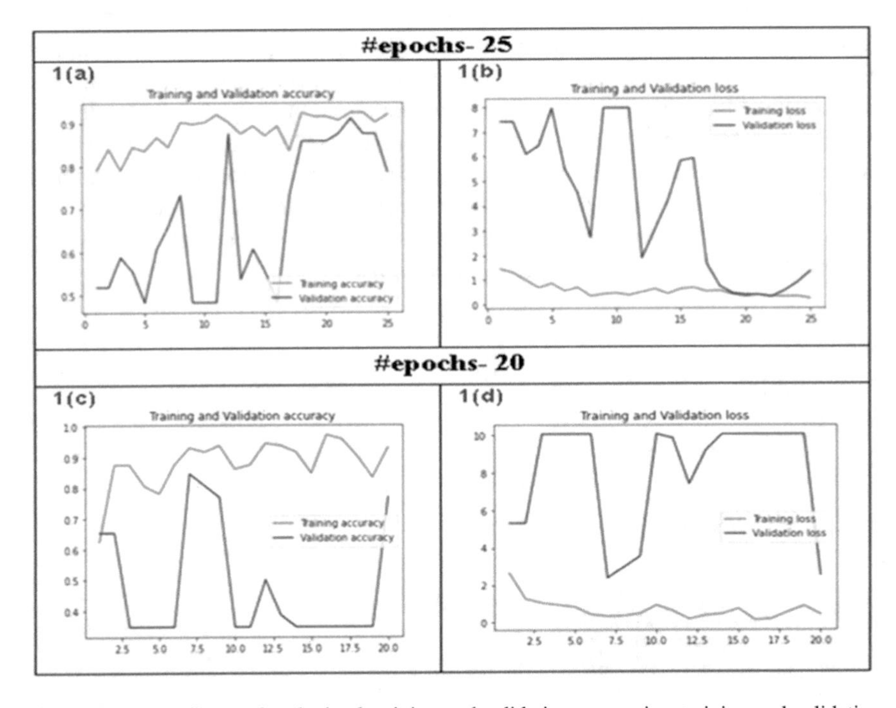

Fig. 6 Corresponding to the obtained training and validation accuracies, training and validation loss plotted as graphs for the train and test split of 80–20% [1(a)–1(d)]

Table 1 Summary of the results obtained

Model	Size of dataset	Train-test split	Accuracy in %	
			Number of epochs-20	Number of epochs-25
CNN with SoftMax classifier	Orange fruit dataset (rotten, healthy), 800 images	80–20	76.92	78.57

Table 2 Comparison of accuracy obtained with different model architectures based on CNN, SVM for different fruits

Type of the fruit	Model name	Methods used for classification	Accuracy (%)
Orange [6]	SVM Classifier	Gabor+LBP+GCH	61.29%
Orange [6]	SVM Classifier	Gabor+CLBP+LTP	64.52%
Orange [6]	SVM Classifier	ColorMoment+GLCM+Shape	67.74%
Banana [12]	CNN	3-Convolutional layers, FFully connected layer, ReLU (activation function), Max Pooling layer	80%
Grapes [13]	CNN	Color Feature Extraction(RGB, HSV), Morphological features such as shape	79.49%
Tomatoes [14]	Convolutional Autoencoder	The encoder is used to generate internal representation of input images, the decor converts those internal representations into outputs	79.09%
Tomatoes [14]	ResNet2 based CNN	Batch Normalization, RELU (activation Function)	87.27%

From Table 2, we can observe that classification of Oranges using SVM resulted in the accuracy that falls in the range from 61.29–67.74%. Whereas, when we try to classify the oranges using CNN, we obtained the accuracy of 78.57%. As we change the ratio of train to test split, the accuracy might vary a bit, with extensive experimentation, we have noted that accuracy always lies above 70%.

5 Conclusion

In the proposed method, we have successfully classified the rotten oranges from the rest and robustly analyzed the same by specifying the accuracy of the work done. The methods which are used here are flexible and potent. The work can also be used to improvise the accuracy by combining with other classification methods. This method can also be applied for the automatic prediction by improvising a bit, which is a great use for the agricultural field to yield better crop in the current era. However, classifying and identifying the defects of the orange fruit by implementing this model in real-time can be done in the future.

References

1. https://apeda.gov.in/apedawebsite/six_head_product/FFV.htm
2. Abdelsalam AM, Sayed MS (2015) Real-time defects detection system for orange citrus fruits using multi-spectral imaging. In: 2016 IEEE 59th international midwest symposium on circuits and systems (MWSCAS), Abu Dhabi, pp 1–4. https://doi.org/10.1109/MWSCAS.2016.7869956
3. Patel H, Prajapati R, Patel M (2019) Detection of quality in orange fruit image using SVM classifier. In: 2019 3rd international conference on trends in electronics and informatics (ICOEI), Tirunelveli, India, pp 74–78. https://doi.org/10.1109/ICOEI.2019.8862758
4. Chen Y, Wu J, Cui M (2018) Automatic classification and detection of oranges based on computer vision. In: 2018 IEEE 4th international conference on computer and communications (ICCC), Chengdu, China, pp 1551–1556. https://doi.org/10.1109/CompComm.2018.8780680
5. Recce M, Taylor J, Piebe A, Tropiano G (1996) High speed vision-based quality grading of oranges. In: Proceedings of international workshop on neural networks for identification, control, robotics and signal/image processing, Venice, Italy, pp 136–144. https://doi.org/10.1109/NICRSP.1996.542754
6. Satpute MR, Jagdale SM (2016) Automatic fruit quality inspection system. In: 2016 international conference on inventive computation technologies (ICICT), Coimbatore, pp 1–4. https://doi.org/10.1109/INVENTIVE.2016.7823207
7. Kumar C, Chauhan S, Alla RN, Mounicagurram H (2015) Classifications of citrus fruit using image processing-GLCM parameters. In: 2015 international conference on communications and signal processing (ICCSP), Melmaruvathur, pp 1743–1747. https://doi.org/10.1109/ICCSP.2015.7322820
8. Najeeb T, Safar M (2018) Dates maturity status and classification using image processing. In: 2018 international conference on computing sciences and engineering (ICCSE), Kuwait City, pp 1–6. https://doi.org/10.1109/ICCSE1.2018.8374209
9. Hussain Hassan NM, Nashat AA (2019) New effective techniques for automatic detection and classification of external olive fruits defects based on image processing techniques. Multidim Syst Sign Process 30:571–589 (2019). https://doi.org/10.1007/s11045-018-0573-5
10. Sri MK, Saikrishna K, Kumar VV (2020) Classification of ripening of banana fruit using convolutional neural networks. Available at SSRN: https://ssrn.com/abstract=3558355 or https://doi.org/10.2139/ssrn.3558355
11. Kangune K, Kulkarni V, Kosamkar P (2019) Grapes ripeness estimation using convolutional neural network and support vector machine. In:2019 global conference for advancement in technology (GCAT), Bangaluru, India, pp 1–5. https://doi.org/10.1109/GCAT47503.2019.8978341
12. Tran, T-T et al (2019) A comparative study of deep CNN in forecasting and classifying the macronutrient deficiencies on development of tomato plant. Appl Sci 9:1601
13. Samajpati BJ, Degadwala SD (2016) Hybrid approach for apple fruit diseases detection and classification using random forest classifier. In: 2016 international conference on communication and signal processing (ICCSP), Melmaruvathur, pp 1015–1019. https://doi.org/10.1109/ICCSP.2016.7754302
14. Sattar N, Ziauddin S, Kalsoom S, Shahid AR, Ullah R, Dar AH (2019) An orange sorting technique based on size and external defects. https://doi.org/10.1109/ICOEI.2019.8862758

Car Comparison Portal Using Real-Time Data Analysis

Archit Nangalia and Ashutosh Mishra

Abstract The system focuses on providing buyers an estimate of the similarities and dissimilarities between two cars. The end-user can select two vehicles and the system will generate a brief report indicating the benefits and defects. Thus, it helps the user to analyze cars effectively and user can make best decision before buying. This system not only provides users details of the cars, but also provides an analysis based on the data mined from Twitter in real-time and gives the users an overview of what the consumers think about the vehicle. This way the user may have a chance to fairly compare cars and come onto a better decision. The tweets are categorized as positive, negative, and neutral. In the past decade, there has been an exponential surge in the online activity of people across the globe. The volume of posts that are made on the web every second runs into millions. To add to this, the rise of social media platforms has led to flooding to content on the internet. Making use of this very content accurate sentiments about a vehicle can be conveyed.

Keywords Real-time · Analysis · Opinion mining · Prediction · Data analytics · Data science · Social mining · Sentiment analysis

1 Introduction

The urge to have your private cocooned space while travelling has led to a surge in car sales across the globe. The automotive market is predicted to see an upward trend in car ownership, rental or lease. Automobiles have tons of options based on features, make type, model, brand, class, size, price range, etc. and people easily tend to get confused to choose one particular. For the aforementioned reasons, we decided to design a portal for comparison multiple cars based on safety features, luxury features,

A. Nangalia (✉)
University of Mumbai, Mumbai MH - 400032, India
e-mail: architnan@gmail.com

A. Mishra
Springer Heidelberg, Tiergartenstr. 17, 69121 Heidelberg, Germany
e-mail: ashu.tech57@gmail.com

N. Chaki et al. (eds.), *Proceedings of International Conference on Computational Intelligence and Data Engineering*, Lecture Notes on Data Engineering and Communications Technologies 56, https://doi.org/10.1007/978-981-15-8767-2_34

necessities, fuel type, range, price, etc. Alongside these basic comparisons the USP of this portal stands at the sentiment analysis capabilities. Car Comparison Portal using Real-Time data analysis is a portal specifically designed to help people compare cars, by taking into account the opinion of fellow owners and potential customers as well as experts of the industry and the specifics they have in mind to filter down to buy a car. The system uses data mining to process data from a database and show results in form of a graph as well as tabulated results for comparison.

2 Literature Survey

The literature review will discuss existing systems that exist and why they are not capable along with the project idea and objective. This study will further advance sentiment research by helping people discover vehicles and emotions related to them.

2.1 Existing System

CarComplaints.com [1] is a little less of a comparison site, since it doesn't have tools to directly compare one vehicle against another, but it does offer a ton of information on virtually any make and model of vehicle, including common owner complaints and issues, recalls, crash tests, active investigations, and any vehicle Technical Service Bulletins (TSBs) the site can get their hands on.

The site is well worth a look even if you're looking at the vehicle you already own to make sure you haven't missed anything, but it's also great if you're looking for the complaints and experiences of people who already own and have had experiences with the vehicle you're thinking about buying. Plus, you can add your own complaints and experiences to the site to give a little back, or search complaints on the site to see if anyone's having the same issues as you are.

2.2 Project Idea and Objective

The attempt is to provide digital platform to manual comparison. The manual comparison of cars using a pen and paper is a tedious job needs a lot of information to be collected. It is always convenient to have a portal to solve your problem of choosing a car, all things in one place and graphs make it better to understand. Any person logging on the website will get to compare two cars at a time. Real-time updating of data helps people get relevant data based on other's opinion and also view their feedbacks posted on Twitter.

3 Proposed System Architecture

This chapter includes a brief description of the proposed system and requirements needed for the portal. The unique, industry first implementation of sentiment analysis of vehicles is achieved by using Twitter's API and python. The Car Comparison Portal using Real-Time data analysis comprises of different components describing different actions performed by the web application. These different components are classified into client connection, the web server, file server and the data layer. The system architecture is depicted in the Fig. 1.

3.1 Client Connection

The client accesses the website. Details from the database will be displayed on the web app. All the queries such as car search is sent to the server for processing. The client side scripts such as JavaScript make it possible to easily interact with various functionalities of the system and to visualize details.

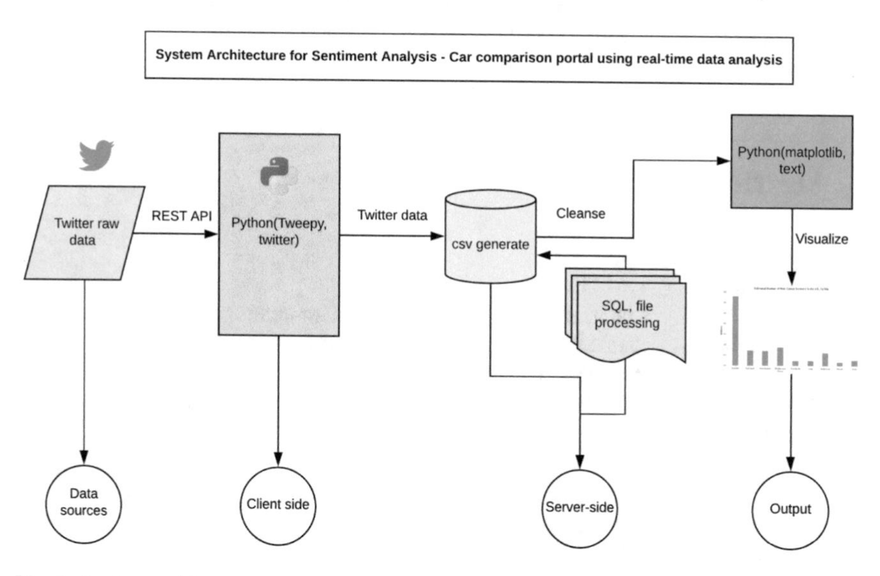

Fig. 1 System architecture

3.2 Web Server

The requests from client application is processed by the web server, where python is run locally to fetch data from the twitter client and store it locally. The file is then passed on to other layers for processing and various different operations.

3.3 File Server

This layer is responsible for processing the different file formats saved locally and prepping them for further processing. The tweets received from API response are stored in a csv file.

3.4 Data Layer

After the files are cleaned, they are stored in the databases and calculations are performed to generate polarity values. Hence, reliability of data layer is of utmost importance.

4 Algorithm

1. Initialize blank list "tweet"
2. Initialize blank list "tweetText"
3. Connect to Twitter's database using the twitter api parameters
4. Download/import tweets using two parameters, "search keyword" and "number of tweets" and store in the "tweets" list
5. Define 7 variables "polarity" "positive" "wpositive" "spositive" "negative" "wnegative" "snegative" "neutral"
6. For each tweet:
   a. Clean tweet by removing all characters that are not alphabets and spaces.
   b. Calculate polarity using nlp
   c. Polarity = polarity + calculated_polarity
   d. If polarity = 0
   Neutral = neutral + 1
   .
   .
   e. Calculate percentage for all the above variables using the number of tweets taken in step 4
   f. Calculate average polarity => polarity = polarity / NoOfTerms
   g. Print all the percentages

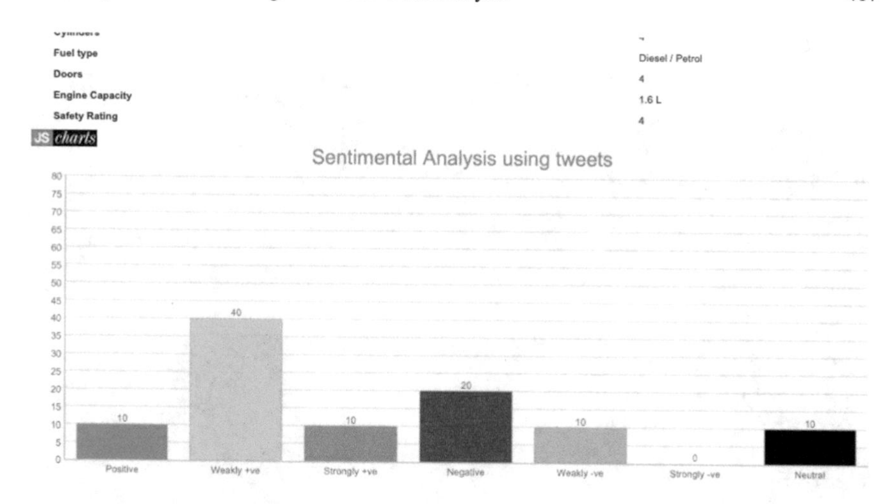

Fig. 2 JS chart generated using tweets

5 Implementation

The system uses python language at its core to process all the tweets fetched from social media platform, in this case Twitter. A token and secret key is passed to Twitter's API and vehicle specific tweets are fetched in real-time and stored on the server. The python script runs locally on the client platform.

Figure 2, depicts illustration of the data after it has been cleaned and removed of all attributes that do not contribute towards the sentiment polarity calculation. Using JavaScript, the charts are then visualized.

The tweets that were used to calculate polarity for sentiment analysis can be viewed in real-time and can also be filtered depending on the emotions. Figures 2 and 3 presents the tweets for "Hyundai Tucson" vehicle in this instance (Fig. 4).

6 Results

The results of the final web portal with actual screenshots to depict the implementation of the functionalities, follow below.

The Fig. 5 shows the image of the first screen when a user visits the website.

Figure 6 is the image of the screen displaying the list of cars with filtered search option.

Figure 7 shows the section where nearby dealers are located for the user on the map using the Google Maps' paid API.

Figure 8 shows a bar graph of the polarity of the words used relating to the car.

The admin can add or remove images of the cars added previously (Fig. 9).

The admin can modify the specifications of the car from this page (Fig. 10).

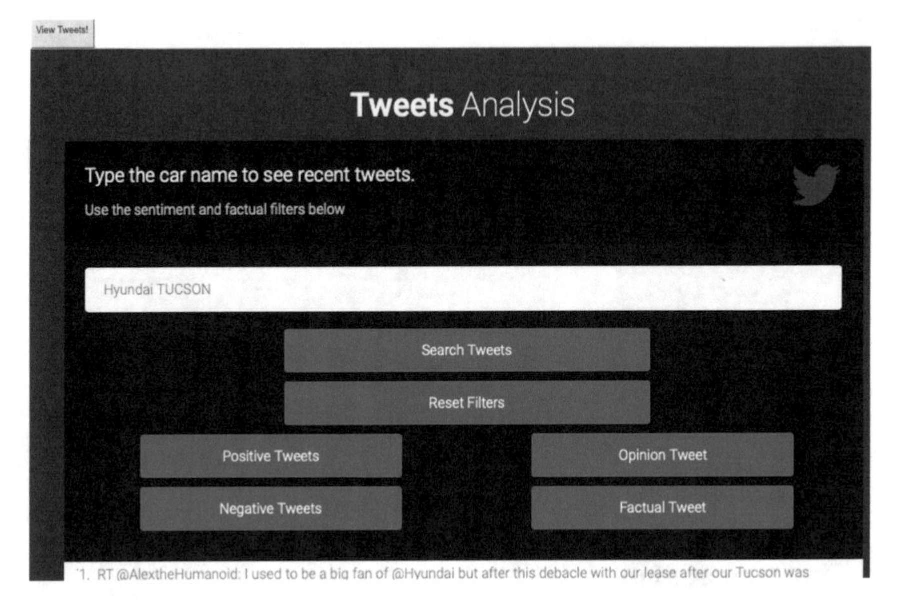

Fig. 3 Categorized tweets analysis

1. RT @AlextheHumanoid: I used to be a big fan of @Hyundai but after this debacle with our lease after our Tucson was wrecked, I will never de...
2. I used to be a big fan of @Hyundai but after this debacle with our lease after our Tucson was wrecked, I will never deal with them ever again. Car was totaled more than 200 days ago, and now you want to mess with my credit cause your team can't get their stuff together?
3. RT @AbassAdisa: Clean Hyundai Tuscon 2008 for sale 4 Cylinders 2.0L Capacity engine Call: 08186069664 for more info Mileage: 121,507km Pric...
4. @RKMac65 The exact same make and model: Hyundai Tucson Active X 2019, which I'm very happy about. I had to give my insurance a copy of the bill of sale so they could see what I'd bought. The car dealer called me yesterday to confirm the purchase of the car and needing my address.
5. NEW ARRIVAL 2017 Hyundai Tucson 1.7 CRDI Sport Edition £13,495 or £245.94 per month with only £99 deposit BUY NOW AND DON'T PAY UNTIL MAY 47,000 Miles Full service history 2 Year Hyundai... https://t.co/hnnPwxTJVJ
6. RT @AbassAdisa: Clean Hyundai Tuscon 2008 for sale 4 Cylinders 2.0L Capacity engine Call: 08186069664 for more info Mileage: 121,507km Pric...
7. oh to be a goat in the trunk of a hyundai tucson https://t.co/DPLpl8awMI
8. Drop in today and check out this beautiful 2013 Hyundai Tucson! $7,495. #houstoncars #houstoncarsales #texascarsales https://t.co/Ha97wOl98t
9. Looking for the right mid-sized vehicle for your lifestyle? The 2020 Hyundai Tuscon is perfect for nights on the town, weekend road trips or just your everyday commute. Shop our inventory here https://t.co/53RMyhbFvY https://t.co/TdJ2VXS1md

Fig. 4 Tweets related to one particular car

Fig. 5 Main page

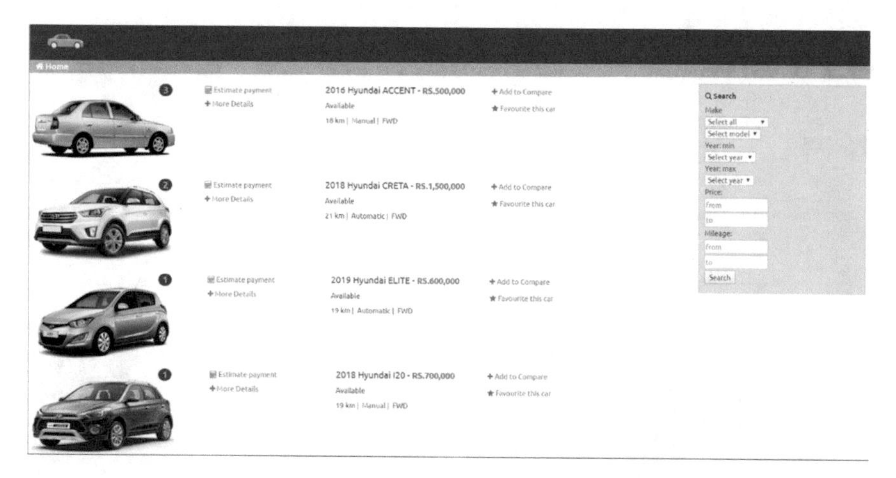

Fig. 6 List of cars

The admin can view the complete database of the inventories used for the website (Fig. 11).

7 Conclusion

The "Car Comparison Portal using Real-Time data analysis" system throughout its inception, development and design has aimed at improving the efficiency and ease of the vehicle purchase decision making for the client while being effective

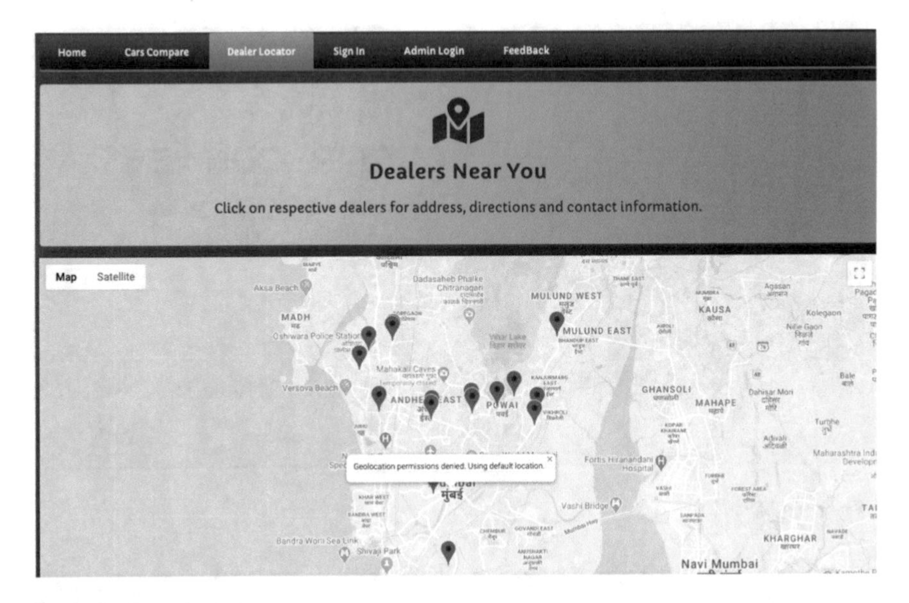

Fig. 7 Dealer locator

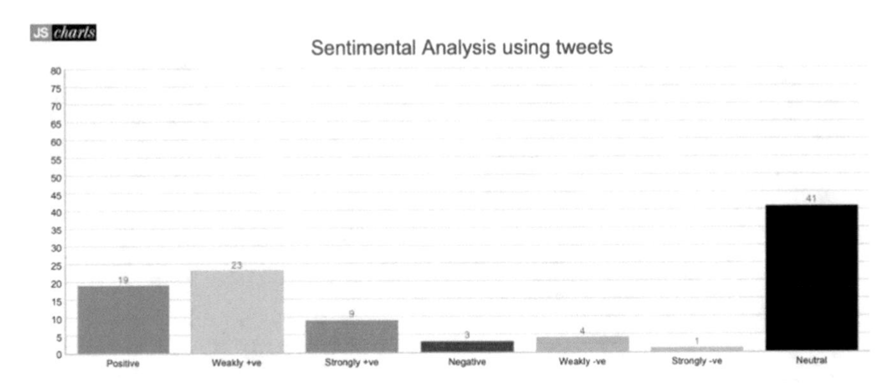

Fig. 8 JavaScript charts

and innovative. The drawbacks of existing systems have been reduced to minimum. Research on the existing system and technology has been achieved with functions that can be useful for different purposes. The system tries to be as simple and efficient as possible using accurate new technologies including data mining in real-time to automate the process of manually finding data. The visualization algorithm is purposefully designed to give detailed results. The proposed system provides almost complete automation for the data analysis of multiple vehicles simultaneously. It will give the client suggestions so as to which vehicle suits their needs and exactly what they are looking for. We hope to help many users across the globe to make right choices and reduce the human strain of manually updating and visualizing data

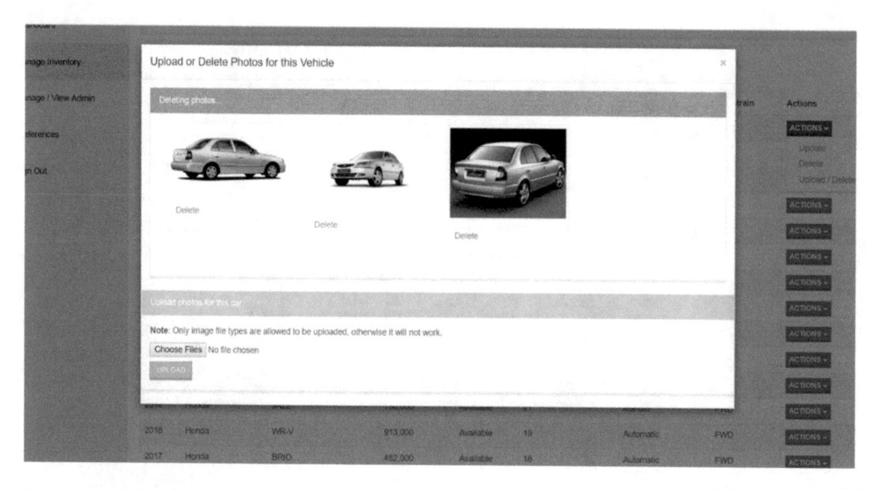

Fig. 9 Image addition/removal

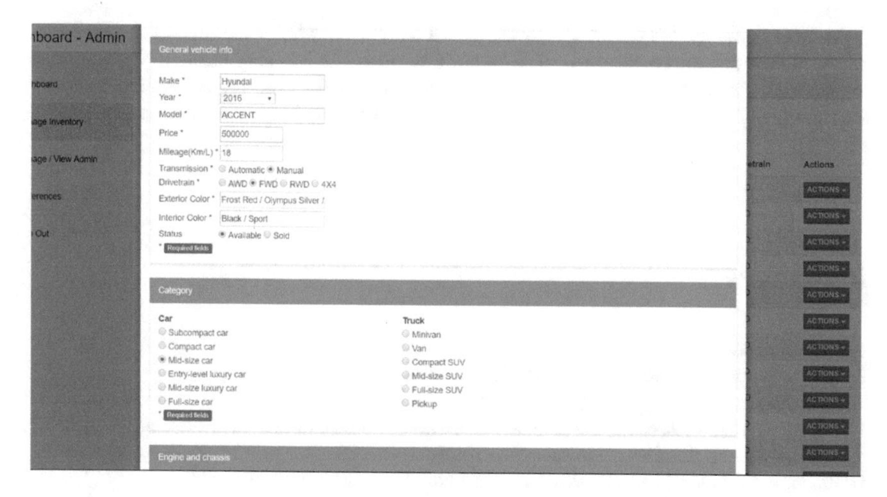

Fig. 10 Car specification modification

for the potential buyers. The proposed system reduces the work time and eliminates hours upon hours of tracking the visualizing the data. The proposed system one of the most efficient form of automating the car data analyzing process.

Dashboard - Admin

- Dashboard
- Manage Inventory
- Manage / View Admin
- Preferences
- Sign Out

Admin Panel > Manage Inventory

Manage Vehicles

Here goes vehicles from the inventory.

Year	Make	Model	Price (Rs.)	Status	Mileage (Km/L)	Transmission	Drivetrain	Actions
2016	Hyundai	ACCENT	500,000	Available	18	Manual	FWD	ACTIONS
2018	Hyundai	CRETA	1,500,000	Available	21	Automatic	FWD	ACTIONS
2019	Hyundai	ELITE	600,000	Available	19	Automatic	FWD	ACTIONS
2018	Hyundai	I20	700,000	Available	19	Manual	FWD	ACTIONS
2017	Hyundai	TUCSON	2,000,000	Available	12	Automatic	AWD	ACTIONS
2008	Hyundai	XCENT	500,000	SOLD	18	Manual	FWD	ACTIONS
2004	Hyundai	GRAND	300,000	Available	17	Manual	FWD	ACTIONS
2005	Honda	ACCORD	2,000,000	Available	14	Automatic	FWD	ACTIONS
2016	Honda	CIVIC	1,200,000	Available	16	Manual	RWD	ACTIONS
2018	Hyundai	ELANTRA	1,200,000	Available	12	Automatic	FWD	ACTIONS
2018	Honda	CRV	3,000,000	Available	12	Automatic	AWD	ACTIONS

Fig. 11 Inventory management page

Acknowledgements We would like to whole-heartedly acknowledge the support received from our institution and the constant guidance by our professors. As honorary mention, it is important to realize Twitter's generosity for providing free access of its API to upcoming developers. Lastly, the tremendous support from friends and family during this long journey of development made it possible.

References

1. https://carcomplaints.com/index.html
2. https://www.slideshare.net/ajayohri/twitter-analysis-by-kaify-rais
3. https://www.geeksforgeeks.org/twitter-sentiment-analysis-using-python/
4. https://www.analyticsvidhya.com/blog/2018/07/hands-on-sentiment-analysis-dataset-python/
5. https://blog.insightsatlas.com/7-benefits-of-sentiment-analysis-you-cant-overlook
6. https://www.digitaltrends.com/cars/10-car-technologies-to-be-thankful-for/
7. https://www.altexsoft.com/blog/business/sentiment-analysis-types-tools-and-use-cases/

Design and Development of an Efficient Malware Detection Using ML

Mathe Ramakrishna, Aravapalli Rama Satish, and P. S. S. Siva Krishna

Abstract Enormous growth and generation of data is happening in every day from various sources. The generated data is presented in various formats, i.e., in structured, unstructured, semi-structured, pdfs, docs, csvs, and raw file formats. All these files are not genuine or pure in all scenarios cause which is generated from identified and unidentified sources. The modern malware is designed with mutation characteristics, that means, it can change its behavior based on the properties of physical file. It is a contraction from malicious software. The tremendous growth of the data is very helpful to the malware designers to execute the malware files such as Virus, Trojans, and Ransomware in any file. The formation of modern malware poses a variety of challenges to the antivirus industries. In this paper, we are going to induce a system with a lightweight model to accurately detect the malware for industrial use with high accuracy. In this, we are identifying nine different types of malwares like Ramnit, Lollipop, Kelihos_ver3, Vundo, etc., on huge amount of data (0.5 TB) that is provided by Microsoft.

Keywords Windows malware · Computer security · Machine learning · Static analysis · Malware classification · Microsoft malware data

M. Ramakrishna (✉)
Research Scholar, School of Sciences, Career Point University, Kota, Kota, Rajasthan, India
e-mail: mathe.ramakrishna@gmail.com

A. Rama Satish
Assistant Professor Sr. Grade 1, School of Computer Science & Engineering, VIT-AP, Amaravati, Andhra Pradesh, India
e-mail: ramsatpm@gmail.com; rama.satish@vitap.ac.in

P. S. S. Siva Krishna
Nannaya University, Rajahmundry, India
e-mail: sivakrishna606@gmail.com

N. Chaki et al. (eds.), *Proceedings of International Conference on Computational Intelligence and Data Engineering*, Lecture Notes on Data Engineering and Communications Technologies 56, https://doi.org/10.1007/978-981-15-8767-2_35

1 Introduction

The term malware has deteriorated as malicious software. The goal of malware is any bit of software that was made with the plan of doing mischief to information, security, gadgets, or to individuals. With the rise of the internet, malware usually infects a system by tricking users; while users are clicking or installing an unknown program or software from the internet, a malware is going to access systems through the emails and other unknown resources. On the other hand, spreading of the malware through USB and other secondary devices. Social engineering and phishing attacks are the most common malwares spreading techniques. By fooling the people, malware developers update their technology to attack the systems. By observing the attacking ways and their characteristics, malwares are basically categorized into Virus, Trojans, Spyware, Worms, Ransomware, and Botnets.

Rise of malware technology challenges anti-malware industries. To produce adequate computer security and protection, anti-malware industries need to update themselves to prevent the attacks. Whenever anti-malware developers update their mechanisms to detect and prevent the malwares, malware writers update the mechanisms to elude from the preventions.

In recent years, traditional signature-based techniques will be the basis of anti-malware vendors to detect the malware. The rise of the darknet is the basis for malware development. Malware coders develop their code by using the genetic and mutation technology to update their malware mechanism. Because of this, mutation technology detection of the malware is tougher since all the present-day malware technologies are, in general, having polymorphic and metamorphic layers to sidestep the detection from the anti-malware mechanisms. Inventors of malware developed it using inventory techniques to escape from the modern malware protection systems; at the same time, anti-malware vendors are developing countermechanisms to perceive and block the malware.

Analyzing the gigantic data to recognize malware families is a time- and cost-utilizing process. Automatic and manual signature extractions are identifying the previously trained malware. A petty solution is there, i.e., by selecting the random data in order to identify whether malware is present in that or not. This method leads to update the signature repositories with enhanced malware. Defect with this problem is the identification of malware with random data had performance issues and low accuracy values.

To viably investigate a huge number of new mischievous code, documents hostile to malware sellers implement the machine learning (ML) technology and integrate this to the signature extraction repositories to update and detect the malware.

2 Literature Survey

Ahmadi et al. [1] in the current paper gave importance to the stages identified with the extraction and determination of the features for the powerful interpretation of malware samples. Depending on the malware behavior, features can be grouped and their merging is accomplished by per class weighting model.

For the computer security group, tagging of malware samples based on their symptoms is very important since they get a tremendous number of malware consistently, and the signature extraction process is typically based on malicious parts tagging. Analysis of this dataset helped in building a novel and viable way of sorting the malware variants into their legitimate family group. With the limited resources, he suggested a model with novel structural features that could gain in performance. Be that as it may, this technique isn't yet been tried for potency against evasion or poison attacks.

Nissim et al. [2] unlike previously, making malware today became simple in light of the fact that mischievous code libraries are shared between attackers. Besides, nowadays invaders have got more advanced technologies, making mischievous code documents that appear to act like normal documents which are harder to identify, for example, the case with Trojan horse. Furthermore, before making mischievous codes, attacker will try to find the susceptibilities already existed or will shadow the recent announcements regarding the susceptibilities. Additionally, the attacker will try to make use of the time delays in delivering the updates by the anti-virus vendors, which allow the virus to attack and spread.

In the process of improving the model, the authors proposed a framework and learning techniques for regular update of anti-virus software by concentrating on marking the file that is destined as malware or benevolent document. Both new and marked document lists must be updated regularly to the anti-virus and detection model. New malware which are not recognized by anti-virus can be detected by using a competent detection discovery model. The detection model is depended on a classifier which is trained by utilizing static examination on both malicious and benevolent documents. It enriches the speculation capacities of the model which in turn boosts up the new malware distinguishing capability at a high positive rate, reducing the possibility of infecting the host PC. The framework shows how we can obtain the details of significant files, benevolent files, and new malware by using learning techniques, which improve the performance of classifier, permitting it to discover and update the anti-virus signature repository with newly discovered files. As with text categorization, static investigation system was introduced for identifying unknown malevolent codes. In static investigation, the analysis will be done without executing the actual program. The detection and classification of unknown malware can only be done if its actual behavior known, which can be obtained by dynamic analysis. In this model, dynamic analysis not performed.

Fan et al. [3] proposed malware detection framework by mischievous sequential pattern identification using an operative sequence mining algorithm and constructing a classifier called All-Nearest Neighbor (ANN) on the identified patterns. The proposed framework can outperform in the detection of malevolent patterns both in sample file as well as newly unobserved malware samples.

Heuristic-based recognition strategy, which uses data mining and ML, is created for malware identification. This methodology discover special patterns that catch the malware qualities. The discovery procedure is done in two stages: primary and secondary. In the primary stage, static or dynamic investigation is done to differentiate features from malware samples. Basing on the primary stage extracted feature, classification is applied to distinguish the malware instinctively.

Since the system proposed in here just spotlight on detection of malware, for example, regardless of whether a sample is malware or not, it cannot classify which specific sorts of malware it is. This shortcoming would limit the technique not being applicable to increasingly broad applications. Another shortcoming of the technique acquires from the the conventional kNN strategy, i.e., the absence of an unequivocal model. This leads our doors open to work on framework in future by consolidating a few techniques, for example, data reduction so as to improve the classification efficiency. These features denote the noteworthy contrast between malicious and benevolent files.

Tian et al. [4] proposes a versatile methodology in which the behavioral features are explored utilizing logs of different API calls to distinguish malicious files and clean files. The work suggests having, in contrast to the conventional technique for recognizing malware records, a robotized order framework utilizing runtime features of malware documents. As a result, the dynamic malware classification framework was designed by using the dynamic API call sequences as features. The whole experiment is carried out in a virtual environment for a limited period of time. The actual behavior of the code is revealed in the form of trace reports. Now distinguishing of cleanware from malware can be done using trace reports of binary files in two ways. (i) malware versus cleanware classification and (ii) malware family classification. By integrating both features that we can increase classification and detection accuracy in malware analysis.

The behavioral feature analysis helps a lot in malware detection. The most essential contributions in the detection are planning of a technique to extract relevant behavioral features of API calls; arrangement of statistical investigation of the API calls from log documents; defining a classification model for recognizing malware from cleanware.

Gavrilut et al. [5] proposed an adaptable system wherein one can utilize distinctive machine learning algorithms to effectively differentiate between malware files and clean files, while limiting the quantity of false positives. In this work, they present the thoughts behind our framework by working right off the bat with cascade one-sided perceptron's and furthermore with cascade kernelized one-sided perceptrons.

In the wake of having been effectively tried on medium-sized datasets of malware and clean documents, the thoughts behind this framework were submitted to a scaling-up process that empowers us to work with enormous datasets of malware and clean files. A problem that occurs when working with large datasets is overfitting caused by the noise appearing in the form of human annotation errors. Not all of the malware-designated samples are actually malware, and not all of the clean samples are clean indeed. That is why, the bigger the database, the more likely is to get misclassified samples in the training set. Because our algorithms aim to reduce the number of false alarms to 0, the detection rate (sensitivity) obtained on a large dataset will be much smaller (due to the misclassification issue).

2.1 Problem Statement

To protect a system against malware, the significant undertaking is to distinguish whether the given piece of document/program is influenced with malware or not. So as to distinguish the malware, we need to analyze and classify the data and construct groups which are called as families and recognize the families.

3 Proposed System

Microsoft develops countermechanisms to detect the malware; for this, it performs an analysis on almost 15 million systems to gather the malware data. And after that, it produces a dataset which is almost half terabyte and it contains nine different malware families. Those malwares are Ramnit (R), Lollipop (L), Kelihos_ver (K3), Vundo (V), Simda (S), Tracur (T), Kelihos_ver1 (K1), Obfuscator.ACY (O), Gatak (G). Nine classes of malware are labeled with the integer values from range 1 to 9, i.e., Ramnit (R) is class 1 and Gatak (G) is class 9.

In this work, we evoke a model that will classify malware based on their characteristics and assign it to the respected families. This approach uses learning-based mechanism to characterize the malware. Learning-based mechanism does not rely on packed and de-obfuscation systems. When the packer is known and extracts more valuable features, it is a cost taking process and performs analysis on customized packers which is complex too. Thus, we aim to perform the classification task without unpacking the samples.

Alongside to this, malware classification system assumes all the samples as if they are malware, and because of this, the packed systems are not analyzed. Finally, rather than analyzing and detecting malware evade techniques, we are more consigned toward the classification of malware.

Malware detection and classification can be considered as two important stages in signature development process for anti-malware products. Malware detection is to detect the malware and classification mechanism classifies the malware and assigns it to the respected families. These two tasks are performed on both dynamic and static investigation. *Dynamic analysis* concentrates on the behavior of the malware and its characteristics while it is executing. Developers use Windows API to access the files, processes, and other systems. At the same time, the development of API's leads to use them rather than using system calls, even applications use APIs to execute themselves apart from the direct system calls. By monitoring the APIs and their

Table 1 Few researches under gone with their features and structure

Year	Authors	Analysis	Type	Features	Structure
2008	Tian et al. [4]	Static and dynamic	Classification	Static and dynamic	–
2009	Santos et al. [6]	Static	Classification	Byte	N-gram
2010	Ronghua et al. [4]	Dynamic	Classification	API	String
2011	Gavrilut et al. [7]	Dynamic	Classification	Binary	Graph
2014	Nissim et al. [8]	Dynamic	Detection	Byte	N-gram
2016	Ahmadi et al. [9]	Static	Classification	Structural	N-gram
2017	Liu et al. [10]	Static	Classification, detection	Texture	Gray scale images

parameters, we can analyze the metadata to identify the malwares. As the dynamic analysis has a limitation of analyzing one malware at a time, programmers using static analysis, by doing it, we analyze the malware without actually executing it. So, in this analysis, we have to identify the patterns of the malware; there are many techniques to analyze the patterns like SAFE, SAVE, Microsoft's header of the PE (Portable Executable), and body of the PE.

All the malware detection and malware classification systems performed by extracting the features from dynamic, static features, both features were same. All the researchers used the same features to discriminate and classify the malware. By employing dynamic analysis, we have to extract the features from it but we already know the classifier assumes that all files were affected with malware. We mainly focus on the classification of malware based on static features to design more accurate and performing system. Allotment of data among the malware shows the statistics of the malware, in that class 3 malware is highly present. The proposed method is to design a novel method that could provide a performance within a limited feature and with low computational task. Rather than using the clustering system, we use multi-threading system to maintain a trade-off between complexity and performance (Table 1).

3.1 System Design

We majorly focus on malware classification by using statistic features, in that the relevant issue is to choose the appropriate features from the analysis. The aim of our model is to achieve more accurate and fast classification results, for this, we will choose the most appropriate features and integrate them. We are going to use both content-based and structural features to achieve the results (Fig. 1).

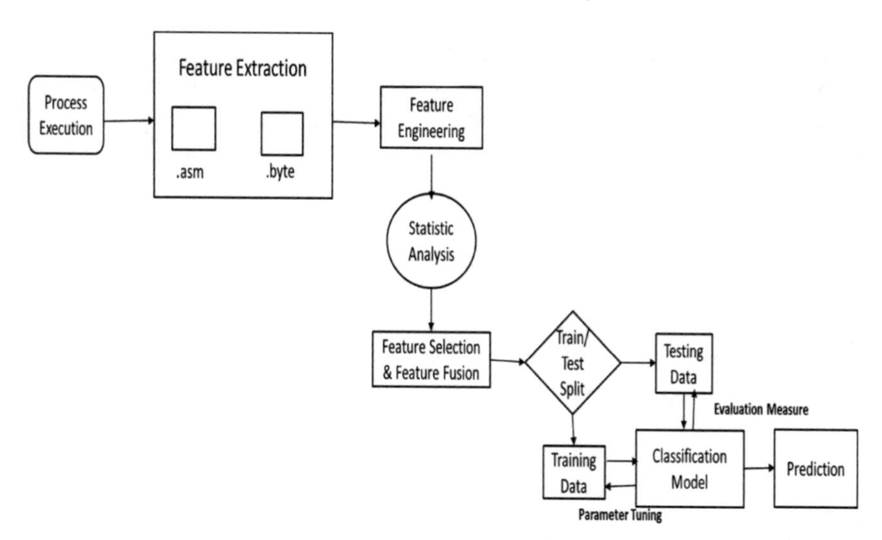

Fig. 1 Proposed system design

3.2 Malware Data Representation

Feature extraction of malware classification is always based on data, so before going to know about the features, we get an overview about the malware which is presented on the data. Every malware sample will contain two different types of files, that is, hexadecimal code (hex) file and disassembled code or byte code. When a program is executing raw code, it is converted to assembly language by compiler and assembler converts it into hexadecimal code. Then malware sample representations are presented by the two views, namely hex view and assembly view. The assembly view of data is an unassembled data and it is the collection of PE Headers, PE body, Instructions, and opcodes.

3.3 Program Code

.asm Files (Assembly View)

```
.text:00419B02 2B CE sub       ecx, esi
.text:00419B07 8D 4C 85 E01    ea ecx, [ebp+eax*4+var_20]
.text:00419B0B 8B 31           mov    esi, [ecx]
.text:00419B0D 8D 3C 16        lea     edi, [esi+edx]
```

The hex view data is presented in the form of digits, which is a collection a 16-bytes words, it is presented in the following representation: the initial value signifies the opening address of these machine codes in memory and each values (byte) bears significant elements for the PE, alike instruction codes or data.

.byte files (Hex-View)

```
00401000 56 8D 44 24 08 50 8B F1 E8 1C 1B 00 00 C7 06 08
00401010 BB 42 00 8B C6 5E C2 04 00 CC CC CC CC CC CC CC
00401020 C7 01 08 BB 42 00 E9 26 1C 00 00 CC CC CC CC CC
00401030 56 8B F1 C7 06 08 BB 42 00 E8 13 1C 00 00 F6 44
```

3.4 Features Extracted from the Byte Files

n-gram: From a given sample of data of the application in a given sequence, the contiguous n items are called n-gram. It is typically collected from text. The representation of malware data is present in byte sequence of words. To effectively analyze this data to get the beneficial information from it and to identify the malware properties, sliding windows of n-gram lengths are used. The n-gram construing was worked on data without decompression and decryption. The element in byte data is taken as a sequence of words, if 100 words are present in that, **1G** delivers 100^1-dimensional vector words that are present there and **2G** delivers 100^2-dimensional vector words present in 2-byte frequency. As our model representation, we are just considered 1-gram features for low computational process.

Classification: Over the last few classifications, many were proposed by the experts. Multiple classification algorithms are proposed to classify the malware; among those, we have selected the best one using the parameters like accuracy, train rate, test rate, and error rate. In total, 179 classification algorithms from 17 families have been selected to classify the data, and they concluded that SVM and Random forest classification algorithms performance is the worthiest. Most of the Kaggle winners are using XGBoost algorithm to increase the model accuracy. By observing all these in this paper, we are using ensembled algorithms, i.e., Random Forest and XGBoost to accurately classify the malware.

Evaluation Measures: Accuracy and log-loss are the two evaluation measures used to estimate the classification performance. Accuracy is defined as the fraction of the correct the predictions and can be presented as the confusion matrix. Log-loss is the delicate estimation of precision that is solidified with the notation of probabilistic confidence. It is called as cross-entropy between the true labels and the anticipated labels distribution. It functions admirably for multi-class classification. Negative log-likelihood of the model can be demosntrated as

$$logLoss = \frac{-1}{N}\sum_{a=1}^{N}\sum_{b=1}^{M} y_{ab} * log(p_{ab})$$

where N is the number of observations, M is the number of class labels, log is the natural logarithm, y_{ab} is 1 if observation a is in class b and 0 otherwise, and p_{ab} is the predicted probability that observation a is in class b.

4 Results

After extracting the features from the represented data, we can, without much of a stretch, measure the malware samples according to their malevolent sequential patterns. By comparing four selective classification algorithms on both *byte* and *asm* file features, the effectiveness and proficiency of our experiment is evaluated. Those algorithms are K-Nearest Neighbor, Logistic Regression, Random Forest, and XGboost. Only ensemble methods are applied on data when combining all features. Reduction of variance and bias is done by K-fold cross-validation. K-values can have between values ranging *n* 5–10. The cross-validation time is directly proportional to chosen fold value. We set k-fold cross-validation value to 5, to find a baseline model for cross-validation.

The below tables are described as follows: The algorithms which we mentioned already, performed on the extracted features from the data. K-NN is a very simple and elegant classifier, it classifies the malware based on their malicious patterns from the extracted features on both *byte* and *asm* data. We already mentioned the features of the data, according to that, K-NN classifies the malware using byte file features with 95.5% accuracy and 97.98% classifies the malware using *asm* file features. Even though logistic regression classifies the malware with less accuracy that is with 77.68% on byte file features, but it gets high accuracy 90.39% when it uses asm file features. Random Forest is an ensemble algorithm which classifies the malware families more accurately than the normal classifiers and it classifies malware with 97.98% accuracy using byte files and get 98.85% accuracy on *asm* data features (Fig. 2).

By using bagging technique, we combined all the features to accurately classify the malware, in that, we are only using Random Forest and XGboost ensemble classifiers in feature fusion technique and XGboost gets a very high accuracy among all the other algorithms, which is 99.83% accuracy along with 0.01 log-loss value (Tables 2, 3 and 4).

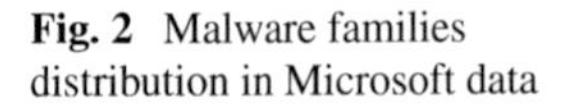
Fig. 2 Malware families distribution in Microsoft data

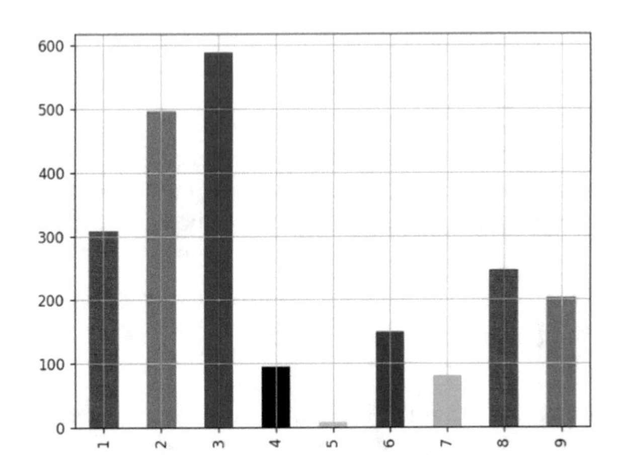

Table 2 Accuracy values of various classifiers on byte file features

Classifier	Test Log-loss (%)	Misclassification rate (%)	Accuracy
K-NN	0.24	4.50	95.50
Logistic regression	0.528	12.32	77.68
Random forest	0.085	2.02	97.98
XGboost	0.078	1.24	98.76

Table 3 Accuracy values of various classifiers on asm file features

Classifier	Test Log-loss (%)	Misclassification rate (%)	Accuracy
K-NN	0.089	2.02	97.98
Logistic regression	0.415	9.61	90.39
Random forest	0.057	1.15	98.85
XGboost	0.048	0.87	99.13

Table 4 Accuracy values of ensemble classifiers on combined features

Classifier	Test Log-loss (%)	Misclassification rate (%)	Accuracy
Random forest	0.031	0.88	99.12
XGboost	0.01	0.17	99.83

5 Conclusion

The goal of this work is to develop model that can detect and classify the known and obscure malware productively. With prediction and classification tools, we can identify the new unknown malware that can be used for sustaining an anti-virus tool.

Anti-virus tools and the detection or classifier model must be refreshed with new and label files to recognize the unknown malware.

In this work, we introduced a malware classification framework characterized by a low intricacy in feature extraction and classification mechanism. For this, we proposed various novel features to define the characteristics of the malware. Specifically, we concentered around features based on content, because of its simplicity to compute, and without the need of dynamic features and unpacking of malware classification of is possible. The principle inspiration driving this work is to make a light framework for the industrial use where the cloud technologies are growing exponentially [11], and is to keep up a trade-off among computational, performance, and classification complexity on unknown data. The ease in the usage of features are designed to understand the industrialists compared with the complex one.

References

1. Ahmadi M, Ulyanov D, Semenov S, Trofimov M, Giacinto G (2016) Novel feature extraction, selection and fusion for effective malware family classification. https://doi.org/10.1145/2857705.2857713
2. Nissim N, Moskovitch R, Rokach L, Elovici Y (2014) Novel active learning methods for enhanced PC malware detection in windows OS. Expert Syst Appl 41:5843–5857. https://doi.org/10.1016/j.eswa.2014.02.053
3. Fan Y, Ye Y, Chen L (2016) Malicious sequential pattern mining for automatic malware detection. Expert Syst Appl 52. https://doi.org/10.1016/j.eswa.2016.01.002
4. Tian R, Islam R, Batten L, Versteeg S (2010) Differentiating malware from cleanware using behavioural analysis. In: Proceedings of the 5th IEEE International Conference on Malicious and Unwanted Software, Malware, pp 23–30. https://doi.org/10.1109/MALWARE.2010.5665796
5. Gavriluţ D, Cimpoesu M, Anton D, Ciortuz L (2009) Malware detection using machine learning. 4:735–741. https://doi.org/10.1109/imcsit.2009.5352759
6. Santos I, Penya Y, Devesa J, Bringas P (2009) N-grams-based file signatures for malware detection, pp 317–320
7. Gavrilut D, Ciortuz L (2011) Dealing with class noise in large training datasets for malware detection. In: 2011 13th International symposium on symbolic and numeric algorithms for scientific computing, Timisoara, pp 401–407. https://doi.org/10.1109/SYNASC.2011.39
8. Nissim N, Moskovitch R, LiorRokach, Elovici Y (2014) Novel active learning methods for enhanced PC malware detection in windows OS. Expert Sys Appl 41(13):5843–5857. ISSN 0957-4174
9. Ahmadi M, Ulyanov D, Semenov S, Trofimov M, Giacinto G (2016) Novel feature extraction, selection, and fusion for effective malware family classification. CODASPY 183–194
10. Liu L, Wang B, Yu B, et al (2017) Automatic malware classification and new malware detection using machine learning. Frontiers Inf Technol Electronic Eng 18:1336–1347. https://doi.org/10.1631/FITEE.1601325
11. Kishore V, Bhavani GGG, Ramakrishna M (2012) Simplifying complex tasks: Cloud computing. Int J Comput Sci Inf Technol 3.5:4988–4996
12. Chowdhury M, Rahman A, Islam MR (2018) Malware analysis and detection using data mining and machine learning classification. 266–274. https://doi.org/10.1007/978-3-319-67071-3_33

Estimating Food Nutrients Using Region-Based Convolutional Neural Network

Ekta Sarda, Priyanka Deshmukh, Snehal Bhole, and Shubham Jadhav

Abstract Good Nutrition and Healthy habits are the keys to achieve a healthy lifestyle for an individual. Changes in working environments and food habits have shown a great impact on the large part of the modern population. Due to such changes, obesity, overweight, and chronic heart diseases are some of the commonly faced issues. Even for people with adequate weight, poor nutrition is associated with major risks that can cause illness and even death. A combination of physical activities and a balanced diet can help to achieve and maintain the ideal weight, reduce risks of diseases like heart attack, cancer, etc. and assist an individual to promote overall health. Food Intake Calories Estimation is an assistive calorie measurement system to help the people to determine the nutritional value in the food that they eat every day. This is an image processing based approach that allows users to capture an image of the food and measure the amount of nutrient content in it automatically. The two commonly known algorithms Faster Region-based Convolutional Neural Networks (Faster-RCNN) and Mask Region-based Convolutional Neural Networks (Mask RCNN) are compared here on various criteria to achieve the goal of estimating the nutrients in the food item in the given image.

Keywords Nutrition · Nutrient and calorie measurement · Faster RCNN · Mask RCNN · Object detection

Supported by organization Ramrao Adik Institute of Technology, Nerul.

E. Sarda (✉) · P. Deshmukh · S. Bhole · S. Jadhav
Ramrao Adik Institute of Technology Department of Computer Engineering, Mumbai, India
e-mail: ektasarda16@gmail.com

P. Deshmukh
e-mail: priyankadeshmukh186@gmail.com

S. Bhole
e-mail: bhole.snehal.16ce1063@gmail.com

S. Jadhav
e-mail: shub13711jadhav@gmail.com

N. Chaki et al. (eds.), *Proceedings of International Conference on Computational Intelligence and Data Engineering*, Lecture Notes on Data Engineering and Communications Technologies 56, https://doi.org/10.1007/978-981-15-8767-2_36

1 Introduction

The life we live in today is much different than what our ancestors have lived. The modern life comes with various advantages like fast-growing technological advancements, higher living standards, development of health facilities, vaccines, preventive measures for the spread of an epidemic, etc. On the other hand, lifestyle patterns have adverse effects on our health physically, mentally, and emotionally. A few of these are poor eating habits, high-calorie intake food, irregular eating time, long working hours, etc. Due to a lack of physical activity with a combination of high-calorie fast food, obesity has become a serious threat. World Health Organization (WHO) [1] determines the criteria for obesity. For adults, if the Body Mass Index (BMI) is greater than or equal to 25 then it lies in the overweight category whereas if BMI greater than 30 is in the obese category.

The fundamental cause of obesity is an imbalance in the amount of calorie input and requirement of the body. Nowadays, people are aware of the fact that there cannot be any alternative to healthy habits. Thus, there are various approaches people consider for maintaining the balance between their hectic working schedule and health. Due to advancement of technology seeking help for online resources is one of them.

Various systems are working for assisting them to maintain this balance. Food Intake Calorie Estimation is also like one of those systems that they can consider. It is a web application where they can know the calories and nutritional content of food by clicking a picture of the food item which can help them to set the control on their eating habits. This could be one of the basic steps an individual can take to live healthily.

1.1 Literature Survey

Many studies have attempted to monitor the number of calories present in the given food dish. In our review, we have found the following papers related to Food Calorie prediction and Convolutional Neural Networks. We have summarized the information of research papers studied as below.

The paper [2] focuses on three existing works and two newly proposed methods for estimating food calories. The first system, Calorie Cam estimates food calories from a single food item image. 120 Japanese food images containing 20 + categories of food were used as a dataset. It uses reference object color pixel-based k-means clustering and grab-cut segmentation for region extraction using which they obtained relative error of 20%. In the second system, region segmentation-based food calorie estimation is done which can be used for multiple dish meals. Third System AR DeepCalorieCam V2 uses the Apple ARKit framework to calculate the actual size of the meal area by the latest Apple phones to calculate actual volumes automatically. As compared to previous methods it improves accuracy. Another newly proposed method is RiceCalorieCam that uses rice grains as reference objects to calculate the real size of the input image for which real-size annotated boiled rice photos were used as a dataset.

In [3], a system that uses feature extractions and a classifier to predict the food calories present in the image are proposed. They used the Food Pics dataset consisting of 568 food images. The calories to be predicted were validated by a trained nutritionist to provide accuracy in the results. Features were extracted using nine image properties and in the classifier part, training and testing of calories were carried out using generalized regression neural network. Some food calories can be closely predicted but the other foods have large calorie prediction errors.

The paper [4] describes the prototype which can identify calories for five types of traditional Malaysian desserts. It uses a color histogram for feature extraction and an artificial neural network for food classification. The dataset used for training consisted of the images taken from publicly available sources. The images from the dataset were inconsistent in terms of background and angles. The accuracy of 80% was obtained by this system.

Nutritrack [5] is an android-based food recognition app that uses Clarifai API In the android environment to detect the food items with specialized ML algorithms. The Nutritrack app utilizes a cloud-based database hosted by an android API to support implementation. These APIs deliver precise results for detecting food and nutrients based on the users input image.

The paper [6] uses multi-task CNN to estimate both food calories and food categories. Dataset consisted of 4877 images on 15 categories, obtained from six different commercial cooking websites. They experimented with both single task CNN and multi-task CNN, which showed that multi-task CNN outperformed single task CNN. By training food calories and food categories simultaneously by multi-task CNN, the performance of both the estimation was boosted compared to the result by independent single task CNN. The average classification accuracy on the top 200 samples with the larger error was 71%.

The paper [7] proposes a method for identification of the different food items present in the plate. The technique used in this paper is purely based on the HSV color model. The hue, saturation, and value of the image are calculated using various equations based on RGB color channels. The calculated values along with image segmentation are then used to make the final detection. The GUI for the system was designed using MATLABR2011a. However, as the detection of food was based only on the HSV values, there were some limitations in this system.

2 Proposed Methodology

In this section, we will give brief description about the proposal of work and methodology to be used.

The Fig. 1 shows the diagrammatic representation of the flow or sequence in which the process takes place. First Step in the process is Image Acquisition in

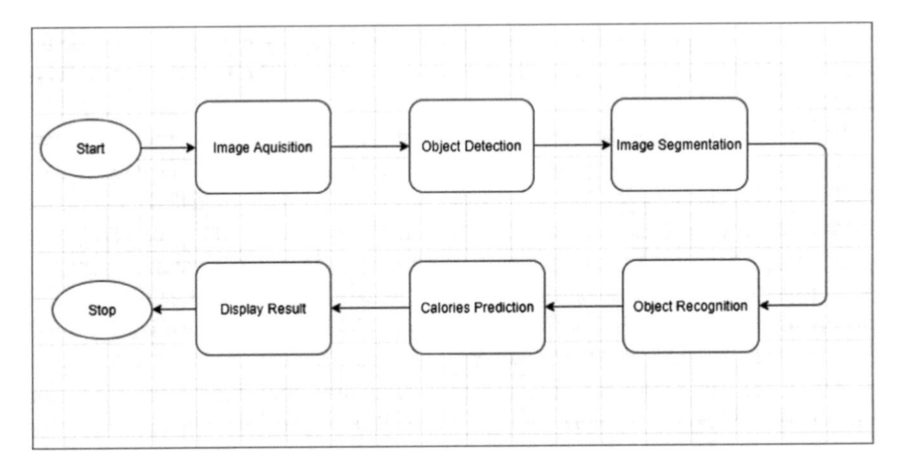

Fig. 1 Flowchart of the proposed system

which image is taken as input to determine the calories of the food. Second step is Object Detection in which Regional Convolution Neural Network uses Region Proposal Network (RPN) to give proposal of various objects present in an image. Third step carries out image segmentation from the detected Objects. In the fourth step, object is recognized from the segments obtained in the previous step. Finally, the calories corresponding to classified objects are displayed [8].

2.1 Faster R-CNN

Faster R-CNN is a state-of-the-art object detection neural network that depends on region proposal algorithm to hypothesize object locations. Faster R-CNN consists of three networks: pretrained image detection network such as VGG, region proposal network (RPN) for creating region proposals, and a network for detecting objects using these proposals.

2.1.1 Feature Network

Deciding which of the network configurations will produce the best performance is debatable. Initially, ZF-net and pretrained VGG were used in Faster R-CNN. We have used ResNet-50, a complex neural network trained on the ImageNet database consisting of a million images. ResNet has the benefit of being faster and bigger than VGG, so it is far more able to learn what is really important for performing tasks such as object detection and object classification.

Fig. 2 Bounding boxes generated for apple and banana

2.1.2 Anchors

Here the goal is to locate rectangular boxes in the picture in different measurements and aspect ratios. Anchors are those fixed bounding boxes that are placed all over an image with distinct aspect ratios and sizes. We can make fine adjustments to the reference boxes by changing the aspect ratio, which is basically width and height of the reference box. Accuracy, as well as the speed of the model, can be increased if the correct set of anchors is configured. These anchors are defined according to the feature maps. In this paper, we have used 256 anchors per image for RPN training.

2.1.3 Region Proposal Network

The RPN takes all the obtained reference boxes (anchors) and successfully offers a set of proposals for objects. This is achieved by generating for each anchor two seperate outputs. The first output is the probability whether an anchor, is an object or not. RPN doesn't really care to what class of object it belongs to, but whether it is an object or not. Then this objectness score used to filter out the bad predictions for the second stage. The second output will be the bounding box regression for adjusting the anchors so they fit the object better than its predicting.

2.1.4 Detection Network

Input from Feature Network and RPN are used in this network, to generate the final class using bounding boxes. It is made up of 4 Dense layers, wherein a classification layer and bounding box regression layer shared by 2 stacked common layers. The features are cropped according to the bounding boxes, to help it identify only the inside of the bounding boxes. The Fig. 2 and Fig. 3 shows Bounding Boxes created for certain images from our Dataset.

Fig. 3 Bounding boxes generated for vada paav and samosa

2.2 *MASK R-CNN*

A deep neural network called Mask Region Based Convolutional Network (Mask R-CNN) is aimed to solve the problem of instance segmentation in machine learning. It is built on top of Faster R-CNN. In addition to the class label and bounding box coordinates for each object, it also returns mask for each object (Figs. 4, 5).

2.2.1 Backbone Model

Many ConvNets are available to serve as a backbone model for feature extraction and classification. After studying popular ConvNets such as VGG, Inception, and ResNet, we found that ResNet has huge memory and computation requirements, especially while training. However, it produces more accuracy. So, in this paper

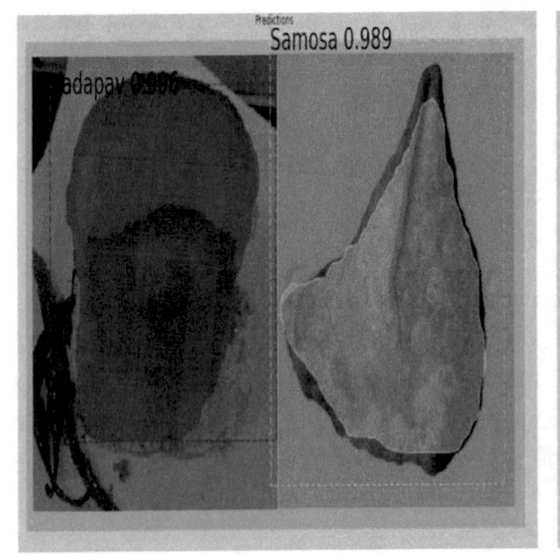

Fig. 4 Masks generated for samosa and vada paav

Fig. 5 Mask generated for banana and mango

to extract feature maps from the image, we used the ResNet 50 architecture. After features are extracted from the image, they are passed as an input for the next Layer. Region Proposal Network (RPN) and obtaining region of interests to predict class labels and bounding boxes are similar to Faster RCNN. In addition to this, Mask R-CNN also generates the segmentation mask.

2.2.2 Segmentation Mask

Once we have the Regions of interests for the given image, we can add a mask branch to the existing architecture which returns the segmentation mask for each region that contains an object. In our model, a mask of size 224 × 224 is returned and then scaled down for inference

2.3 Dataset

In this section, we will give brief description of the food dataset along with various image annotation tools used for preparing the dataset. The dataset we have used is a custom dataset containing about 350 images belonging to 5 different classes. All the images used in the dataset are scraped from Google and each image is labeled manually. Table 1 shows different types of food categories present in the dataset.

2.4 Pre-Processing

Fist step in implementation process is **pre-processing**. In pre-processing, we used two bounding boxes annotation tools to annotate dataset images.

Table 1 Food type categories

Category	Type of food
1	Apple
2	Banana
3	Mango
4	Vadapav
5	Samosa

1. LabelImg-LabelImg is a graphical image annotation tool. We used this tool to manually bound input bounding boxes on the training images for performing classication using Faster RCNN. This created an XML per image that includes the lename, path and the coordinates of the bounding boxes so that the model can understand where the object of interest is present. Then each xml is converted into csv so that all the features are in eld format and it is easier to visualize and use the data for training.
2. VGG Image Annotator-VGG Image Annotator is a simple manual annotation software for image, audio, and video. We used VIA tool to manually create the masks on each training image for performing classication using MASK RCNN. Each mask is represented as a set of polygon points. Annotations are then saved in JSON le format that includes lename, class ids, region attributes, and shape attributes.

3 Result Analysis

In this project, we have implemented both Faster RCNN and Mask RCNN, over the five mentioned categories. To compare these two algorithms we have used Mean Average Precision, wherein we used images from test set and the IOU threshold was set to 0.5, respectively. Finally, the mAP values obtained for Mask RCNN and Faster RCNN are, 0.85 and 0.78, also shown in Table 2. Thus, we can conclude that Mask RCNN works better for this project. The Fig. 6 displays the results generated on an interface, for input image of Vada Pav which includes its prediction probabililty, total amount of calories present in the image, and other nutrients associated with it. Learning is slow in Faster RCNN as compared to Mask RCNN. As we train the model, learning becomes more slower process and it takes more time to learn and reduce loss. Mask R-CNN, in addition to the class label and bounding box coordinates for each object, will also return the object mask. Mask RCNN uses Instace segmentation to get the object mask. Thus Mask RCNN generates better and accurate results than Faster RCNN with less training.

Table 2 mAP values for faster RCNN and mask RCNN

Algorithm	mAP(Iou≥ **0.5**)
Mask R-CNN	0.85
Faster R-CNN	0.78

Result: Vada Paav Probability : 0.9999062

Total Calories: 262 Kcal

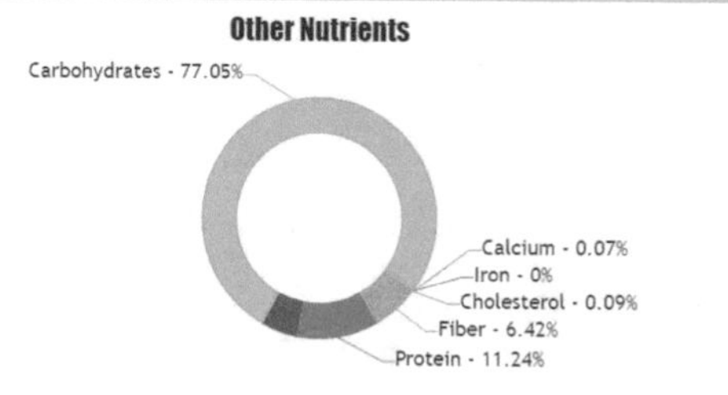

Fig. 6 Result generated for vada paav

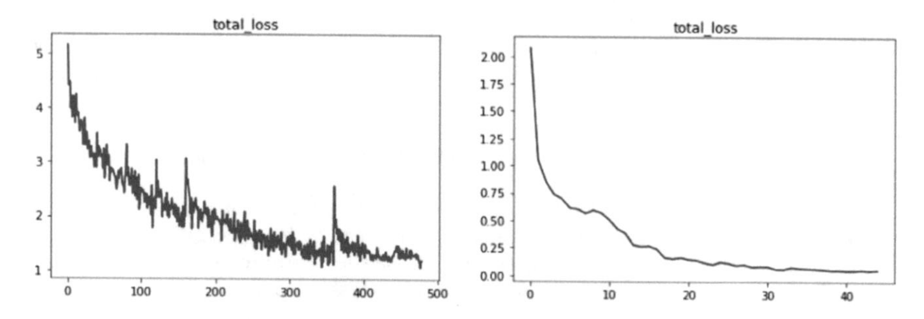

Fig. 7 Graphs generated while training **a** Faster RCNN **b** Mask RCNN

Figure. 7 shows the graphs generated by plotting the statistics generated while training. X-axis denotes number of epoch and Y-axis denotes total loss. By Looking at both the graphs, we found that Mask RCNN model produced better and more accurate results with less amount of training whereas Faster RCNN required more training to produce acceptable results.

4 Conclusion

We implemented two algorithms, namely, Faster RCNN and Mask RCNN for object detection and nutrients were displayed based on the results from their output. We developed a system that uses an input image from the user and determines the food item on the plate and shows its nutrients in a graphical format that shows the distribution of percentages of nutrients using Nutrition tables. In the comparative study between these algorithms, we found that Mask RCNN produced better results as compared to Faster RCNN. RCNN networks use a different approach than the one used in traditional algorithms for object detection thus both these algorithms are faster than other object detection approaches. Mask RCNN is built on Faster RCNN and is thus better than Faster RCNN and appears to be promising approach.

References

1. WHO-Obesity and Overweight, https://www.who.int/en/news-room/fact-sheets/detail/obesity-and-overweight
2. Ege T, Ando Y, Tanno R, Shimoda W, Yanai K (2019) Image-based estimation of real food size for accurate food calorie estimation. IEEE
3. Gunawan FT, Kartiwi M, Abd Malik N, Ismail N (2018) Food intake calorie prediction using generalized regression neural network. IEEE Conference
4. Muhammad NA, Lee CP, Lim KM, Razak SF (2017) Malaysian food recognition and calorie counter application. In: IEEE 15th student conference on research and development (SCOReD)
5. Ocay AB, Fernandez JM, Palaoag TD (2017) Nutritrack: Android-based food recognition app for nutrition awareness. In: 3rd IEEE international conference on computer and communications
6. Ege T, Yanai K (2017) Simultaneous estimation of food categories and calories with multi-task cnn. In: Fifteenth IAPR international conference on machine vision applications (MVA) Nagoya University, Nagoya, Japan
7. Trambadia S, Mayatra H (2016) Food detection on plate based on the hsv color model. In: Online International Conference on Green Engineering and Technologies (IC-GET)
8. Nutrition information for raw fruits, vegetables, and sh. https://www.fda.gov/food/food-labeling-nutrition/nutrition-information-raw-fruitsvegetables-and-sh

Comparative Study of Routing Protocols in Vanets on Realistic Scenario

Rajiv Chourasiya and Mohit Agarwal

Abstract Vehicular Ad Hoc Network (VANET) has been emerged out as one of the most popular areas of research which gathers the attention of the research community. VANET is basically deployed to integrate the responsiveness of widely active wireless-based networks to the vehicles. The basic idea here is to provide the abundant connectivity to the vehicles either with the help of efficient vehicle-to-vehicle or vehicle-to-infrastructure communication link which permits the adoption of Intelligent Transportation Systems (ITS). In order to design and develop a right and capable routing protocol for the VANET, a rigorous study of popular existing VANET routing protocols is always required. In this work, some of the most widely used routing protocols like AODV, AOMDV, DSR, DSDV are taken into the consideration for the comparison purpose in terms of routing performance based on a set of parameters chosen. The purpose of this work is to explore the simulation of two real-world-scenarios of wireless systems in VANET. The routing protocols are compared with the help of network simulator—2 (NS 2), MOVE, SUMO, etc. The evaluation metrics used for the comparison purpose include PDR (Packet Delivery Ratio) and NRL (Normalized Routing Overhead). The outcome of simulations suggests that AODV and AOMDV are much suitable out of the four protocols for the real-time scenario which was taken into the consideration.

Keywords Vehicular ad hoc network (VANET) · Wireless network · Transportation system · Packet delivery ratio · Normalized routing overhead

R. Chourasiya
Department of Information Technology, KIET Group of Institutions, Delhi, India
e-mail: rajiv.chourasiya@kiet.edu

M. Agarwal (✉)
Department of Computer Science & Engineering, School of Engineering & Technology, Sharda University, Greater Noida, India
e-mail: mohit.agarwal1@sharda.ac.in; rs.mohitag@gmail.com

N. Chaki et al. (eds.), *Proceedings of International Conference on Computational Intelligence and Data Engineering*, Lecture Notes on Data Engineering and Communications Technologies 56, https://doi.org/10.1007/978-981-15-8767-2_37

1 Introduction

Recent development and advancement in the field of wireless communication and technology really helping the general masses in every dimension of their life. One of such areas where change is clearly visible is the maintenance of transportation system. With the advent of better technologies and the availability of cheaper transport facilities, there has been a rapid increase in the movement of vehicles on the road [1]. However, passage of vehicles on the allowed path is governed by situations/circumstances like congestion of traffic, speed zones, prevailing weather conditions, maintenance work of roads, etc. Such conditions result in formation of the group or cluster of moving vehicles in order to manage traffic flow in all possible directions.

Under such situation and constraints, sometimes it is impossible for the vehicles to maintain a direct communication link with each other by taking help of single hop, which is related to the specified area of coverage. Such situations force the researchers to think for the creation of internetwork by considering the node from the different clusters. In order to manage the communication line among the out of ranged nodes or vehicles, different routing protocols are used. It is also observed that the available and proposed routing protocols which proved their efficiency in ad-hoc networks are not very much compatible with the scenario exists in VANET due to above underlying constraints. Therefore, existing protocols require certain changes and improvements so that they can prove efficiency with the mentioned conditions for the existing network and this makes it an area of focus for many researchers.

Till date, most of the comparisons done between protocols have been limited to their individual characteristics and specification, and as such very less research and analysis is done in a Real Map Scenario. By generating simulations in Real Map Scenario, we can actually determine their productivity and efficiency under various circumstances and find how they will actually behave in the real world.

The purpose of this research work is to highlight the importance of routing protocols in two different kinds of VANET scenarios which we have taken into the consideration. In order to achieve the same, we performed a number of simulations using the NS2 simulator for various combinations of parameters.

Rest of this work is organized as Sect. 2 used to present the survey of the area and protocols; VANET and its architecture is discussed in Sect. 3; concept of various routing protocols is presented in Sect. 4. Sections 5 and 6 is used to present our proposed work and results, respectively, while conclusion of this work is present in Sect. 7.

2 Literature Survey

Authors in their work [2] state the concepts of VANET precisely and explore the inspiration behind their design and analyze the development of these protocols along with the comparison of some different kind of routing protocols in VANET, their advantages, and shortcomings of these different routing protocols, and finally pointing out some issues with possible record of future research related to VANET routing. Apart from these, some other issues discussed in the paper are

It explains the various distinguishing features of VANET that make it different from MANET along with the brief discussion of different types of routing protocols.

In work presented by the author [3], Authors analyzed DSR, AODV, and STAR protocols using the GloMoSim simulation environment. After studying this paper, we can conclude that it is very difficult to determine that, which three protocols are suitable when comparison are done in ad hoc network. None of the protocol is ideal for each scenario. Size and expected traffic load might be the good criteria to choose a protocol for a selected network. There is a scaling problem with on-demand routing protocol which means the network overhead increases linearly as the number of nodes increases.

In the paper [4], two tools are characterized by authors such as SUMO, MOVE that permits users to provide real-world models to stimulate Vanet. The tool MOVE is basically above SUMO that is open source micro-traffic simulator. The output of MOVE is a model of real-world mobility and can be utilized by NS-2 and qualnet simulator. This tool can also be utilized to understand the simulation of VANET.

In the paper [5], there is an analysis of Ad-Hoc routing protocol which is for real time situations of VANET. After a comparative study of AODV, OLSR, and DSR in realistic scenario of VANET we got to know that execution of AODV in terms of PDR is well established approximate 97% in both situations.

In the paper [6], authors came up to deal with a characterization of ad hoc routing protocols, and further it provides some specified rules in consideration to the properties. The protocols presented here are identified in accordance with their entities by this paper.

The paper [7] gives the performance evaluation results of three reactive protocols, namely AODV, AOMDV, and DSR in comparison with a proactive routing protocol DSDV. The simulation has been done by using different velocities of vehicles moving in different locations by changing maps.

In paper [8], authors state that the dropping rate of packet for DSR is few OLSR, DSDV, and AODV giving back its highest efficiency. Both AODV and DSR give better performance under high flexibility of traffic vehicles than DSDV. The high mobility of the traffic vehicles is seen due to the recurrent link failures and also the overhead which is foreseen in improving all the nodes within the new routing details as present in DSDV.

DSR which mainly uses source routing and route caches is basically not dependent on any periodic or time-based activities. AODV is a mechanism to prevent loops and to determine triggering of paths that uses routing tables, one route per destination, and the ultimate destination sequence numbers.

3 VANET

This section contains the comparative views on conceptual fundamentals and issues of VANET [2] structure and their properties. VANET network is primarily a form of Mobile ad-hoc Networks, to furnish communication in between the fixed equipment of the vehicles and the vehicles nearby, i.e., equipment's on the roadside. VANET or Intelligent Vehicular Ad-Hoc Networking furnishes a significant and an enlightened path of using vehicular Networking [9]. VANET is basically a technology that primarily uses moving vehicles comparing them with the nodes in a network in order to create a mobile network.

Every vehicle is connected with VANET device and will be a node in the Ad-hoc network. The main functionality is that it can receive and transfer other fragments throughout the wireless network. Driving has become more dangerous as well as challenging due to major increase in vehicles on daily basis. Roads are confined. There are no such protocols that are to be followed for safety distance and reasonable speeds. Efforts have been made so that VANET is to enhance and improve road safety. In order to match this, the vehicles have to act as sensing tools and had to transfer and exchange warnings of obstacles or—we can say—telemetric details which allow the drivers to be more active and react fast to any dangerous situations that might occur due to traffic cramps (Figs. 1 and 2).

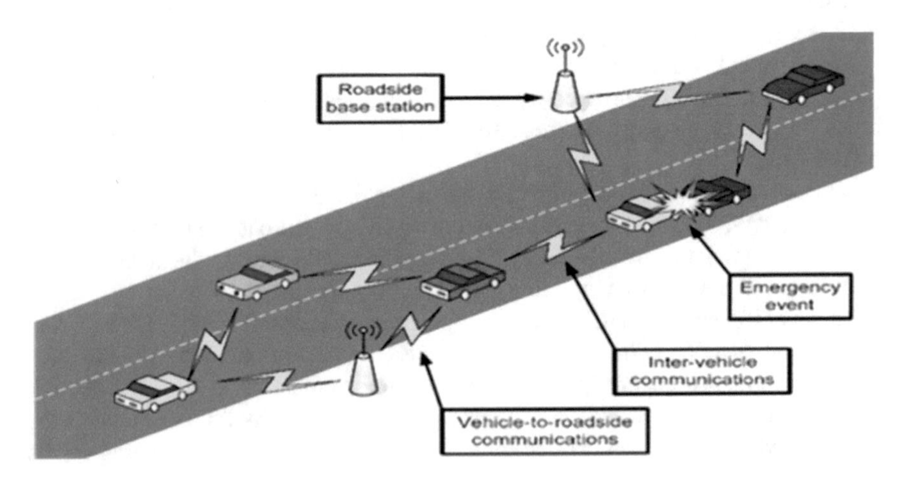

Fig. 1 VANET in on-road scenario

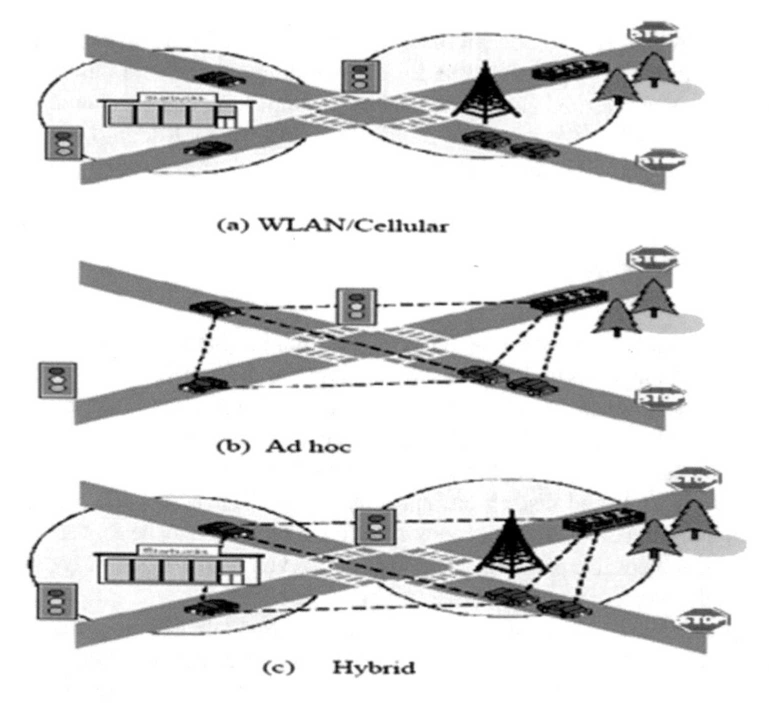

Fig. 2 Architecture of VANET

4 Routing Protocols

Routing in network may be defined as the procedure of selecting a path in a concerned network data or traffic can be send. In packet-switched networks, routing puts packet forwarding (the transit of logically addressed packet from a given source to their ultimate destination) through intermediate nodes [10]. These routing protocols use already present information that exist in the network to perform packet transferring in the forward direction. Routing protocols can be further categorized into proactive and reactive protocols [11]:

- ***Proactive routing protocols***: These routing protocols infer that the information related to routing being transferred, like the next hop being forwarded and is maintained in the background irrespective of any kind of communication requests. One of the advantages of the proactive routing protocol is there is no discovery of routes as the route for the destination is stored and kept in the background, but the disadvantage of this protocol is that low latency rate is provided for real-time applications in this protocol. The various types of proactive routing protocols are as follows: Fish-Eye Routing Protocol (FSR), Optimized Link State Routing (OLSR), Destination Sequenced Distance Vector Routing (DSDV), etc.
- ***Reactive/*Ad hoc*based routing***: Reactive routing flashes the route, opens only when it is matter of urgency to communicate with one another. Reactive routing

consists of many phases, one of them is the discovery of route; in which network is flooded with the query packets for the searching of path and this phase will get over when desired route is available. Examples of reactive routing protocols includes AODV (Ad-hoc On-Demand Distance Vector Routing), Temporally Ordered Routing Algorithm (TORA), and Dynamic Source Routing (DSR).

5 Proposed Work

VANET in terms of experimental analysis have no designed mechanism and thus still in conceptual design. Therefore, we are creating a VANET simulator to meet our requirements since the already existing tools do not meet our requirements. Our proposed simulator will be able to create its own maps to produce distinct situations of traffic for a real understanding of the road system. This situation consists of vehicles having Wi-Fi tools and distinct attributes that may either be created manually or randomly. Special events like the roads or different blocked may be activated. Many security techniques like pseudonym changes or mix zones are triggered. The process is extremely distinct and fully consistent at any time.

Our main aim is to simulate VANET routing protocols namely—Ad-Hoc On demand Distance Vector routing (AODV), Dynamic Source Routing (DSR), Ad-Hoc On Demand Multipath Distance Vector routing (AOMDV), Destination Sequenced Distance Vector routing (DSDV) protocol with real map scenario in Network Simulator 2 (NS-2) and make a relative study actually following the results of the simulations [12–14]. By producing simulations in these two cases, we can actually find their productivity and efficiency under various activities and can identify what they will believe in the real time scenario.

This research work mainly highlights the significance of routing protocols in VANET under distinct factors (especially through pragmatic scenarios) and keeping a check in two other scenarios. It will be our endeavor to investigate a complete analysis of these protocols under various parameters by means of rigorous simulation test cases and the comparative analysis of different scenarios.

6 Result and Discussion

This paper contains a full simulation criterion which is quite mandatory for taking into the consideration for the completion of specified objectives and also to understand the behavior of routing protocols in Vehicular Ad-Hoc network (VANET) [15]. For the same, we consider the realistic vehicular traces, i.e., by simulating with real map scenario.

The map considered is of Dilshad Garden, New Delhi, India and Indirapuram, Ghaziabad, India which are shown in the figure (Fig. 3).

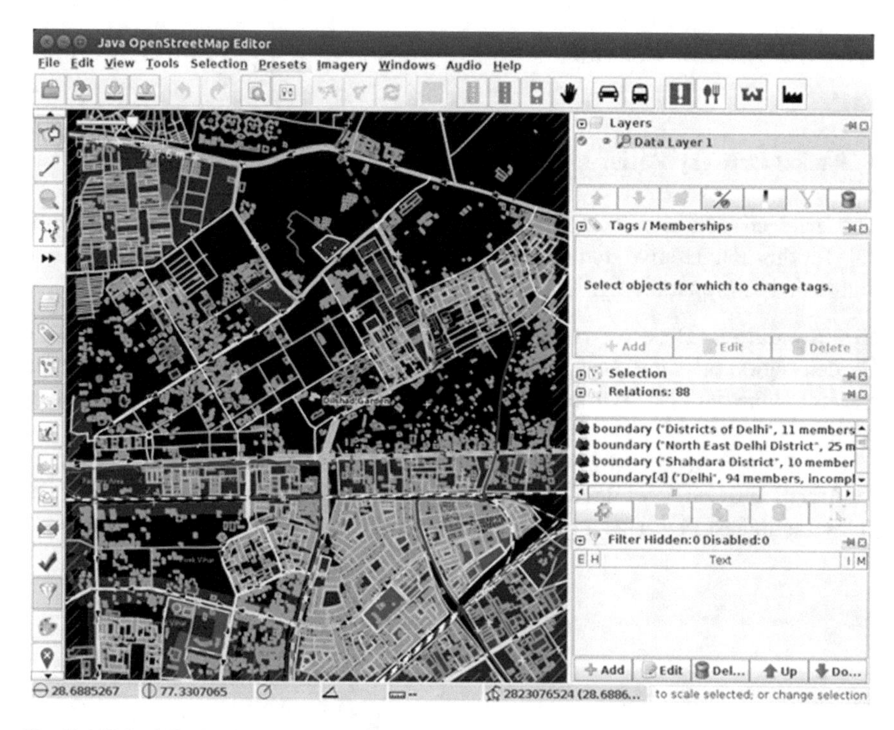

Fig. 3 Dilshad Garden Map

In order to generate the instances for the simulation work, we need to define certain variables which are required to be used within the simulation script for further action. For this work, required common variables for two different locations are shown in Table 1.

In the above table, variables with their optimal values are highlighted which are going to be used for the simulation purpose for both the locations. According to this, a typical simulation running time is 1000 ms. Routing protocols which are deployed individually for the analysis purpose in each simulation are AODV, AOMDV, DSDV, and DSR [16]. At last, UDP agent is used to define the traffic type for nodes communication which is more reliable than TCP.

Table 1 Common variables

Variable	Value
Simulation time	1000 ms
Routing protocols	DSDV, DSR, AODV, AOMDV
Type of traffic	UDP

6.1 Simulation Results

- **Packet Delivery Ratio:**

The analyzed metrics of PDR and the required values for the specified routing protocols for this comparative study is presented by the help of Fig. 4 for Dilshad Garden region and Indirapuram region (Table 2).

Result Analysis: Observed percentage of PDR (Packet Delivery Ratio) in Dilshad Garden region are quite acceptable in case of AODV, AOMDV, and DSR but the results obtained for DSDV routing protocol is significantly lower when compared to others.

And it has been observed that the Packet Delivery Ratio dropped very drastically for AODV and DSDV while AOMDV and DSR are offering very high PDR and DSR gives the highest PDR as compared to all other protocols for Indirapuram region.

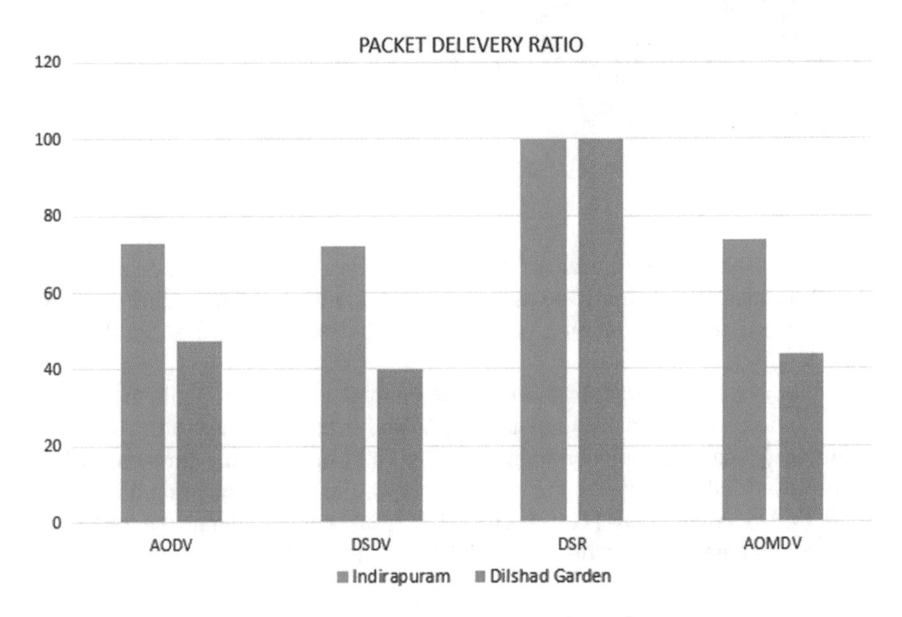

Fig. 4 Packet Delivery Ratio of protocols in Dilshad Garden and Indirapuram region

Table 2 Analyzed data for Packet Delivery Ratio

Packet Delivery Ratio				
	AODV	DSDV	DSR	AOMDV
Indirapuram	72.87	72.22	99.9	73.62
Dilshad Garden	47.3	40.14	99.91	44.077

Table 3 Analyzed data for Normalized Routing Load

Normalized Routing Load				
	AODV	DSDV	DSR	AOMDV
Indirapuram	1.064	1.097	1.001	1.262
Dilshad Garden	1.483	1.92	1.001	2.014

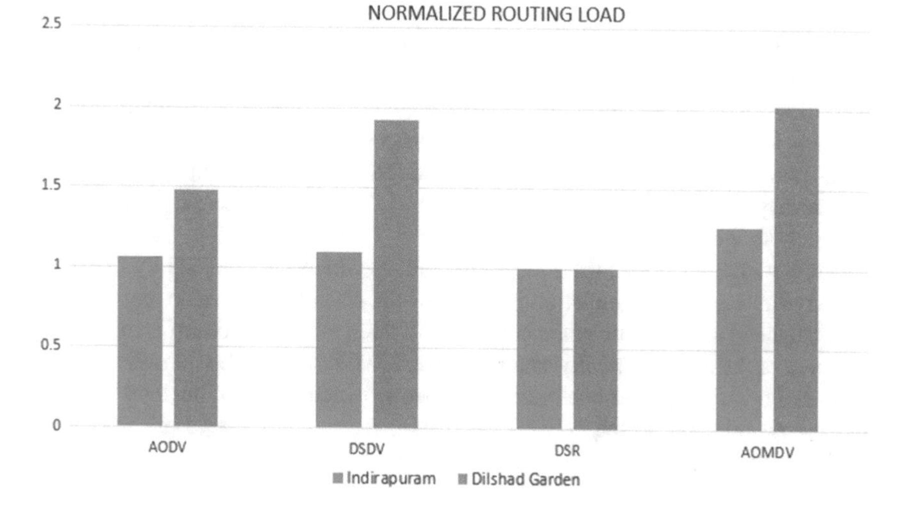

Fig. 5 Normalized Routing Load of protocols in Dilshad Garden and Indirapuram region

- **Normalized Routing Load:**

The analyzed numerical results for the locality of Dilshad Garden and Indirapuram, scenario from the generated trace files for individual routing protocols are given in Table 3 (Fig. 5).

From the above given charts following inferences can be drawn out:

Result Analysis: Normalized Routing Load value obtained for DSR is less when compared to the values obtained for other protocols. AOMDV gives the highest NRL so it is considered worse than other protocols. The protocol which has less routing overhead (load) is considered to be an efficient one. Here after recording the Normalized Routing Load values for each of the protocol, and it has been observed that the Normalized Routing Load for AOMDV is highest, so its performance is worst when it comes to send bulky data. While the Normalized Routing Load for AODV and DSDV decreases comparatively, then DSR offers the minimal Normalized Routing Overhead which makes it the best protocol to be used for sending bulky data.

A comparative and detailed analysis for the popular routing protocols which is based on the investigative metrics is thoroughly discussed. In a brief, this study comprises two decisive factors: "Packet Delivery Ratio" and "Normalized Routing

Overhead". Both the factors directly depend upon the method of the formulation of the respective models in two different scenarios.

7 Conclusion

Due to the variety of challenges in VANET especially for real life scenario we came up with the problem of finding the better protocols that works on real life scenario. The results obtained for the real map models outwardly remains competitive in case of routing protocols: AODV and AOMDV. Values are obtained for PDR (Packet Delivery Ratio) for both ranges in between 96 and 99% approximately, while, the values obtained for Normalized Routing Load (NRL) for all participating protocols were nearly within the same range except for DSR, which had exceptionally low NRL in both the regions.

Consequently, in short, the end results' values which are obtained by means of extensive and rigorous simulation process for the specific deployable test beds are quite balanced and realistic when compared to the real roads and traffic conditions. Outcome of the analysis clearly establish that AOMDV and AODV clearly emerged out as the two efficient ad hoc routing protocols, which are also found as the most appropriate selection at the network layer of given cases, i.e., real map model in VANET with varying traffic concentration in different regions.

References

1. Ho YH, Ho AH, Hua KA (2008) Routing protocols for inter-vehicular networks: A comparative study in high-mobility and large obstacles environments. Comput Commun 31(12):2767–2780. https://doi.org/10.1016/j.comcom.2007.11.001
2. K., S., Deshmukh, A. R., & Dorle, S. S. (2016) A survey of routing protocols for vehicular ad-hoc networks. Int J Comput Appl 139(13):34–37. https://doi.org/10.5120/ijca2016909541
3. Jiang H, Garcia-Luna-Aceves JJ (n.d.) Performance comparison of three routing protocols for ad hoc networks. In: Proceedings tenth international conference on computer communications and networks (Cat. No.01EX495). https://doi.org/10.1109/icccn.2001.956320
4. Khairnar VD, Pradhan DSN (2010) Mobility models for vehicular ad-hoc network simulation. Int J Comput Appl 11(4):8–12. https://doi.org/10.5120/1573-2103
5. Kumar Singh P, Lego K, Tuithung DT (2011) Simulation based Analysis of Adhoc Routing Protocol in Urban and Highway Scenario of VANET. Int J Comput Appl 12(10):42–49. https://doi.org/10.5120/1716-2302
6. Nadia T, Mourad A, Hamouma M, Hamoudi K (2019) A Survey on vehicular ad-hoc networks routing protocols: classification and challenges. J Dig Inf Manage 17(4):227. https://doi.org/10.6025/jdim/2019/17/4/227-244
7. Araghi TK, Zamani M, Mnaf ABA (2013) Performance analysis in reactive routing protocols in wireless mobile ad hoc networks using DSR, AODV and AOMDV. In: 2013 International Conference on Informatics and Creative Multimedia. https://doi.org/10.1109/icicm.2013.62.
8. Malhotra R, Sachdeva B (2016) Multilingual evaluation of the DSR, DSDV and AODV routing protocols in mobile ad hoc networks. SIJ Trans Comput Netw Commun Eng 04(03):07–13. https://doi.org/10.9756/sijcnce/v4i3/0103550101

9. Chandrasekar A, Vinnarasi FSF (2019) VANET routing protocol with traffic aware approach. Int J Adv Intell Paradig 12(1/2):3. https://doi.org/10.1504/ijaip.2019.10017739
10. Samara G (2018) An intelligent routing protocol in VANET. Int J Ad Hoc Ubiquitous Comput 29(1/2):77. https://doi.org/10.1504/ijahuc.2018.094399
11. Maan F, Mazhar N (2011) MANET routing protocols vs mobility models: a performance evaluation. In: 2011 third international conference on ubiquitous and future networks (ICUFN). https://doi.org/10.1109/icufn.2011.5949158.
12. Perkins C., Belding-Royer E, Das S (2003) Ad hoc On-Demand Distance Vector (AODV) Routing. https://doi.org/10.17487/rfc3561
13. Yuan YH, Chen HM, Jia M (n.d.) An Optimized Ad-hoc On-demand Multipath Distance Vector (AOMDV) routing protocol. In: 2005 Asia-Pacific conference on communications. https://doi.org/10.1109/apcc.2005.1554125
14. Tuteja A, Gujral R, Thalia S (2010) Comparative performance analysis of DSDV, AODV and DSR routing protocols in MANET using NS2. In: 2010 international conference on advances in computer engineering. https://doi.org/10.1109/ace.2010.16
15. Khan S, Alam M, Fränzle M, Müllner N, Chen Y (2018) A traffic aware segment-based Routing protocol for VANETs in urban scenarios. Comput Electr Eng 68:447–462. https://doi.org/10.1016/j.compeleceng.2018.04.017
16. Madhanmohan DR (2019) A study on Ad-Hoc On-Demand Distance Vector AODV Protocol. Int J Trend Sci Res Dev 3(4):1019–1021. https://doi.org/10.31142/ijtsrd24006

An Efficient Algorithm for Prawn Detection and Length Identification

Chaladi V. N. Koushik, Rampilla V. N. Kamal, Channagiri Tarun, Koya Dinesh Teja, and Suneeth Manne

Abstract Prawn fishery has been gaining vast popularity in the aquaculture industry. But farmers fail to know health status of the prawn. Length and weight are parameters for assessing the health of the prawn. It is usually easier to measure the length of the specimen than the weight, and weight can be predicted using the length–weight relationship. In this application, Faster Region-Based Convolutional Neural Network (Faster RCNN) algorithm is used for the detection of the prawn and to draw a bounding surrounding the specimen. Faster RCNN returns coordinates in the form of [ymin, xmin, ymax, xmax] which can be used to localize prawn in the image. Pixel length is achieved using the above coordinates, and pixel per metric is used to derive length in centimeters from pixel length. The weight is achieved by applying length–weight relation. This weight is stored in the database. A graph with weight analysis is returned. In this work, we considered 100 images as test set and got an accuracy of 95% and ∓ 2 cms approximation in length.

Keywords Faster RCNN · Region proposal network (RPN) · Length–Weight relationship · Region of interests (ROIs)

1 Introduction

Prawn fishery is growing rapidly. Prawn production in India is about 15% of world prawn production (indianAgro.net). It earns foreign exchange for our country. Prawn fishery is playing a significant role in the Indian economy. Generally, a prawn takes 180 days to gain weight of 40 gm, i.e., it grows 0.23 g per day. But, farmers fail to

C. V. N. Koushik (✉) · R. V. N. Kamal · C. Tarun · K. D. Teja
B.Tech, Department of Information Technology, VR Siddhartha Engineering College, Vijayawada, India
e-mail: vnkoushik.ch@gmail.com

S. Manne
Department of Information Technology, VR Siddhartha Engineering College, Vijayawada, India
e-mail: Suneethamanne74@gmail.com

N. Chaki et al. (eds.), *Proceedings of International Conference on Computational Intelligence and Data Engineering*, Lecture Notes on Data Engineering and Communications Technologies 56, https://doi.org/10.1007/978-981-15-8767-2_38

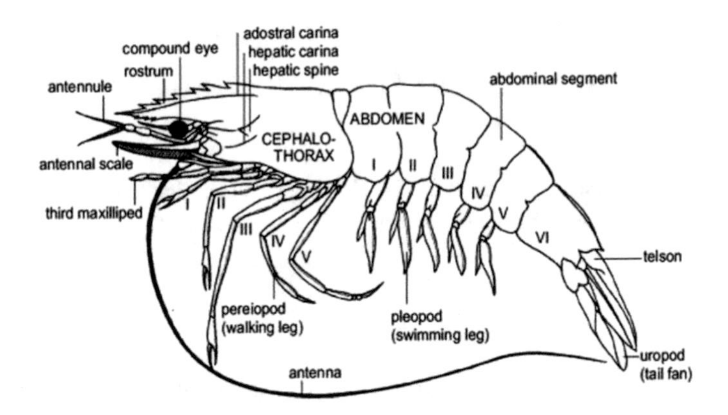

Fig. 1 Prawn (Sucharita, 2013)

assess the health of the prawn with the naked eye. It is difficult for farmers to afford regular supervision of doctors.

Based on the above constraints, here we have developed an application to find the length and weight of the prawn which are important factors for health. Here we used the Faster Region-Based Convolutional Neural Network for the detection and localization of the prawns. Length is identified using localization coordinates and weight is calculated using a length–weight relationship. The prawn body structure is shown in Fig. 1.

A few works [1, 2] done by researchers are restricted to classification with less accuracy compared to our technique. There is no scope for localization and length calculation. They used edge detection, Fuzzy logic methods, and Support Vector Machine (SVM). In [3], length is calculated using logistic regression of pixel area and pixel length which is extremely sensitive to outliers. Our method is robust to outliers with increased accuracy.

A prawn length is measured from its head to tail. Many assume that curved length is taken but length is measured straight from head to tail. Prawn body has three parts: head, carapace, and abdomen.

Cephalothorax: Cephalothorax is again divided into two portions:

Head: Head is the front part of cephalothorax. In head, there are five pairs of segmented appendages, which are a. Antenulle b. Antenna c. Mandibles d. Maxillula and e. Maxilla.

Thorax: Behind the head thorax is situated. Head and thorax together form the cephalothorax. The first three pairs of appendages are maxilla like and known as maxillipeds.

Abdomen: The elongated portion of the body after the cephalothorax is abdomen. It is round dorsally and a bit compressed laterally. The abdomen consists of six segments.

2 State of the Art

Early work on the length of the prawn is done by Alf Harbiz [4] which determines the length of the carapace using regression of pixel area and pixel length. The location of the prawn is identified using the red component present in the image. This approach would be useful only when the object in the image is a prawn and any object other than a prawn would lead to wrong result. In [1, 2, 5, 6], methods for the detection of prawn are specified but did not discuss about the length of the prawn. Those papers mainly focused on prawn species detection. In [1, 3], detection is done based on histogram values of the image which has less accuracy, a time taking process, and robust to outliers which may cause a serious problem in the detection of the prawn. In [2] method, detection was based on edge detection. In this method initially, noise is removed based on wavelet transform, and edge detection is made based on eight feature vectors. Work in [7–9] is the research of prawn length and weight relationship and the conditional factor of prawns; this paper did not specify any methods of detection of length measuring techniques but only had the research on the length and weight.

3 Proposed Approach

This chapter consists of design methodologies used in the project. The proposed Architecture Diagram is shown in Fig. 2.

Farmer (Client-Side): Upload Image: Farmer upload image of prawn for identification of length.

Receive Length Analysis: Farmer receives length analysis based on this data present in the database.

Server Side: Trained algorithm: Uploaded image is sent to the R-CNN algorithm which is trained using the prawn dataset.

Localizing prawn: After the detection of prawn in the image, it is localized using **bounding boxes regressors**. Bounding Boxes coordinates in the form [ymin, xmin, ymax, xmin].

Calculation of Length and Weight: The length of the prawn is calculated using image using pixel per length metric using a reference object in the image such as a coin.

$$\text{pixels\_per\_metric} = \text{object\_width}/\text{know\_width} \tag{1}$$

Weight Calculation

$$\mathrm{W} = 0.0108 * \mathrm{L}^{2.6978} \tag{2}$$

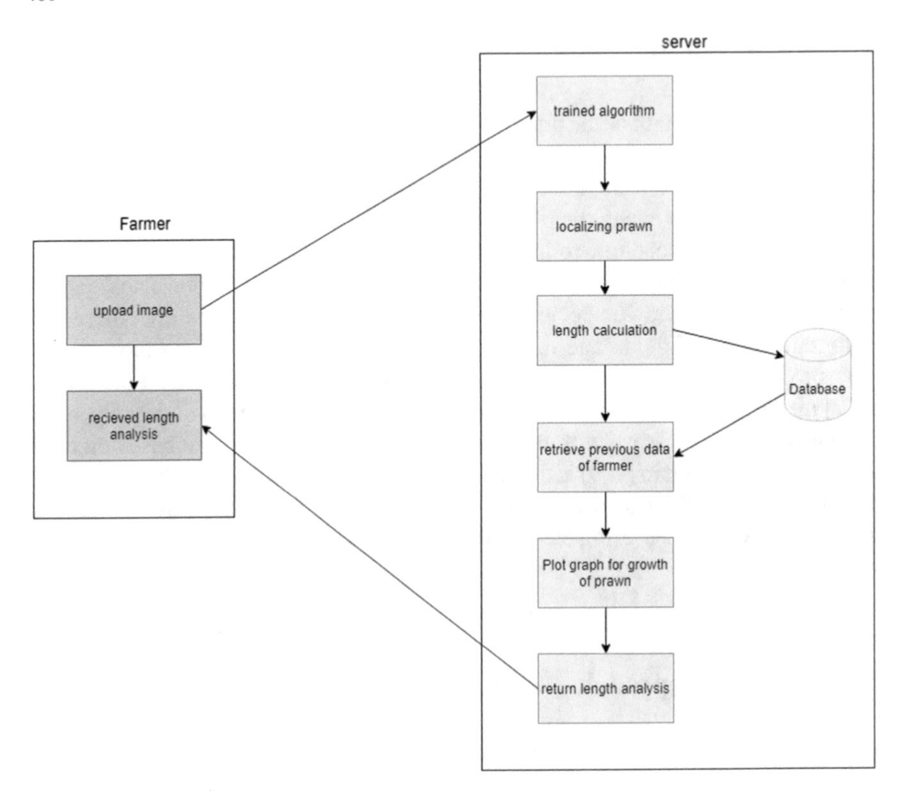

Fig. 2 Proposed architecture diagram

Retrieve Previous Data of farmer: Previous data of framers is retrieved to analyze trends of growth.

Plot graph for growth of prawn: Trends in growth of prawn are plotted using matplotlib package in Python.

Return length analysis: Analysis of length is returned to the farmer.

The algorithm used in the application is **Faster Region-Based Convolutional Neural Network (Faster RCNN)**. Using this algorithm, one can localize the object in the image. An abounding box around a trained object is returned as output. Faster RCNN is one among the foremost distinguished object detection neural networks [1, 2]. It is additionally the idea for several derived networks for segmentation, 3D object detection, the fusion of measuring instrument purpose cloud with image, etc. The Faster CNN flow diagram is shown in Fig. 3.

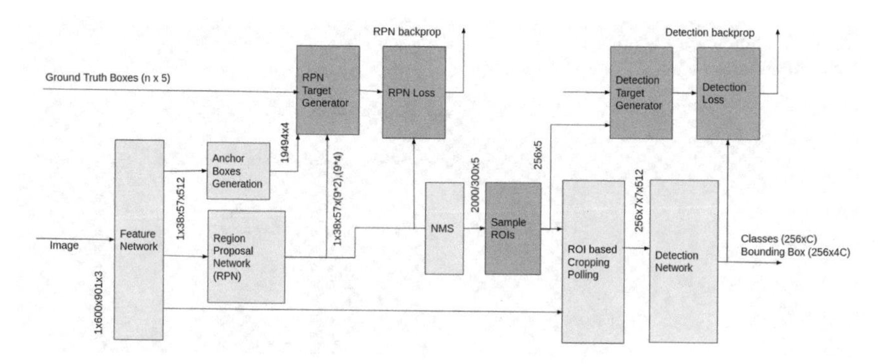

Fig. 3 Faster RCNN architecture diagram

3.1 Components of Rcnn

Feature Network

The Feature Network is typically a well-known pre-trained image classification network like VGG minus some last/top layers. The performance of this network is to come up with sensible options from the pictures. The output of this network maintains the form and structure of the first image.

Region Proposed Neural Network

The RPN is typically an easy network with three convolutional layers. There is one common layer that feeds into 2 layers—one for classification and another for bounding box regression. The aim of the RPN is to come up with many bounding boxes referred to as Region of Interests (ROIs). The output from this network is many bounding boxes known by the element coordinates of 2 diagonal corners.

Region of Interests (ROIs)

Region of Interests (ROIs) are regions that have a high probability of containing any object. Faster RCNN Architecture diagram is shown in Fig. 3.

Anchor Boxes

Anchors play a vital role in quicker R-CNN. Associate degree anchor may be a box. Within the default configuration of quicker R-CNN, there are nine anchors at a foothold of a picture. The anchor boxes default configuration is shown in Fig. 4.

Non-Maximum Suppression

NMS removes boxes that overlap with different boxes that have higher scores (scores are not normalized possibilities, e.g., before softmax is applied to normalize).

ROI Pooling Layer

Converts into fixed-length feature map. The RPN tries to tighten the middle and also the size of the anchor boxes around the target. This is often referred to as the bounding

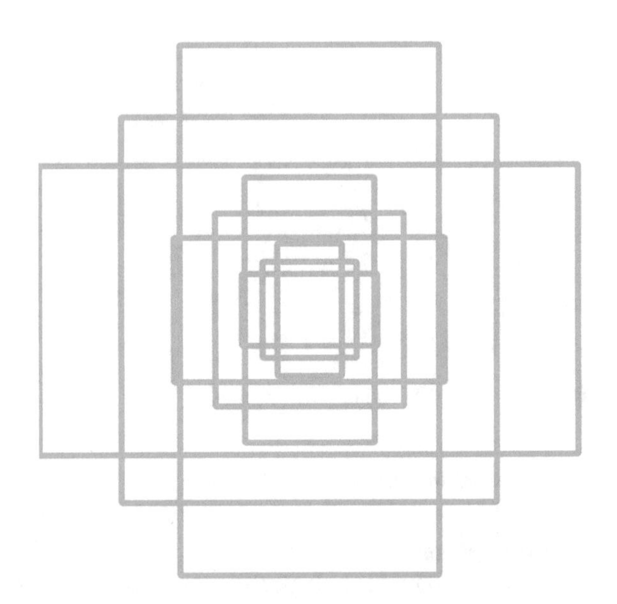

Fig. 4 Anchor boxes default configuration

box regression. For this to happen, targets have to be compelled to be generated, and losses have to be compelled to be calculated for backpropagation. The space vector from the middle of the bottom truth box to the anchor box is taken and normalized to the scale of the anchor box. That is the target delta vector for the middle. The scale target is that the log of the quantitative relation of the scale of every dimension of the experimental results and analysis.

3.2 Description of Datasets and Tools Used

Dataset: It is the collected prawn pictures from Google images. It preprocesses images using wavelet transforms and divides 80% of images to train and 20% into the test set.

Sample Training set: For training of this model two classes are taken as class-1 that contains prawn images whereas class-2 contains images other than prawn such as frog, crab, and snake. The train set is annotated using the LabelImg tool which in turn gives an XML file that consists of coordinates of the object in the image.

Sample Test set: For testing of this model, two classes are taken: class-1 contains prawn images whereas class-2 contains images other than prawn such as frog, crab, and snake. The training and test sets are shown in Figs. 5 and 6.

Tools Used: LabelImg is a tool programmed in python to label images for training the model. LabelImg tool returns an xml file which consists of boundary coordinates of the object in the image. Xml file consists of annotations.

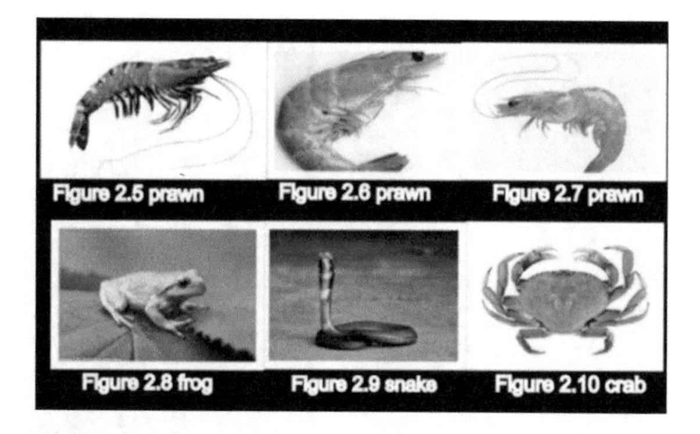

Fig. 5 Train set

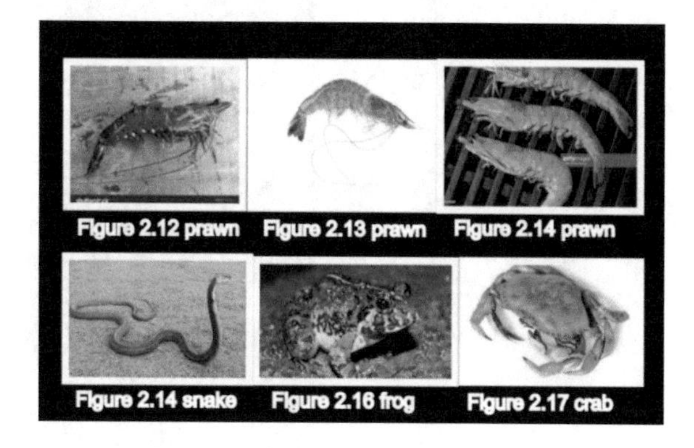

Fig. 6 Test set

4 Experimental Results and Observations

This chapter consists of results and observations obtained by the project. The result comprises the prawn pictures with the localization of the prawn in it. Additional to this, we also get a graph showing the growth of prawn given and also the optimal growth of the healthy prawn. So the result showed the weight of the prawn along with the graph plotting its weight to the number of days.

The algorithm detects the picture and identifies whether the picture consists the prawn, if it detects that the prawn is there then it draws a bounding box along its boundaries. Then it finds the length of the prawn with the help of the boundaries it detected from the bounding box. If the algorithm detects that there is no prawn in the picture, it displays "NO PRAWN" and does not go to next steps of finding the length by drawing the bounding box. The test set results are shown in Figs. 7 and 8.

Unknown Images

In Fig. 7, the image is not a prawn but it is a snake, so the bounding box localizes the snake and gives the label as unknown with the confidence interval of 98%. As the image is other than a prawn image so the outcome does not result in showing the length. In Fig. 8, the image is not a prawn but it is a frog, so the bounding box

Fig. 7 Snake

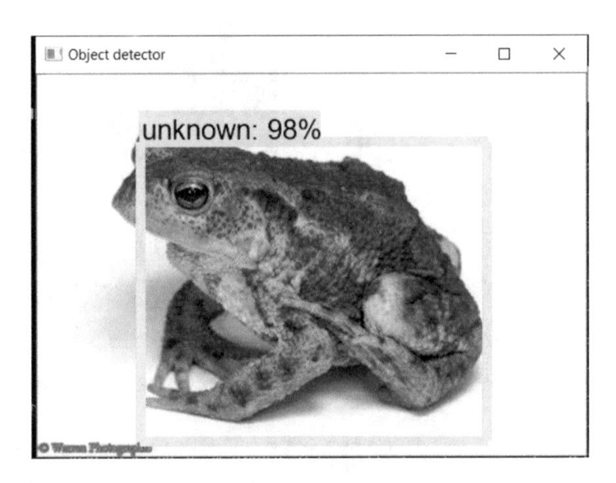

Fig. 8 Frog

localizes the frog and gives the label as unknown with the confidence interval of 98%. As the image is other than a prawn image so the outcome does not result in showing the length.

In Fig. 9, the image is not a prawn but it is a crab, so the bounding box localizes the crab and gives the label as unknown with the confidence interval of 98%. As the image is other than a prawn image so the outcome does not result in showing the length. In Fig. 10, the image is a prawn, so the bounding box localizes the prawn and gives the label as prawn with the confidence interval of 96%. As the image is a prawn image so the outcome results in showing the length along the boundaries.

```
17.05631983548403
Prawn Detected
Weight of your Prawn 22.465024422262896 Optimal Weight in mg 4.4
```

Fig. 9 Crab

Fig. 10 Prawn

```
10.5502410364151
Prawn Detected
Weight of your Prawn 6.160003813894443 Optimal Weight in mg 4.4
```

In Fig. 11 the image is a prawn, so the bounding box localizes the prawn and gives the label as prawn with the confidence interval of 99%. As the image is a prawn image so the outcome results in showing the length along the boundaries. The sample final output is shown in Fig. 12.

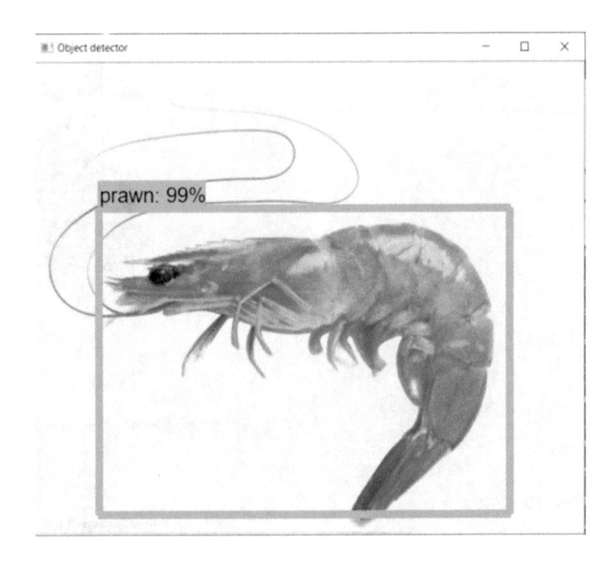

Fig. 11 Prawn

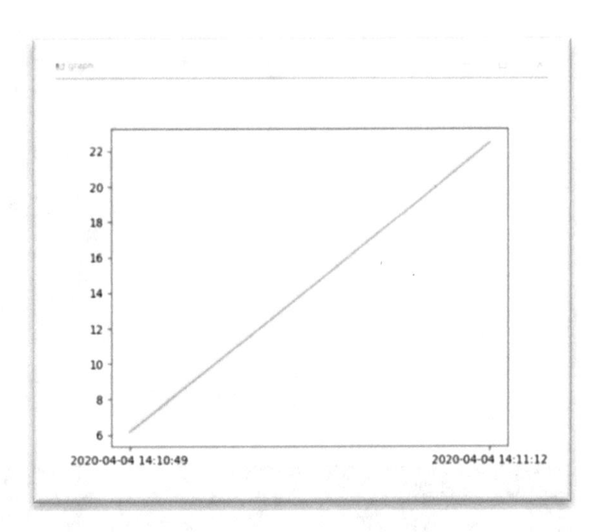

Fig. 12 Sample final output

5 Conclusion

In this application, we used Faster RCNN algorithm, and using this algorithm we got an average accuracy of 95% with 100 prawn test results and ∓ 2 cms approximation in length. In this research, we got better accuracy when compared to traditional algorithms. Algorithm used in the project gave good accuracy than the algorithms which were used earlier for detection of the prawn. Identification is also precise and accurate. Accuracy can be increased by increasing the training dataset and adding more features. Detection of diseases from prawn images can be also done. Need to do processing of water images to determine quality of water. Automatic food provision using health metric is also needed to be done and providing suggestions to the framers based on the previous images uploaded by farmers.

References

1. Sucharitha V, Jyothi S (2013) An identification of penaeid prawn species based on histogram values. Int J Adv Res Comput Sci Softw Eng. 3(7). ISSN: 2277 128X
2. Nagalakshmi G, Jyothi S Image acquisition, noise removal and edge detection methods in image processing using matlab for prawn species identification. ICETET 29th–31st
3. Kasmir Raja SV, Shaik Abdhul Khadir A, Riaz Ahamed SS (2009) Moving toward region-based image segmentation techniques: a study. J Theor Appl Inf Technol
4. Harbitz Alf (2007) Estimation of shrimp carpace length by length analysis. CES J Mar Sci 64(5):939–944
5. Djerouni A, Hamada H, Berrached N (2011) MR imaging contrast enhancement and segmentation using fuzzy clustering. IJCSI Int J Comput Sci Issues 8(4)(2)
6. Nagathan A, Manimozhi J, Mungara J (2014) Content based image retrieval system using feed-forward backpropagation neural network. IJCNS 14(6)
7. Udoinyang EP, Amali O, Iheukwumere CC, Ukpatu JE (2016) Length-weight relationship and condition factor of seven shrimp species in the artisanal shrimp fishery. Int J Fish Aqua Stud
8. Fuller PL, Knott DM, Kingsley-Smith PR, Morris JA, Buckel CA, Hunter ME, Hartman LD (2014) Invasion of Asian shrimp, Penaeus monodon Fabricius, 1798, in the western north Atlantic and Gulf of Mexico. Aqua Invasions 9(1):59–70
9. Hadil R, Albert GG (2001) Resource assessment of the tiger shrimp, Penaeus mondon of Kuala, miri-sarawak. Malaysian Fish J 2(1):221–237
10. Khademzadeh O, Haghi M (2017) Length-weight relationship and condition factor of white leg shrimp Litopenaeus vannamei. Int J Fish Aqua Stud

Author Index

N. Chaki et al. (eds.), *Proceedings of International Conference on Computational Intelligence and Data Engineering*, Lecture Notes on Data Engineering and Communications Technologies 56, https://doi.org/10.1007/978-981-15-8767-2

Printed in the United States
by Baker & Taylor Publisher Services